Finite Element Analysis for Satellite Structures

Gasser F. Abdelal · Nader Abuelfoutouh
Ahmed H. Gad

Finite Element Analysis for Satellite Structures

Applications to Their Design, Manufacture and Testing

Springer

Gasser F. Abdelal
School of Mechanical and Aerospace Engineering
Queen's University Belfast
Belfast
Northern Ireland, UK

Ahmed H. Gad
Egyptian Space Program, National Authority for Remote Sensing and Space Sciences
Cairo
Egypt

Nader Abuelfoutouh
Aeronautical and Aerospace Engineering Department—Faculty of Engineering
Cairo University
Cairo
Egypt

ISBN 978-1-4471-4636-0 ISBN 978-1-4471-4637-7 (eBook)
DOI 10.1007/978-1-4471-4637-7
Springer London Heidelberg New York Dordrecht

Library of Congress Control Number: 2012950393

Printed on acid-free paper

Springer is part of Springer Science+Business Media (www.springer.com)

Preface

This book is about finite element analysis of spacecraft structures. I tried to reflect the importance of numerical simulation not only in design stages but also in manufacturing simulation and testing. This book covers a wide range of applications using finite element method. This includes linear and nonlinear analysis, which makes this book very useful for many engineers in different areas and not only in spacecraft structures design.

This book is the outcome of my academic and industrial experience. After I got my Ph.D. degree in Aerospace Structures and did a post doc at West Virginia University in USA, I got the chance to work with YUZHNOYE Design Office in Ukraine through the Egyptian Space Program at NARSS; later I worked for Bluewater Energy in Netherlands, then a consultant for a number of aerospace and mechanical companies in UK, while working for the Virtual Engineering Centre, University of Liverpool, UK. Currently, I work as a Lecturer in Aerospace and Manufacturing, School of Aerospace and Mechanical Engineering, Queen's University in Belfast, Northern Ireland, UK. My research is in the area of Multiscale modeling and manufacturing simulation, in an effort to produce better products and reduce cost.

I would also like to thank my family, my father and Prof. Ever Barbero who always supported me.

I would like to express my gratitude to Allah and his prophets Moses, Jesus, and Muhammad, whose teachings have always guided me through all my life steps.

Gasser F. Abdelal

Contents

Chapter 1
Literature Review

Abstract The purpose of this chapter is to review several basic concepts from the areas of satellite design. It begins with defining the basic phases of structural design. Then a literature review of the categories of structures is presented. This is followed by a literature review of some previous satellite structural designs. Conventional spacecraft design methods and materials are included. Finally, the literature review chapter is concluded with a review of research on structural optimization methods.

1.1 Structural Design Phases

Structural design consists of three phases: conceptual design, preliminary design, and detailed design.

Conceptual design is the phase of establishing feasibility and estimate cost and risk for one or more spacecraft configurations or sets of derived requirements to support system trade studies or proposals. It also contains deriving requirements, identifying candidate types of structures, materials, and attachments. Finally, it develops the designs far enough to estimate and compare weight, cost, and risk; and select from options.

Preliminary design is the time for looking closer at the winning candidates and identifying the best arrangement, shape, and sizes of structural members. During this phase, types and forms of materials are selected; and design of attachments is also developed. Finally, it includes the beginning of manufacturing plan, and development of testing.

Detailed design is the time for final dimensions and manufacturing tolerances, identifying fastener sizes and installation torques, designing tertiary structures such as cable-support brackets, and doing all analyses necessary to justify

G. F. Abdelal et al., *Finite Element Analysis for Satellite Structures*,
DOI: 10.1007/978-1-4471-4637-7_1,

decisions. At the same time, the product team develops manufacturing processes and plans verification tests. Detailed design ends when the last engineering drawing for manufacturing is released.

All three phases are important. One overlooked detail during detail design can lead to expensive drawing changes, a test failure that drives a redesign, or even mission failure. Decisions during preliminary design and how these decisions can be documented will determine how well problems are avoided during detailed design and budgeted cost and program schedules are met. Conceptual design heavily influences all that follows. The requirements and ideas carried forward from conceptual design will affect performance and cost far more than anything done later.

1.2 Categories of Structures

A space mission typically requires a wide array of structures [1]. Structures support every physical part of a satellite. In fact, every physical part is a structure, because anything that has mass must carry structural loads due to weight or inertia forces when accelerated by the launch vehicle. Structures are categorized into three categories: primary structure, secondary structure, and tertiary structure. Structures are not categorized because one is more important than another, but because different requirements typically drive their designs.

Primary structure is the backbone, or the major load path, between the spacecraft's components and the launch vehicle. It carries shear, bending moments, axial loads, and torsion. For satellite, it consists of the spacecraft main body structure and the LV adapter. The primary structure takes its shape from the stowed packaging of spacecraft components and becomes integral to the configuration.

Secondary structure includes support beams, trusses, antenna dishes, and solar panels. Most of the considerations for primary structures also apply to secondary structures.

*Tertiary structure*includes component housing, mounting brackets, cable-support brackets, and connector panels.

1.3 Previous Satellite Structures

Aerospace structures generally require lightweight designs. The main goal is to optimize the strength per weight, or efficiency of the design. Satellite structural design has evolved greatly over the past four decades. Traditionally, efficiency has been accomplished using a combination of various structural designs and materials. This section is begun by discussing basic primary structural designs, and concludes with presenting traditional materials used in spacecraft and methods of attachment.

1.3.1 Conventional Structural Types

Primary structures are designed using many criteria that depend on the mission requirements [2]. Conventional spacecraft integrates four basic primary structural types:

- Skin-frame structures;
- Truss structures;
- Monocoque cylinders;
- Skin-stringer structures.

A *skin-frame structure* uses an interior skeletal network of axial and lateral frames to mount exterior skin panels using fasteners or rivets [2]. The frames support bending, torsion, and axial forces. The skin reinforces the structure by supporting the shear forces introduced by the interior member connections. To be stable, the ends of a skin-stringer structure must be closed with the skin. The skin is sometimes minimized to save mass, even though the thin skin leads to some structural instability. When the skin buckles due to shear, it transfers all additional shears loading to in-plane tension forces at 45° which must be supported by the connections. The buckling modes of the skin exhibit large deformations that make it insufficient for exterior mounted components such as solar cells. Intermediate frames are used to mount equipment or increase the buckling strength of the skin or stringers. Internal access can be difficult; therefore, removable panels may be needed. Sandwich and isogrid panels can be used to mount components.

A *Truss structure* uses an array of members that can only support axial loads. Truss members are produced independently and arranged typically in arrays of triangles for stability [2]. The members are manufactured using extruded tubes made of composite, metallic, or sheet metal materials. A stable truss is statically determinate and has no excess members to introduce alternate load paths. Trusses are generally mass-efficient when the members are configured into rectangular or triangular cross-sectional assemblies. However, they become less efficient as the cross-section becomes more circular or hexagonal. Buckling is typically the critical failure mode for trusses. Also, the design of the structure creates stress concentrations at interface mounting points, such as separation systems. Components may be mounted both internally and externally, and the absence of shear panels enables easy access to a payload. However, this absence of shear panels is not helpful to spacecraft requiring body mounted solar cells. Machining a full side of the truss from a single piece of metal is usually more economical than fabricating and assembling individual members.

A *Monocoque cylinder* is an axi-symmetric shell that does not contain any stiffeners or frames [2]. The shells are manufactured using metallic or sandwich panels with curved sections formed by rolling. Typically, two or three curved sections are fabricated and assembled into the cylindrical configuration. The strength of monocoque cylinders is usually limited by its buckling strength. The shells are most efficient when the loads are distributed evenly throughout the structure. Components are typically mounted to the walls using fasteners; however, care must be taken not to overload the shell and cause local failure. To achieve a uniform load distribution, the mating structure must be either another monocoque cylinder or a stiff transition adapter. The monocoque cylinder design is applicable to spacecraft with body mounted solar cells and relatively lightweight components. A solid-skin cylinder made of sheet or plate metal is the simplest and least expensive structure, but solid-skin cylinders are normally suitable only for stiffness-critical designs because of low buckling stress. The use of sandwich construction results in a light structure, and isogrid shells can also be made at relatively low weight.

Cylindrical skin-stringer structures are designed using axial and lateral frame members attached to an outer skin [2]. These designs are similar to skin-frame structures; however, this class of structures refers to circular cylinder configurations. The skin is sometimes minimized to save mass, even though the thin skin leads to some structural instability. The post-buckling behavior of the skin transfers the additionally applied shear loads to torsion by the diagonal tension phenomenon described above. The skin and members must attach uniformly to enable the assembly to act as a continuous structure. Typical connection methods include fasteners and/or rivets. Internal access can be difficult, so it may need removable panels or framed cutouts. Interior components are usually mounted to the walls at locations along the stringer assembly. This method is more efficient than monocoque cylinder component mounting in introducing local loads. The skin must be designed sufficiently stiff to enable mounting of exterior entities such as body mounted solar cells.

1.3.2 Materials

Satellite structural designs use several different materials [2]. Materials are chosen based on their performance characteristics, cost, and complexity. There are two typical materials used in space applications: metal alloys and advanced composite materials.

1.3.2.1 Metal Alloys

Many available metal alloys are suitable for spacecraft; each has desirable characteristics for a specific design. Aluminum alloys are light and strong, and are used virtually for every type of structure, including skins, truss members, and brackets. But titanium, beryllium, and stainless steel often appear somewhere in a spacecraft because of their unique properties.

Aluminum alloys are the most commonly used metallic materials in spacecraft manufacturing. The advantages are high strength to weight ratios, high ductility, ease of machining, weldability, and availability at low cost. The stiffness to weight ratio is comparable to steel; however, the strength to weight ratio is typically higher. The disadvantages are low hardness, low strength to volume ratio, and a high coefficient of thermal expansion (CTE). Aluminum alloys are typically tempered to increase the material strength. Two typical alloys used in manufacturing are 6061-T6 and 7075-T7. Aluminum 6061-T6 contains silicon and magnesium which strengthens the alloy during tempering. This alloy has good machinability and corrosion resistance. Aluminum 7075-T7 contains zinc and trace amounts of magnesium. The alloy exhibits higher strength than 6061-T6, but is more difficult to machine.

Titanium and titanium alloys are used for applications requiring very high strength materials or high strength at high temperatures. The materials have high strength to weight ratios, low coefficients of thermal expansion, and excellent corrosion resistance. However, they are hard to machine, and some alloys have poor fracture toughness. Ti-6Al-4V, which contains 6 % aluminum and 4 % vanadium, is the most popular titanium alloy used in aerospace applications. The alloy has heritage in wings and missile bodies. Perhaps its most famous applications are the castings used to connect the external fuel tank to the space shuttle and its boosters.

Beryllium is used for very high-stiffness aerospace applications. Its specific stiffness, which is the ratio of Young's modulus to density, is about six times that of most metals. The material is nonisotropic due to its grain alignment, and therefore has low ductility and fracture toughness in its short-grain direction. It is commonly used in lightweight optics and mirrors because it performs well at cryogenic temperatures (i.e., low coefficient of thermal expansion and high thermal conductivity). However, beryllium is expensive, difficult to machine, toxic, and requires special machining equipment. Beryllium parts are machined in a controlled environment because its powder is a known carcinogen when inhaled. The parts may be safely handled once machined.

Steel is mainly used in aerospace applications where low-volume strength and stiffness are important. It has a broad range of strength, hardness, and ductility. Steel provides high wear resistance; is generally easy to machine, and wieldable. However it is not efficient for structural stability because it provides low buckling strength vs. weight. Steels are combined with many trace elements to address a wide range of needs. Austenitic stainless steel is by far the most abundant steel alloy used in spacecraft. It contains 12 % chromium which results in a tough chromium-oxide coating that protects parts from corrosion. Stainless steels are generally used for fasteners and mechanisms whereas many heat-resistant alloys are used for heat shields, rocket nozzles, and other high-temperature applications.

1.3.2.2 Advanced Composite Materials

Composite materials consist of a matrix and reinforcement. The matrix (metal, epoxy) binds the reinforcing fibers (carbon, graphite) together into a continuous system. The efficiency of composite structures is due their high specific modulus and unique load path. The flexural shear loads are transferred from the matrix to axial loads on the high-strength fibers, creating a stiffer and lighter structure. These fibers may either be discontinuous or continuous entities. Discontinuous reinforced composites comprise ceramic or fiber particles that are randomly distributed throughout the matrix. Aluminum reinforced with silicon carbide particles is the most widely used discontinuous composite.

The majority of continuous fiber composites are generally called laminate composites. Laminate composites are manufactured from several layers of woven fibers called laminae. The laminae are composed of several parallel fibers arranged in sheets. The sheets themselves are anisotropic and have few structural applications. However, stacking several of the laminae with fibers aligned at different angles, called lamina angles, creates a more stable laminate composite structure. The laminate may be customized for individual applications by varying the fiber type and the lay up. For example, some graphite/epoxy laminates are modified to have a nearly zero coefficient of thermal expansion, and others may be laid up to exhibit extraordinary specific stiffness properties.

Polymer-matrix composites (PMCs) are the most widely used continuous fiber composites in spacecraft. The matrices consist of two polymers: thermoplastics, and thermosets. Thermoplastics may be melted and solidified multiple times, whereas thermosets are not reusable after curing. These properties enable a multitude of bonding techniques and lay up procedures, many of which are currently proprietary.

The downside of fiber composite structures is the large development cost required for reliable manufacturing. The large cost is due to the sensitivity of adhesive bonding to process variables. This sensitivity makes each part a unique entity which must be tested to verify strength. Extensive testing is required for fiber composite flight hardware to verify its structural integrity. The requirements typically create the need to fabricate a protoflight structure dedicated to qualification testing followed by acceptance testing of the light article. This qualification procedure presently prohibits the use of fiber composite structures on small-budget programs.

1.3.3 Methods of Attachment

To attach structural modules, mechanical fastening hardware (such as bolts and rivets), welding, and adhesive bonding are used. In selecting from these options and designing the attachments, the following considerations should be considered [1]:

- Type of structure and application
- Materials of the joining parts
- Joint strength
- Joint stiffness
- Fatigue life
- Alignment
- Access

- The need for disassembly
- Cost (materials, parts, tooling, manufacturing labor, verification)
- Schedule (availability of materials and parts; manufacturing time)
- Quality assurance and reliability

1.3.3.1 Fastening

Types of fasteners are countless, but it can be divided into broad categories. Most consist of a load-bearing shaft and a head or nut on each end. This type includes threaded fasteners and rivets. Mechanical fasteners are readily available and often inexpensive. They have standard sizes and thread geometry. They can also be installed in almost any facility without expensive tools or process controls. They add damping to a structure. However, they have a big weight and loss of stiffness is hard to be avoided. Premature failure in brittle materials can appear due to stress concentration introduced at fastener locations. Most fasteners can be easily disassembled, and are good for dissimilar materials. They are suitable for most structures and mechanisms made of ductile materials.

1.3.3.2 Welding

Welding is the process of joining two materials—usually with a filler material—by applying intense heat and sometimes pressure. With the right materials and suitable conditions, welding reliably achieves a strong, stiff joint. Often, welding is the most economical method of attachment, although it is limited to joining similar, weldable metals. The key disadvantages of welding are that it is a permanent form of attachment. It can cause distortions and cracks in some materials especially aluminum. Welding quality varies with the process, so workmanship must be developed and tested.

1.3.3.3 Adhesive Bonding

Adhesive bonding is widely used for joining structural parts, particularly for bonding face sheets to honeycomb, and joining polymer-matrix composites. It is also used to electrically isolate solar cells from support structure, and separate dissimilar metals that are susceptible to galvanic corrosion. A key advantage of adhesive bonding is that loads distribute over the entire joined region instead of locally at fasteners. As a result, it can often save weight, and the lower stress concentrations can lead to longer fatigue life. Adhesive can add structural damping, and join dissimilar materials that cannot be welded. However, it cannot be disassembled and has limited shelf life. Some adhesives are toxic and require good ventilation.

1.4 Methods of Enhancing Structural Efficiency

Several methods are available to enhance the structural properties of spacecraft [2]. The most suitable method may vary depending on the design task. Three of the most widely used methods are presented: sandwich structures, multifunctional structures, and isogrid. All of the technologies and their benefits for satellite structural designs are examined.

1.4.1 Sandwich Structures

Sandwich structures are often used in skin-frame designs and solar panels. A sandwich structure consists of two thin face sheets attached to both sides of a lightweight core. The design of sandwich structures allows the outer face sheets to carry the axial loads, bending moments, and in-plane shears, while the core carries the normal flexural shears. Sandwich structures are susceptible to failures due to large normal local stress concentrations because of the heterogeneous nature of the core/face sheet assembly. Component mounting must therefore use potted inserts to distribute the point loads from connections. Sandwich panel face sheets are usually manufactured using aluminum or graphite/epoxy composite panels. The core is typically manufactured using a honeycomb or aluminum foam construction.

Honeycomb sandwich paneling is the lightest option for compressive or bending loading in specific applications. Honeycomb sandwich cores are manufactured using thin strips formed into honeycomb cells. The honeycomb geometry is nonisotropic, with greater stiffness in the longitudinal direction. However, the core acts nearly isotropically for in-plane loads when assembled in a sandwich configuration. The disadvantages of using honeycomb cores are the potted inserts required for mounting and the thermal inefficiencies. These inefficiencies stem from the low thermal conductivity of the adhesive layers used in construction, and make honeycomb prohibitive in optical and mirror aerospace applications.

Aluminum foam sandwich panels use a porous aluminum foam material for the core. The flexural shear stiffness dominates the overall panel stiffness for relatively small panels (i.e., less than 50 inches). Therefore, the core design is an integral part of the sandwich panel design for small spacecraft. The shear stiffness of foam core sandwich panels is generally less than that of honeycomb core sandwich panels of equal mass. However, radial ribs and shear rings may be embedded in the core to overcome the low shear stiffness. A major benefit of aluminum foam construction is an increase in thermal efficiency because the core may be brazed to aluminum face sheets rather than epoxies. Brazing provides a continuous thermal path through the material, which benefits applications such as cryogenic mirrors and solar arrays.

Fig. 1.1 Iso grid Structure [2]

1.4.2 Multifunctional Structures

Multifunctional structure (MFS) technology includes several functions into the primary structure of a spacecraft. The main objective of these members is to minimize parasitic mass by incorporating chassis, cables, connectors, and thermal control components into the satellite primary structural walls. The walls are typically constructed out of fiber composites or sandwich panels, and the electrical components are embedded during manufacturing. The traditional ground plane/ printed circuit board design is performed by copper/polymide (CU/PI) patches, multi-chip modules (MCMs), and the current cabling functions are performed using CU/PI flexible jumpers. The design allows for an easily accessible, removable, and modular electrical system. The benefits of this technology include a 70 % reduction in electronic enclosures and harnesses, a 50 % reduction in spacecraft volume required for these conventional components, a reduction in labor required for spacecraft assembly, and an extremely robust system with wide applicability to several missions. Lockheed Martin has recently proven the technology as an experiment aboard the Deep Space 1 mission.

1.4.3 Isogrid Structures

Isogrid uses an array of equilateral triangle cutouts to increase the stiffness per weight of a structure. The pattern may be manufactured by machining a metallic panel, or it may be constructed using fiber composite materials. The concept began in the early 1960s using metal structures and development continues today with research focusing primarily on composite applications (Fig. 1.1).

References

1. Sarafin TP, Larson WJ (eds) (1995) Spacecraft structures and mechanisms—from concept to launch. Microcosm Press and Kluwer Academic Publishers, Torrance
2. Stevens CL (2002) Design, analysis, fabrication, and testing of a nanosatellite structure. Master thesis, Virginia Polytechnic Institute and State University, Virginia

Chapter 2
Satellite Configuration Design

Abstract This chapter discusses the process of integration of the subsystem components and development of the satellite configuration to achieve a final layout for a satellite; the process will be applied on a test case and it is called "Small Sat". The Small Sat structural configuration is designed to accommodate all of the mission components. All mechanical requirements are derived from the satellite's configuration. The process used to create the satellite configuration of Small Sat is described. It begins with mission definition, launch vehicle selection, and subsystem identification. This is followed by a description of the satellite composition, and the major design constraints that guide the configuration design. Then a configuration development process is presented to create the preliminary configuration. Finally, the issued layout drawings and the calculated mass properties for the developed satellite are presented.

2.1 The Process of Configuring a Satellite

The first step in designing a satellite, once its top level requirements are identified, is to define (at least roughly) the orbit and the payload's function, field of view, required power, mass, and size. From the payload's features, the satellite's total mass and volume can be estimated based on the data collected from previous missions. This information allows us to select a launch vehicle, which dictates the allowable physical envelope of the stowed satellite.

Before we have a preliminary configuration, identifying and trading options are begun to answer many questions related to the design process, like the method of satellite control, the communication system, the need for a propulsion system, and the total power estimated, which determines the solar panel surface area and the battery size. These and many other questions in designing a satellite are not

G. F. Abdelal et al., *Finite Element Analysis for Satellite Structures*,
DOI: 10.1007/978-1-4471-4637-7_2, © Springer-Verlag London 2013

straightforward. The answer of one depends on several or all of the solutions to the others. Often, we cannot find the best answers to the above questions until we try to configure the satellite. However, to start developing the initial configuration, answers can be estimated to the above questions, so that the key components and their critical characteristics may be identified. By doing so, a preliminary equipment list, which includes information such as quantity, size, mass, and the required power for each component, is generated.

Using this list, the launch vehicle's payload envelope, identified fields of view for sensors and antennas, and basic packaging guidelines, arranging the components, and tying them together with structural load paths can be begun. The resulting configuration is just a starting point for a string of iterations. The process of developing a preliminary satellite design is summarized in Fig. 2.1.

The information needed to begin developing a satellite configuration is concerned with all major design elements which have an effect on configuration. The first significant element is the payload, which is the starting point for satellite design and usually the heaviest components. It is characterized by its size, weight, power, data rates, field of view, thermal interfaces, and other constraints. It determines the satellite attitude, and most probably uses a lot of power. Another element having great effect on a satellite configuration is the mission, which is distinguished by its orbit, reliability, design life, operations concepts, and mission constraints. Orbit defines satellite environments and power-gathering capabilities, while reliability and design life influence the number of components and component size.

Launch vehicle has very important effect on satellite configuration design. It is characterized by environments and constraints which contain envelope, mass properties, fundamental frequencies, and access. The stowed envelope can derive the need for complex deployment mechanisms. Data relay and communications also affect configuration design. They specify the frequency, data rate, hardware losses, and receiver station characteristics. Antennas may need special locations for fields of view, and the transmitter typically must be near the antenna. Another element is attitude control approach, which is categorized into spin-stabilized, 3-axis, and gravity gradient. The control types require different types of actuators and affect the configuration in different ways. Subsystems have great influence on satellite configuration design. Key components must be defined early, and minor components can be added as the configuration matures. Schedule and cost limit the development of technology, so risks, schedule, cost, and technical function must be considered.

Table 2.1 describes a general process for configuring a satellite [1]. Because of unique requirements and equipment, no single process applies to all satellites, but this one should be effective for most programs. The products from this process are:

- Layouts of stowed and deployed configurations, showing the arrangement of equipment and the main structural load paths
- An equipment list that summarizes quantity, size, mass, and power for each component

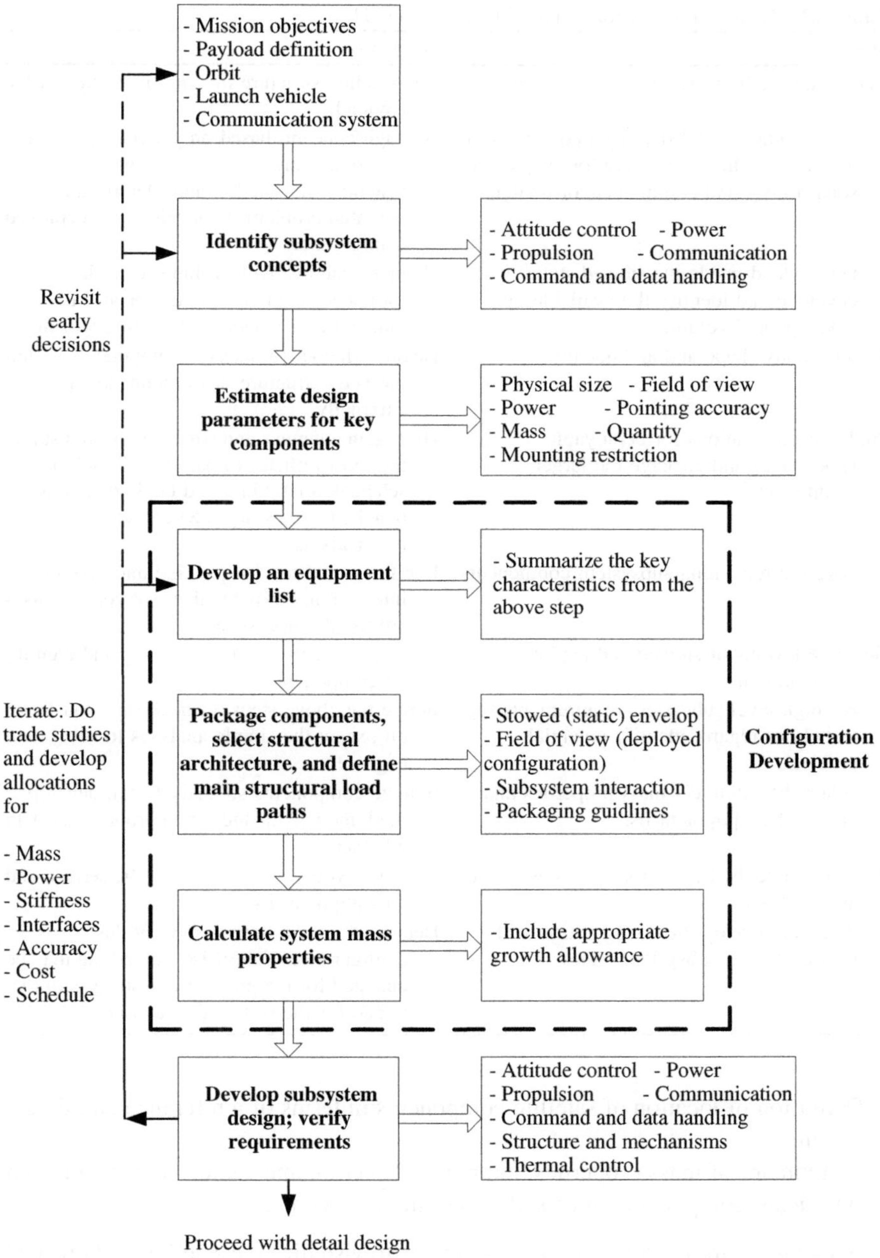

Fig. 2.1 The process of developing a preliminary satellite design [1]

Table 2.1 General process for configuring a satellite [1]

Step	Discussion
Determine the best location for the payload	The satellite structure function is to support the payload
Sketch a "quick-look" deployed configuration based on the fields of regard for the payload, solar arrays, and communications antenna	A rough concept, based on general guidelines for component locations, allows us to visualize the satellite and identify any potential problems in developing a deployed configuration
Fit the payload inside the stowed static envelope and identify the available bus envelope and volume	Compare the available volume with the estimated required volume for an early indication of whether everything will fit
Select a body shape and architecture	Decide whether to package components within the body structure or to mount them externally
Find stowed locations for deployable appendages and package the larger components	High-gain antennas and solar arrays are usually the most difficult to package. Develop schemes for folding and deploying solar panels, if necessary. Identify any needed mechanisms
Package the remaining subsystem components	Use the guidelines for packing and system integration, but recognize that compromises are usually necessary
Generate layouts of stowed and deployed configurations	Make reasonably detailed drawing and identify all components
Assess high-level subsystem requirements such as field of regard; identify potential problems.	Iterate the above steps, as necessary, but leave all except the simple analyses to subsystem engineers
Calculate the satellite's mass properties and update the equipment list	Itemize components so analysts can develop math models. Include an appropriate growth allowance
Release the configuration for subsystem trades and analyses	Provide layouts, tabulated mass properties, and an equipment list
Continue to develop the configuration with feedback from subsystem trades	Decide as a team how to modify the configuration. Otherwise, something may be changed for the good of one subsystem that is bad for the rest of the satellite

- Definition of location of satellite components in terms of a reference coordinate system
- A summary of mass properties, moments of inertia, and center of mass for each significant component, and for the satellite as a whole.

This information allows program designers to visualize the satellite and proceed with subsystem sizing and trade studies. Usually, a program develops more than one configuration to enable trade studies. Developing a satellite configuration has no right answer. With multiple iterations and by considering requirements, cost, and schedule, a capable design team will converge to a configuration that is best for the program. This always results in compromises: for the best system, each subsystem may not be ideal. Reliability and cost are two key considerations in this process, which means we strive for simplicity, the fewest parts, the use of previously qualified components and proven technology, and producible design.

2.2 Mission Definition

The design and size of any satellite are highly dependent on the mission goals. Small Sat satellite is intended for earth observation missions. The results of the Earth remote sensing missions are used to find solutions for many problems in several fields. The most informative remote sensing methods are related to an observation by optical unit. Space images with high resolution are of a great interest for national economy and science, because they make possible to compose the detailed maps and track the slightest changes taking place on the Earth. Data acquisition of Earth optical-electronic observation is useful for information support of economic activity, which include agricultural problems, land use, construction activity, environment pollution monitoring and estimation, and manufacture of digital locality maps. It is helpful also for finding solutions for scientific problems.

Most earth–observation missions require low-earth orbits. The payload for Small Sat satellite is a very precise optical unit to image the earth's surface. Mission is intended to cover all the area of Egypt by taking images. To develop a conceptual configuration for Small Sat, mission requirements are identified according to objectives and purposes. Table 2.2 summarizes preliminary mission requirements for Small Sat, which are typical of the information available at the start of the conceptual design.

2.3 Satellite Functions

To perform the mission requirements, the satellite performs the following functions:

- Acquisition and transmission of telemetry and signal information and data files to the ground control station
- Reception of command-program information from the ground control station
- Pointing the satellite optic-electronic equipment to certain Earth's surface areas
- Imaging of certain Earth's surface areas
- Coding of information of images obtained and transmission to the ground station

Table 2.2 Small Sat preliminary mission requirements

Mission Related Orbit:	668 km at 98.085° inclination
Design life:	5 years
Communication relay:	Ground station in Egypt
Coverage:	Local area of Egypt
Payload Instrument:	Multi band earth imager
Size:	0.45 m diameter by 1.1 m length
Weight:	45 kg mass
Power:	100 watts when operating
Resolution:	2.5 m
Payload instantaneous field of view:	Nadir viewing with a half angle of 2°
Payload field of regard:	half angle of 80 ° from Nadir
Pointing accuracy:	±0.25 °
Position knowledge:	±1 km
Launch Vehicle: Dnepr	
Small Sat allowable mass band (includes launch vehicle adapter):	200–300 kg
Spacecraft Derived Requirements Control:	3-axis (because of off-nadir viewing)
Payload duty cycle:	Approximately 12 min per orbit
Programmatic considerations:	Low cost with minimal development

2.4 Launch Vehicle Selection

At present, the following methods of orbital injection for small satellites are employed in world practice:

1. Single (solitary) launching with the help of a small launch vehicle
2. Series (group) launching of several satellites with the help of one launch vehicle:
   - Launching as the additional payload together with the main satellite
   - Series launching of the satellites of the same class, "cluster launch"
3. Separation from the main satellite, "baggy back"

In the process of selection, it should be taken into account that the small satellite under development will function on a circular sun-synchronous orbit with altitude of 668 km and mass band of 200–300 kg. Therefore, satellite launching from the main satellite is not acceptable, as the disadvantages related to the latter can affect the launching latency (waiting) time. In addition, the orbit of the main satellite specifies the small satellite orbit. Single launching by using small launch vehicle, like Pegasus, is also not accepted, because costs of the launching services are thoroughly included in the satellite launching costs.

The best way to minimize launch costs is using a launch vehicle which deals with series launching. The most famous launchers in this category are Arian 4, Arian 5, Delta 2, Delta 4, Taurus, and Dnepr. Using Delta 2, Delta 4, and Taurus

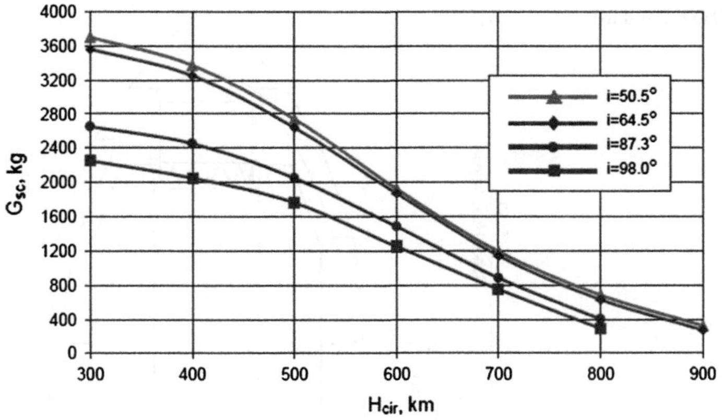

Fig. 2.2 Mass of LV payloads to be injected by Dnepr into sun-synchronous orbit [2]

launch vehicles require a modification in their interface configuration to provide the possibility of launching of a satellite 200–300 kg, which is unacceptably costly. Arian 4, Arian 5, and Dnepr provide the possibility of launching a satellite of 200–300 kg without modification. All launch vehicle types, except Taurus, assure appropriate orbiting accuracy.

Figure 2.2 [2] shows the total mass of the launch vehicle payload to be injected by Dnepr LV into sun-synchronous orbit at different inclination angles. From the figure, the allowable payload mass of Dnepr LV at an altitude of 668 km with inclination 98° slightly exceeds 800 kg, which is suitable to launch a series or group of small satellites. After comparison based on the above discussion and cost criteria, Dnepr Launch Vehicle is found to be the suitable one to launch Small Sat.

For Dnepr Launch Vehicle, the spacecraft "Small Sat" is installed inside the space head module (SHM). The SHM is composed of the fairing, cylindrical intermediate section, adapter, protective membrane, and gas dynamic shield (GDS) or encapsulated payload module (EPM). Layout schematic of the standard length SHM (with both GDS and EPM) is shown in Fig. 2.3. SHM design allows for multi-tier spacecraft layout. One of the options for such layout is shown in Fig. 2.4.

2.5 Satellite Composition

A satellite consists of a payload, which is the mission-specific equipment, and a collection of subsystems [1]. A subsystem is a group of components that support a common function. There is a difference between the payload and the rest of the satellite subsystems, because the payload is typically unique for a given mission, whereas the other subsystems may be able to support different missions. In the next section, a closer look is provided at essential subsystems, focusing on features and

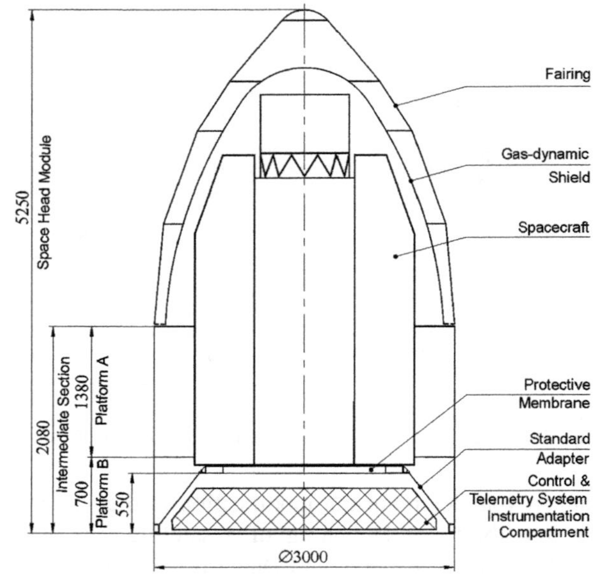

Fig. 2.3 SHM standard length [2]

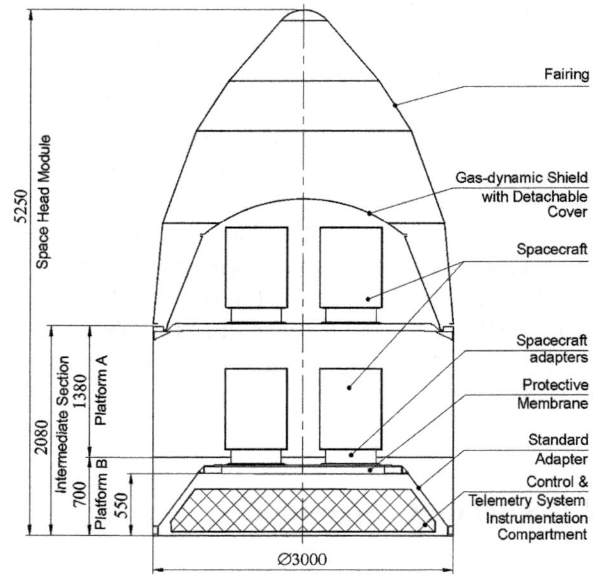

Fig. 2.4 SHM with 2-tier Layout [2]

components that most influence Small Sat's configuration. Satellite consists of the following subsystems:

1. Payload
2. Attitude determination and control subsystem (ADCS)

3. Communications Subsystem
4. Platform command & data handling subsystem (PCDHS)
5. Power Subsystem
6. Thermal Subsystem
7. Structures and Mechanisms Subsystem

Table 2.3 shows the initial equipment list for Small Sat. Quantity, physical size, and mass in (kg) of each component are included. The selection of each component depends on the previous discussion of satellite functions and subsystems identification. The mass shown for the launch vehicle adapter is an estimate. Satellite structural modules include the primary (body) structure, brackets of equipments, and mechanical fastening such as bolts, nuts, and rivets. Their estimated mass is (41 kg), which is about 20 % of the satellite total mass not including the launch vehicle adapter. This is a reasonable estimate based on historical averages.

2.6 Mounting Restrictions and Integration Constrains

This section provides guidelines for arranging a satellite's components, and explains how subsystems affect the satellite configuration. These guidelines can be considered as requirements, so they should be taken into consideration during the configuration process of Small Sat.

2.6.1 Payload

The payload of Small Sat is a multi-band earth imager (MBEI), which is a high precision electromechanical optical unit. This type of payload needs key requirements; often include field of view, pointing accuracy, stability, and thermal isolation. From the previous data mentioned in Table 2.3, MBEI is heavy and large, thus it is the main component affecting the configuration design. Because MBEI requires a field of regard, the most common location for it is the forward end of the satellite, opposite the interface to the launch vehicle. Although MBEI is heavy, this location is often chosen because

- It is easier to provide a clear field of view at this end
- It is sensitive to shock, so it is kept away from ordnance at the LV separation interface
- Structural load during launch is highest at the LV interface, and it is hard to keep large and sensitive payload out of the primary load path

Table 2.3 Small Sat initial equipment list

Subsystem and component	Quantity	Size (mm)	Total Mass (Kg)
Payload			
Multi-band earth imager	1	D 450 × 1100	45
Payload CDH unit	1	230 × 220 × 160	7.2
MEI signal processing unit	2	300 × 200 × 120	3.7 each
ADCS			
Star sensor	1	375 × 215 × 184	4
Angular velocity meter "Gyro"	4	D 150 × 47	1 each
Interface unit for each gyro	4	101 × 81 × 56	0.92 each
Magnetometer	1	150 × 90 × 90	1.5
Magnetorquer	3	D 22 × 170	0.38 each
Reaction wheel	4	195 × 195 × 89	3.3 each
Communications subsystem			
X-band equipment			
X-band electronic module	1	380 × 315 × 60	3.8
X-band antenna	1	D 243 × 120	1.6
S-band equipment			
S-band electronic module	2	380 × 315 × 30	2.2 each
S-band conical antennae	2	D 100 × 112	0.27 each
S-band dipole antenna	1	D 100 × 44	0.13
GPS receiver			
GPS electronic module	1	380 × 315 × 30	1.1
GPS antenna	2	D 70 × 55	0.15 each
Platform CDHS			
On-board digital computing complex	3	380 × 315 × 38	3.7 each
Telemetry module	1	210 × 155 × 95	2.8
Power subsystem			
Battery cell module	1	430 × 280 × 130	16.5
Power-conditioning unit (PCU)	1	380 × 315 × 65	3.6
Cells leveling unit (CLU)	1	380 × 315 × 40	1.9
Solar array panels	4	3.2 m^2 total area	6.8
Cabling	set	–	1.5
Thermal subsystem			
Heat shields	–	TBD	3.6
Insulation, coatings, and sensors	set	–	1.5
Structure and mechanism subsystem			
Satellite structural modules	–	TBD	41
Rotation mechanism	4	TBD	1.7
Locking and releasing mechanism	4	TBD	0.5
Separation transducer	2	TBD	0.1
Satellite total mass			205
Launch vehicle adapter	1	TBD	20
Total mass (including LV adapter)			225

All objects must stay out of the payload's field of view. The only practical way to orient the payload to its target is to rotate the satellite. This is usually the simplest approach, because fixed mounting of the payload is more easier than using a gimbaled mechanism. A high precision MBEI has requirements for accurate pointing. This means the mounting structure must be stiff and provides direct load path into the satellite's primary structure. Structural distortions between the payload and the ADCS sensors must be minimized. Distortions can result from on-orbit structural vibration, on-orbit thermal effects, and any yielding or joint shifting during launch or ground operations. Making the mounting structure stiff avoids problems from on-orbit vibration and lunch effect. Thermal deformation can be controlled by selecting the right materials and by controlling temperatures.

2.6.2 *Attitude Determination and Control Subsystem*

The selected method of control drives the satellite's shape. The satellite configuration, in turn, can derive the types and sizes of actuators. Small Sat is preferred to be symmetric, this will reduce aerodynamic drag and solar radiation pressure, hence a net torque. To minimize this torque, the Small Sat's center of mass should be as close as possible to its center of pressure, which is the centroid of the satellite's projected area. This is provided also by creating a symmetrical front area, so four solar arrays, symmetrical about the satellite's center of mass, will be used. Symmetry also reduces gravity-gradient torques, as does a compact shape.

The configuration of tree-axis control satellite, like Small Sat, is the most severe constraint for ADCS and structural design. Making appendages of Small Sat as short as possible makes it easier to keep natural frequencies above the control system's bandwidth. This will avoid resonance phenomena which lead to structural fracture.

The star sensor of Small Sat requires a narrow field of view, so it must be protected from any obstacles. Bright sunlight can damage the star sensor or causes it to shut down. Therefore, the star sensor mounting will be turned by a certain angle to protect it from sunlight. The Magnetometer must be installed at enough distances from high magnetic field components like ADCS actuators, reaction wheels, and magnetorquer. Alignment is very important for ADCS sensors, so they are grouped on one platform, which is stiff and thermally stable to reduce errors from distortions.

For reaction wheels, a common approach is to align them with the satellite axes and add a wheel at the critical axis to provide redundancy. If any one wheel fails, the redundant wheel can compensate. The Y-axis shown in Fig. 2.5 is the most critical one for the stability of Small Sat, so the redundant wheel is installed on the Y-axis. The same approach is followed for the angular velocity meters "gyros", but the redundant one is added at a skewed axis. The configuration of Small Sat

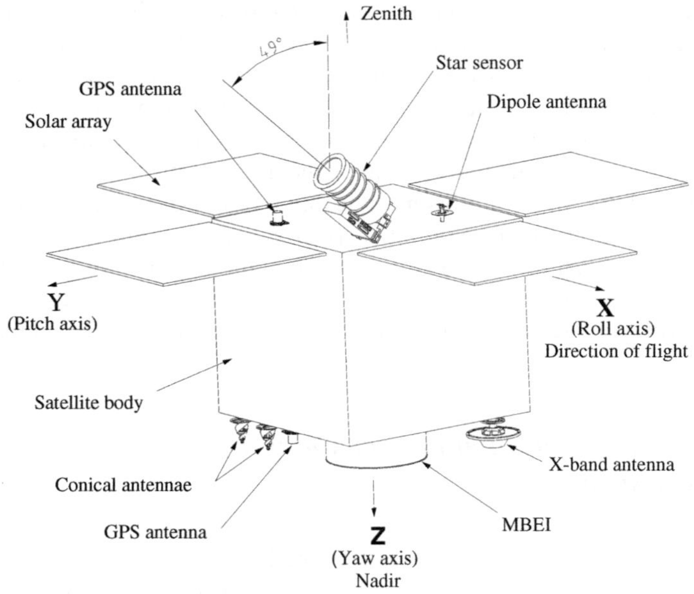

Fig. 2.5 Quick-look for on-orbit configuration of Small Sat

must be developed with a proper mass distribution to provide stability conditions. Therefore, the moment of inertia about the critical Y-axis must be greater than their about the velocity direction axis "X-axis", which is also greater than their about the nadir "Z-axis".

2.6.3 Communications Subsystem

The communication components important to the configuration designer are antennas and power amplifiers. All antennas of Small Sat require a clear field of view. The S-band omni antenna consists of one conical antenna and a dipole antenna, and is used to ensure initial ground communications regardless of the satellite's orientation. So one of them is mounted at the aft end and the other is at the opposite side. The second S-band conical antenna is mounted at the forward end to provide in-orbit communications with the ground station. A high gain antenna of Small Sat "X-band antenna" is mounted also at the forward end of the satellite. The GPS receiver antenna consists of two similar antennas; one of them is installed at the aft end and the other at the opposite side.

Another key consideration is the proximity of the power amplifier to the communications antenna. The amplifier of each antenna in Small Sat is mounted at the related electronics module. To reduce signal losses, each electronics module is installed as close as possible to its antenna. This also leads to minimize the length of the coaxial cables. Brackets are used to mount all antennas except the X-band antenna, because the wave pattern is affected by the distance between the antenna and its mounting surface.

2.6.4 Platform Command & Data Handling Subsystem

The electronic modules of PCDHS in Small Sat, especially on-board digital computing complex (ODCC), are important for the configuration designer. These modules are dense and therefore heavy, so the best location for mounting them is near the aft end. PCDHS equipment will be electrically connected to virtually all of the satellite's nonstructural components. By grouping electronics, cabling losses and mass can be minimized.

2.6.5 Power Subsystem

Small Sat configuration is strongly influenced by the power subsystem components, especially the solar arrays. The design of solar arrays is based on the satellite's power requirements, the orbit altitude, sun-angle conditions, the method of attitude control, and mission and payload requirements. For Small Sat, fixed solar panels mounted on the satellite body surfaces are not used, because Small Sat needs relatively high power with respect to the available surface area. Heat rejection can be another problem of using fixed solar panels. Therefore, four deployed-fixed solar panels are used to supply power for Small Sat. A deployed-fixed solar panel is one that is stowed in one location for launch, and then deployed to a fixed position in space. Rotation mechanisms are used to rotate solar panels and provide fixation into specific positions in space. Locking and releasing mechanisms are needed to fix the solar panels during launch, and then release them at space.

In defining deployed locations for solar panels, shadows from other components should be avoided. Therefore, in Small Sat, solar panels and rotation mechanisms are mounted at the aft end. This also reduces the overall structural loading by keeping the mass of both solar panels and rotation mechanisms near the launch vehicle interface. This minimizes the cable runs to battery, which is also mounted near the aft end of the satellite. Flat solar arrays made of lightweight honeycomb sandwich are the most common and easiest to manufacture. Solar arrays are major contributors to a deployed satellite's modes of vibration, so these should be very light and stiff, with natural frequencies high enough to avoid interaction with the control system. During launch, acoustics combined with transient loads usually cause the highest loads in the solar panels and mechanisms.

The best location for the battery is dictated by weight, temperature sensitivity, and cabling. The battery of Small Sat is heavy, so it should be packaged as near as possible to the launch vehicle interface. The battery also needs a location with temperature that is uniform and somewhat low (5–20°) to maximize the depth of discharge. Thus, it must be protected from direct exposure to the sun or earth. Because battery generates heat during use, it needs a lot of radiator area to maintain low temperatures. The battery is mounted near large power consumers and near the solar arrays to minimize cabling losses and weight. The power subsystem electronic components in Small Sat are the power-conditioning unit (PCU) and cells leveling unit (CLU), which control and distribute power. They are typically dense and heavy, so the aft end near battery is the best location to mount them. Cabling of all satellite subsystems is rather heavy. The main target of reducing cable mass can be achieved during the configuration process by mounting the interact components as close as possible in a compact space, and by co-locating items with many interconnections. The configuration should provide access for installing cabling and connectors. When locating components, free spaces must be provided for the necessary bends of cables and mate electrical connectors.

2.6.6 Thermal Subsystem

Designing the thermal control subsystem begins with the satellite's configuration. Our goal is to use passive thermal control. Doing so requires proper location of powered satellite components and effective use of radiators, insulations, and coatings. The design of Small Sat configuration aims at achieving that goal. The best location for heat-generating components and radiators is the side of the satellite with the least sun exposure. Also for low earth orbit, like Small Sat's, heating can be minimized by shading components from planetary emissions and facing radiators away from earth. Therefore, heat shields are used in Small Sat to cover and protect the internal components from environmental effects.

2.6.7 Structures and Mechanisms Subsystem

The configuration of a satellite's primary structure can be characterized by its architecture, type, and the packaging scheme. This section introduces alternate architectures and packaging approaches. Chapter 1 describes types of structures, materials, and attachments. The shape of the body's cross-section characterizes the body architecture, which is characterized also by whether the body is open or closed. Cylindrical, square, rectangular, hexagonal, and cruciform cross-sections have all been used for satellites.

Open-architecture configurations, which include frames and trusses, have satellite equipment mounted externally on structural members or panels. Closed-architecture configurations enclose the equipment within the body structure. The best type of body architecture depends on the mission and the available packaging volume. Mechanisms are also a major consideration in configuring a satellite. They must be designed to perform their functions under hostile conditions without maintenance. Mechanisms add complexity and risk, so their number should be reduced and they should be kept as simple as possible.

2.6.8 Systems Aspects of the Satellite Configuration

The system requirements and constraints that influence a satellite's configuration are reliability, design life, maintainability, cost, schedule, and environments. To satisfy reliability requirement, which is specified from customers, the program allocates higher reliability values to the subsystems and key components, such as mechanisms. The target reliability can be achieved by using high-grade (space) components and providing redundant or backup components. Redundancy will at most influence the configuration simply because of the extra components.

Satellites have a range of design lifetimes, which depends on the satellite mission and orbit. As design life increases, solar arrays area and battery capacity must grow. Design life also affects structures and mechanisms, but usually more in details than in features that affect the satellite configuration. The maintainability of a satellite is the ability to access or service its components during integration and test. This requirement should be taken into account during configuration development, as well as cost and schedule.

Finally, launch and space environments drive the sizes of structural members and strongly affect the satellite configuration. Sometimes satellite configurations appear to be ideal from the nonstructural subsystems point of view, but it is very difficult to design a structure for these configurations which withstand launch loads without being too heavy. For Small Sat, many of the guidelines mentioned above in this section will conflict with one another. Therefore, subsystem concerns must be compromised to optimize the satellite or the system, which means finding the best design given all program considerations. The goal is to arrive at a cost effective design with compromises that do not affect or risk mission objectives.

2.7 Configuration Development Process

In this section, a conceptual configuration for Small Sat will be developed. To perform this, Fig. 2.1, which summarizes the general process of developing a preliminary satellite design, should be followed. Section 2.2 through Sect. 2.6 discuss the initial data and requirements needed to begin developing Small Sat configuration. Table 2.2 summarizes the preliminary mission requirements for Small Sat, and Table 2.3 summarizes the initial equipment list. Now the process is to package components, select suitable structural architecture, and define main structural load paths. This will be done generally by following the steps in Table 2.1, which describes a general process for configuring a satellite. Products of this phase will be layouts of stowed and deployed configuration. The calculation of mass properties will be discussed in Sect. 2.8. Normally, the conceptual design phase results in several configurations, but only one will be presented to limit work efforts.

From Table 2.2, there are no outstanding requirements that will dictate a revolutionary design or new technology. From Table 2.3, the total predicted mass of Small Sat is 225 kg, including launch vehicle adapter, which is within the allowable mass band (200–300 kg) and leaves a high margin, based on the Dnepr's payload capability of 800 kg for Small Sat selected orbit. The initial equipment list indicates that there are some assumptions already made regarding the satellite's deployed configuration. The 3.2 m^2 of solar-array area is based on the assumption of deployable-fixed solar arrays. Thus, rotation mechanisms are needed to deploy and fix solar arrays in space.

2.7.1 A Quick Look at On-Orbit Configuration

Using this information and the payload requirements, a "quick-look" can be sketched for on-orbit configuration, as shown in Fig 2.5. Because the MBEI is heavy and bulky, it is located at the middle of the satellite and directed to the earth "Nadir", which provides a clear field of view. This location makes the mass distribution as symmetric as possible. Moreover, it enables mounting the payload directly along the primary load path, which reduces the shock effect and distributes structural loads uniformly during launch.

Since the high gain antenna (X-band antenna) communicates through a ground station, it needs to be fixed at the forward end and directed to the earth "Nadir". A dipole antenna of the S-band equipment and one of the GPS receiver antennae are mounted at the aft end to be directed to "Zenith", which is the opposite direction of "Nadir". The other GPS receiver antenna and two conical antennae of the S-band equipment are mounted at the forward end to be directed to "Nadir". Using symmetric solar arrays about the satellite's center of mass minimizes environmental disturbances. They will be most efficient if they protrude from the satellite near the aft end along the axis perpendicular to the orbit plane. Determination of how many solar array panels should be used depends on the configuration shape, method and location of stowed panels, and mass properties of the final configuration. Four solar arrays with 3.2 m^2 total area are assumed to be mounted on the initial configuration. To provide symmetrical shape, each two solar arrays located at opposite sides are identical.

The star sensor requires a narrow field of view to identify the relative location of certain stars, so it is located at the aft end and directed toward the horizon. The star sensor mounting is turned by 49° from Zenith direction in the positive Y-axis to protect it from sunlight. This quick-look configuration establishes only the general placement of major external components. It does not address structural load paths, the shape and size of the solar panels, or the satellite's physical dimensions and its internal arrangement.

2.7.2 Packaging Envelope

The equipment list (Table 2.3) reflects the need for redundancy of certain items in order to achieve the required design life with high reliability. The MBEI is relatively bulky and large, and the solar arrays require considerable surface area. All these factors indicate that packaging volume in the Dnepr launch vehicle will probably be a driving consideration. Thus, the stowed configuration should take the first attention.

Dnepr launch vehicle is designed to perform series launching for several small satellites. Hence, Small Sat will be mounted inside the Dnepr fairing envelope with several other satellites. The main goal during packaging the satellite is to minimize its volume and design it as compact as possible. For Dnepr launch vehicle, the payload "satellite" envelope is a volume within the SHM, which is designed for accommodation of spacecraft. Spacecraft dimensions (including all of its protruding elements) must fit within the specified payload envelope, given all possible deviations and displacements from the nominal position during ground testing and flight phases. The size of the payload envelope within the standard SHM is shown in Fig 2.6.

2.7.3 Body Shape

The main considerations in selecting a body shape for Small Sat are [1]:

Packaging consideration:

- Enough volume to contain the subsystem components
- Ability to package appendages as well as the body within the fairing

Structural considerations:

- Efficient structural load paths between the payload and launch vehicle
- Compatibility with the payload and launch-vehicle mechanical interfaces

In general, a body with a large cross-section is better for equipment packaging, whereas a narrow body makes it easier to stow the solar arrays and simplifies the design of the launch-vehicle adapter. For Small Sat, a large cross-section is selected, because it is more effective in fixation of payload, which is heavy and bulky. In addition, it reduces the bending loads at the launch vehicle interface. Moreover, it improves the fundamental frequencies and the mode shapes of the

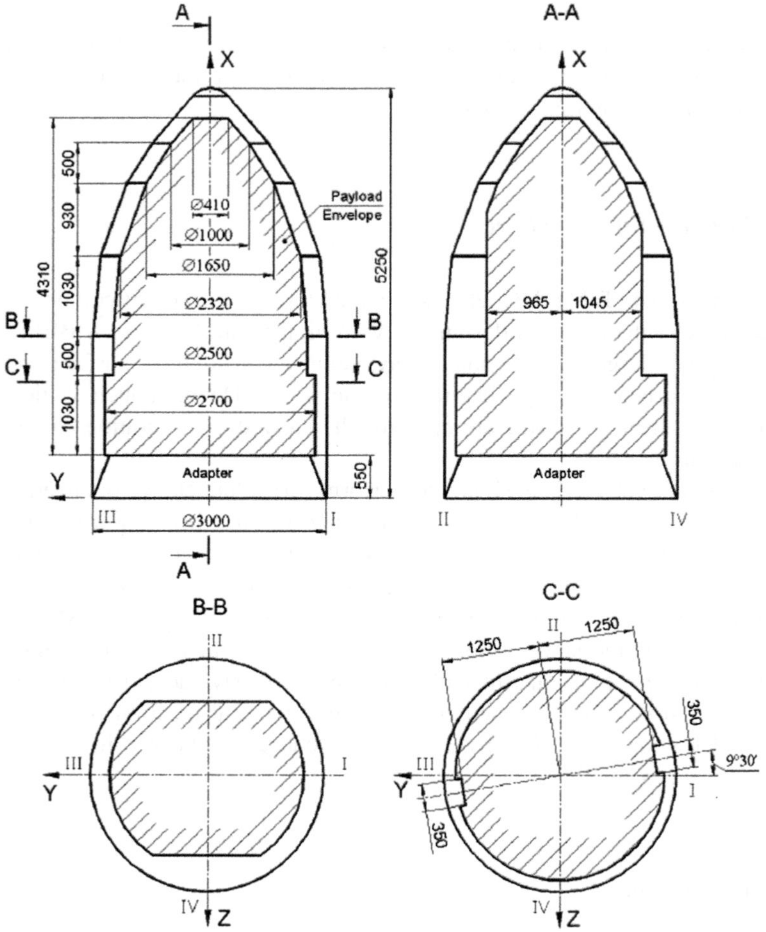

Fig. 2.6 Payload envelope available within SHM with standard adapter [1]

satellite primary structure. Because the packaging volume is tight, a combination between open and closed architecture will be used for the body structure, which will more efficiently use volume. This type of architecture combines the advantages of both open and closed one. It provides greater bending stiffness for Small Sat because of its wider cross-section. Moreover, components can be mounted internally and externally on structural members to provide the best arrangement with minimum volume.

Several possible body shapes can be used as a packaging envelope. Circular, square, rectangular, hexagonal, and cruciform cross-sections have all been proposed or used for satellites. The first criterion for selection is that the shape must be able to contain the largest packaged components, which for Small Sat are the MBEI, Battery, and the electronics modules. All options except cruciform pass this test.

A circular shape will also be a difficult choice because components require flat mounting surfaces. In addition, it will be more difficult to package flat solar arrays on a cylindrical body. The hexagonal shape is reliable, but is more complex in configuration design. Moreover, it cannot provide the minimal volume criterion for Small Sat, because it produces relatively large unused spacing inside the configuration envelope. Thus, only square and rectangular shapes can be considered to provide Small Sat packaging in minimal envelope. However, they present structural problems at the adapter interface where launch loads are highest. These problems can be solved during structural design phase by designing a suitable launch vehicle adapter with sufficient number of fixation connections. The selection between square and rectangular shapes depends on the packaging approach.

2.7.4 Packaging Approach

The next step is to find the packaging approach that will provide the most surface area for mounting components. From the previous discussions, the Small Sat primary structure is a combination between open and closed architecture with square or rectangular shape, so components can be mounted internally and externally on structural members. The best packaging option is enclosing the MBEI, which is the largest component, within the primary structure, while other components can be mounted externally on structural members. The primary structure for Small Sat consists of the main load path structure, which is covered by two plates at the aft and forward ends. Figure 2.7 illustrates the primary structure of Small Sat.

The main load path structure encloses the payload, so it should take suitable shape and dimensions to provide mounting the payload inside it and the rest of equipment outside. The square shape is the best choice for the main load path. The two plates covering the main load path provide enough surface area to mount the external components. The first plate, which connects the main load path structure to the launch vehicle adapter, is called the base plate, while the other plate at the forward end is called the mounting plate. This plate should contain a suitable hole to pass the MBEI forward end. Packaging the rest of equipment on the main load path structure decides the final shape and dimensions of the two plates. The shape can be square or rectangular, while the outer in-plane dimensions should be the same with different thicknesses. The outer surface of the mounting plate carries the components that are directed to the earth. These components are the X-band antenna, one of the GPS receiver antennae, and two conical antennae of the S-band equipment. On the other hand, the outer surface of the base plate, which is connected to the launch vehicle adapter, carries the other GPS receiver antenna and the dipole antenna of the S-band equipment. The inner surfaces of both base and mounting plates are suitable areas for mounting other components.

As mentioned before, the total area of the required solar arrays is 3.2 m^2, which is divided into four solar arrays. Each solar array is connected to one side of the

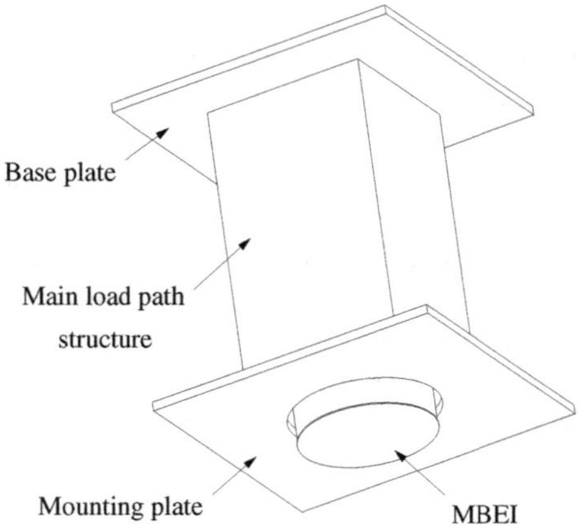

Fig. 2.7 Primary structure of Small Sat

satellite body by a single rotation mechanism, which is fixed at the outer surface of the base plate. Rotation mechanism provides a fixed position for one deployed solar array where the angle between each solar array and the satellite body equals 90°. Stowage of a solar array is done by a locking mechanism, which is mounted at the outer surface of the mounting plate. This is done during transportation and launching process.

At this time, rough concepts have become clear for the payload, satellite body, communications antennae, and solar arrays. Packaging the remaining components is the next step, which is done by using the guidelines presented in Sect. 2.6. Mounting restrictions and system integration constraints should be taken into account during this step too.

A good starting point is to establish locations for the sensitive equipment, which are usually the most difficult to fit. This equipment is the ADCS sensors and actuators, which require accurate mounting positions. Control sensors must be mounted on a stiff, thermally stable platform, and as close as possible to the payload. Thus, a basis block case is used as a stiff and thermally stable platform to group all ADCS equipment and the payload. The components mounted on the basis block case are MBEI, star sensor, four angular velocity meters, four interface units, magnetometer, three magnetorquers, and four reaction wheels. The best location to mount the basis block case is at the middle of the main load path structure. This location provides fixing the payload directly on the main load path. In addition, the star sensor can be directed toward the horizon and protected from sunlight. A proper mass distribution for Small Sat configuration will be provided, which assists stability conditions. The design of the basis unit block should provide mounting requirements and mechanical interfaces with the components. Therefore, it should contain enough surfaces to mount the components on three perpendicular planes. The basis block case consists of the basis plate and four walls connected together to produce an assembled structure as shown in Fig. 2.8

The next step is to present the optimum arrangement of the equipment that should be mounted on the basis block case. The total assembly produced from

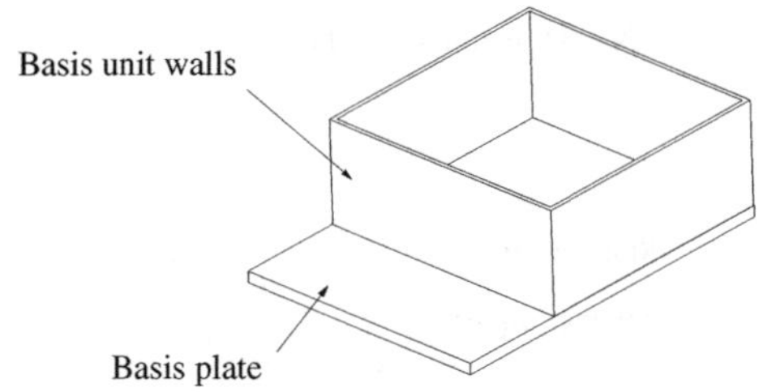

Fig. 2.8 Basis block case of Small Sat

grouping the basis block case and its equipment is called the basis unit block. Computer aided design can be invaluable for identifying interferences and trying various arrangements for equipment, so Mechanical Desktop computer package (MDT) is used to issue the current configurations.

Before starting the configuration process of the basis unit block, there are some constraints which should be taken into account. The star sensor must be turned by 49° from Zenith direction in the positive Y-axis. One of the angular velocity meters (gyro) is redundant, which requires to be mounted at a skewed axis. Therefore, two brackets are designed to provide mounting constraints for both star sensor and the redundant gyro. To minimize the required surface area for mounting equipment, another bracket is used to collect the three pieces of magnetorquers in three perpendicular axes. Each angular velocity meter must be connected to one of the interface units, so each pair is located as close as possible to each other to reduce cabling lengths. The magnetometer must be installed at sufficient distances from high magnetic field components, so a distance not less than 0.6 m must separate it from magnetorquers, and not less than 0.3 m from the nearest reaction wheel. Figure 2.9 shows the final packaging arrangement for the basis unit block of Small Sat.

By reviewing Fig. 2.9, the basis plate carries the star sensor with its bracket, magnetorquers with their bracket, and the Z-direction reaction wheel on one side, while the other side carries MBEI, Z-direction gyro, skewed gyro with its bracket, and interface unit of skewed gyro. The first wall located at the positive Y-axis carries two reaction wheels in the Y-direction, one of them acting as a redundant. The second wall located at the positive X-axis carries the X-direction reaction wheel, X-direction gyro, and the interface unit of the Z-direction gyro. The third wall located at the negative Y-axis carries the Y-direction gyro, and both interface units of Y-direction gyro and X-direction gyro. The magnetometer can be mounted directly on the basis plate or on the third wall with the help of a bracket. The first idea is more reliable because using a bracket will decrease mounting accuracy of the magnetometer. Therefore, the basis plate should be designed to provide high accuracy mounting for the most critical equipments like MBEI, star sensor, and magnetometer. It is clear that there is no equipment mounted on the fourth wall located at the negative X-axis. The reason for this is to produce a free space for other equipment which is relatively big and need special constraints on their locations.

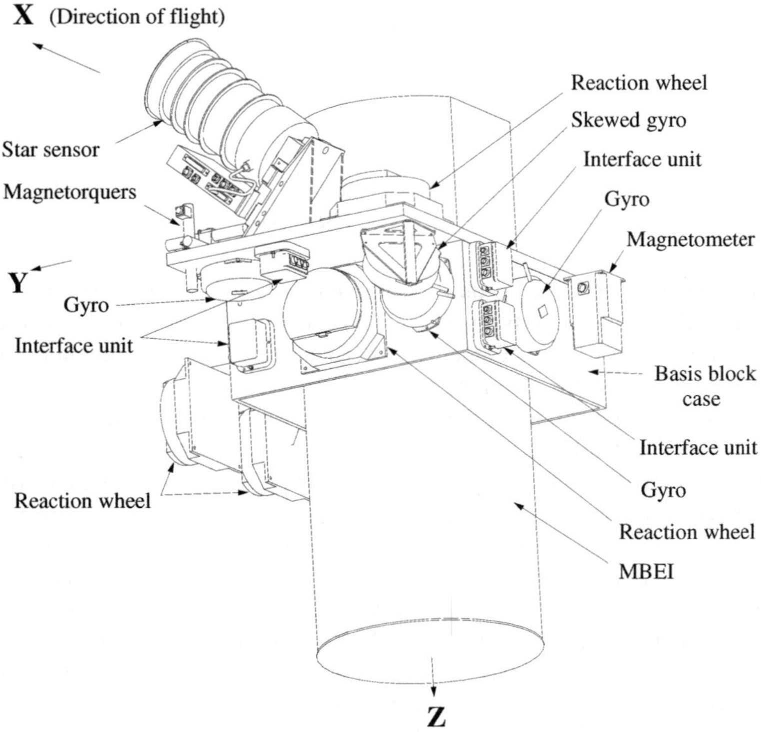

Fig. 2.9 Final packaging arrangement for the basis unit block of Small Sat

The next step is to complete the main load path structure and find suitable surfaces to mount the remaining equipment. Two frames can be used to connect the basis block case with both base and mounting plates, so they have suitable shape and size to enclose the payload inside. Each frame consists of four plates connected together to form a square cross-section, which provides integrity with the basis block case. The first frame that connects the basis block case with the base plate is called the upper frame, while the other that connects the basis block case with the mounting plate is called the lower frame. Both upper and lower frames provide enough area on their external surfaces to mount some of the remaining equipment. In addition, the inner surfaces of both the base and mounting plates can be employed to mount the rest of the equipment. Figure 2.10 illustrates the location of both upper and lower frames.

The battery is considered one of the most difficult equipment to fit inside the satellite because it is heavy, large, and needs a special location protected from direct exposure to the sun or earth. Moreover, it should be packaged as close as possible to the launch vehicle interface and near large power consumers and the solar arrays. The power subsystem electronic component, PCU and CLU, should be mounted as close as possible to the battery. Therefore, Small Sat battery will be

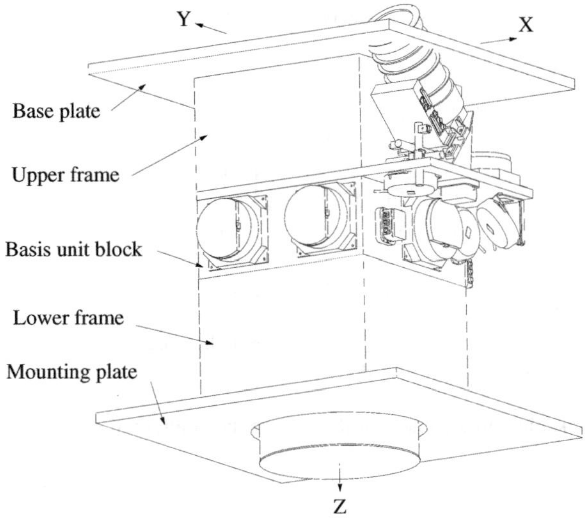

Fig. 2.10 The location of both upper and lower frames

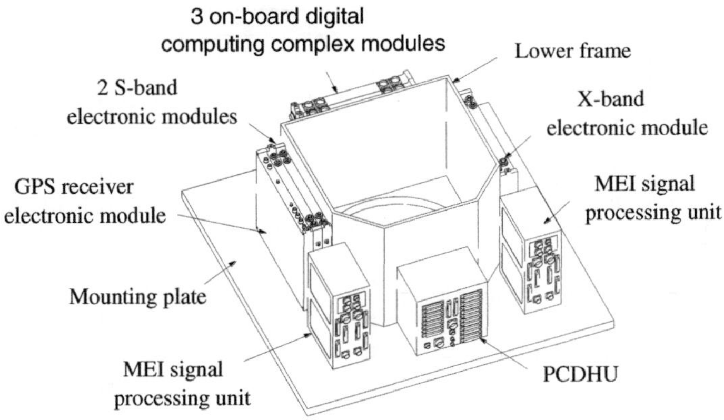

Fig. 2.11 The modified square cross-section lower frame

mounted on one of the external surfaces of the upper frame. By reviewing Fig. 2.10, there are only three faces of the upper frame which can carry the battery because the fourth one is occupied by the star sensor. From the point of view of thermal control, the best location for the battery is on the negative X-axis face of the upper frame. This location provides uniform and low temperature and a lot of radiator area to maintain this condition. The remaining two faces of the upper frame in Y-axis are preferred to be free to make room for the MBEI connectors. To minimize cabling length, PCU and CLU are mounted at the same side of the battery. They are packaged together to save mounting surfaces and provide

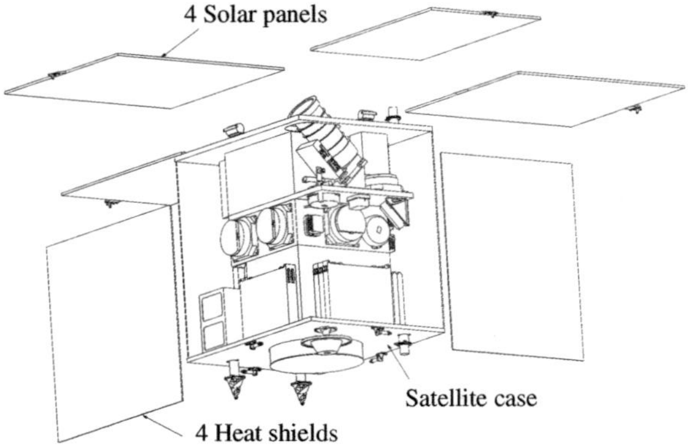

Fig. 2.12 The packaging arrangement of the conceptual configuration

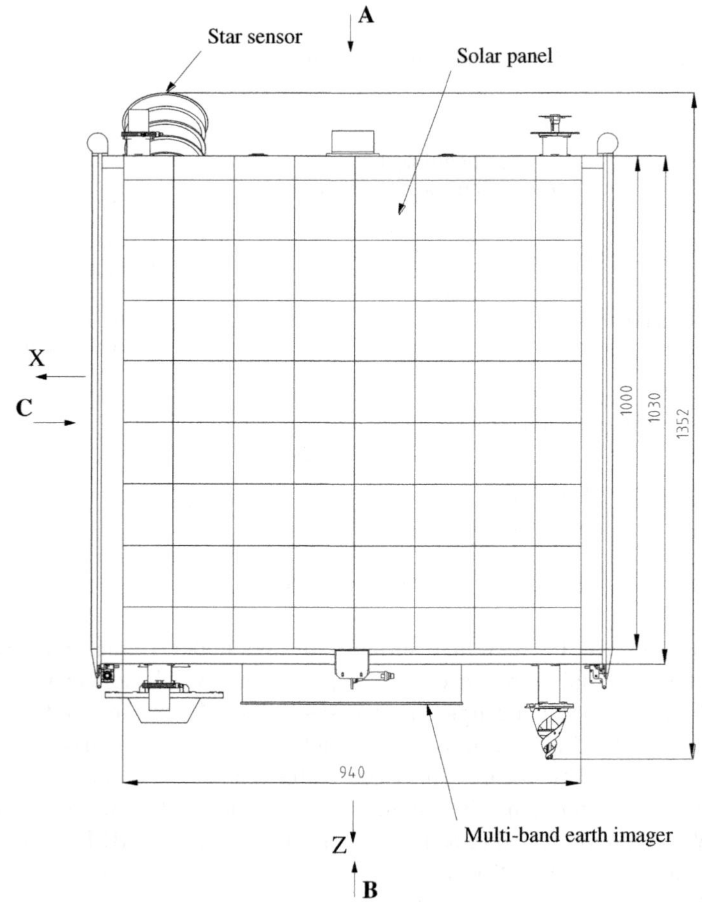

Fig. 2.13 General view of the satellite in stowed configuration in the Y-direction

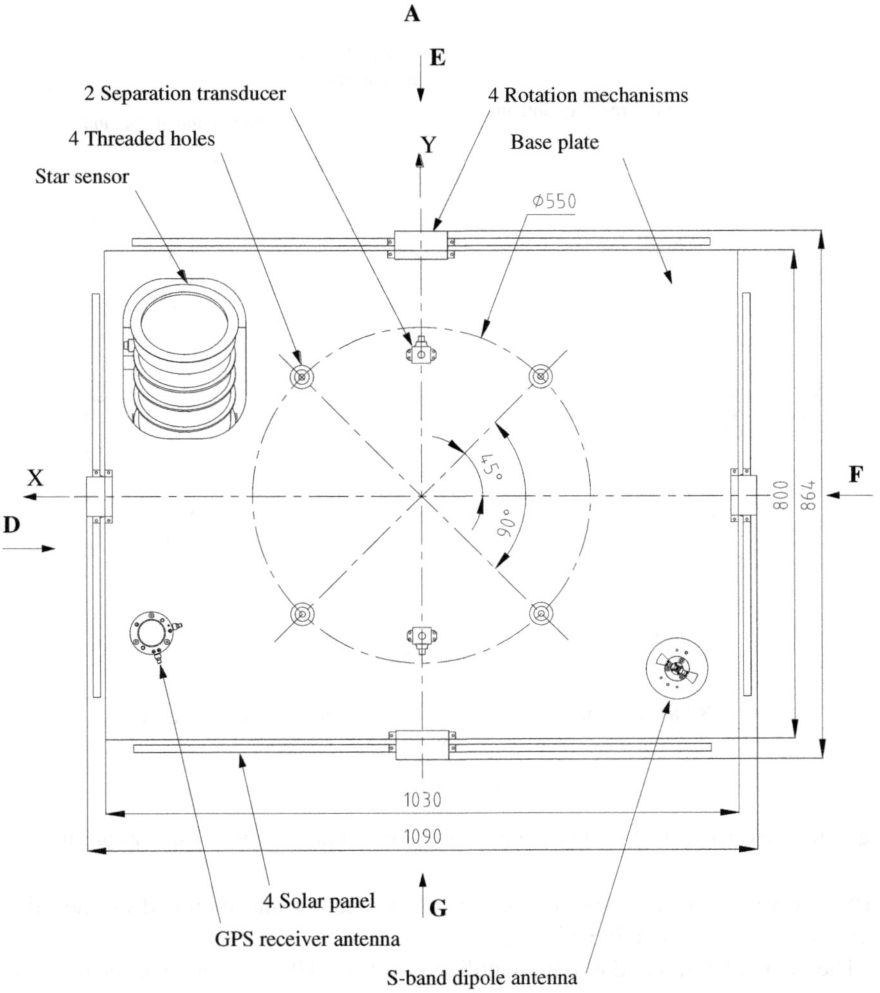

Fig. 2.14 General view of the satellite in stowed configuration in the Z-direction

efficient volume usage. So they are fixed jointly on the basis block case and the lower frame. Hence, the fourth wall of the basis block case is employed.

One of the main objectives in the configuration process is to make the design and mass distribution as symmetrical as possible, so the arrangement of the remaining equipment should follow this concept. Three faces of the lower frame have already been used as mounting surfaces, while the fourth one in negative X-axis is partially occupied by the power subsystem electronic modules, PCU and CLU. The first face located at the positive Y-axis carries the X-band electronic module, while the second face located at the positive X-axis carries the ODCC, which consists of three identical modules. Two S-band electronic modules and

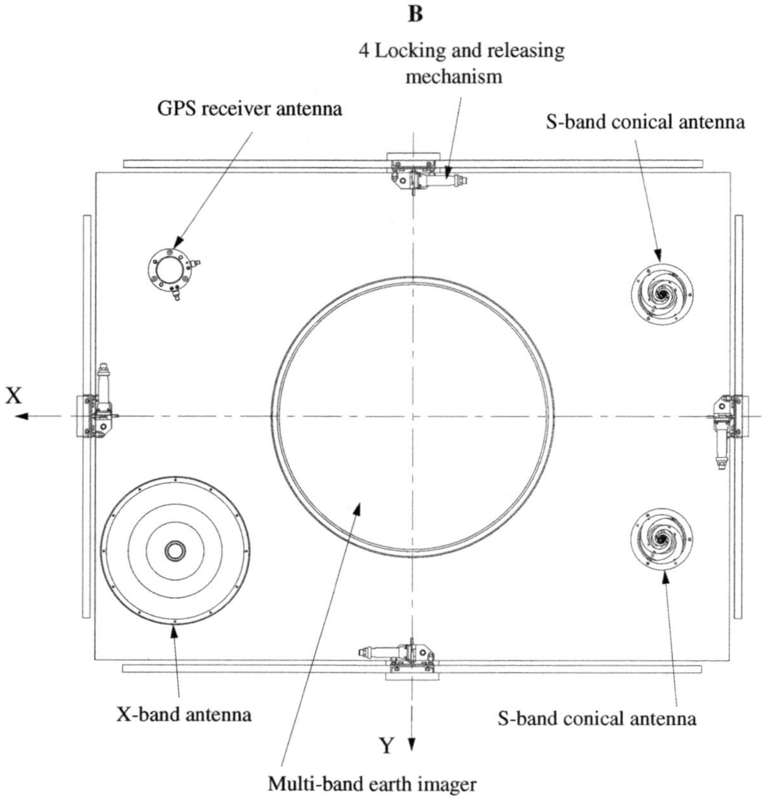

Fig. 2.15 General view of the satellite in stowed configuration in the opposite Z-direction

GPS receiver electronic module are packaged together and mounted on the third face located at the negative Y-axis.

The payload command & data handling unit (PCDHU) and two identical multi-band earth imager signal-processing units should be located as close as possible to each other to minimize cabling length, so they are mounted at the inner surface of mounting plate. They are located at the negative X-axis area of the mounting plate, thus they occupy the remaining area of the fourth side of the lower frame. To provide symmetry, PCDHU is located at the middle between the two MBEI signal-processing units. The shape of the lower frame will be changed from square cross-section to a modified square one to provide minimal occupied volume. Figure 2.11 illustrates the modified square cross-section lower frame. Telemetry module is mounted at the inner surface of the base plate at the positive X-axis area.

After mounting all main equipment inside the satellite, both base and mounting plates have a rectangular shape with the same outer dimensions. Final dimensions of the four solar panels should be selected to be suitable for packaging. Therefore, there are two groups of solar panels; each one of them consists of a pair of panels located at opposite sides. The dimensions of each panel of the first group are

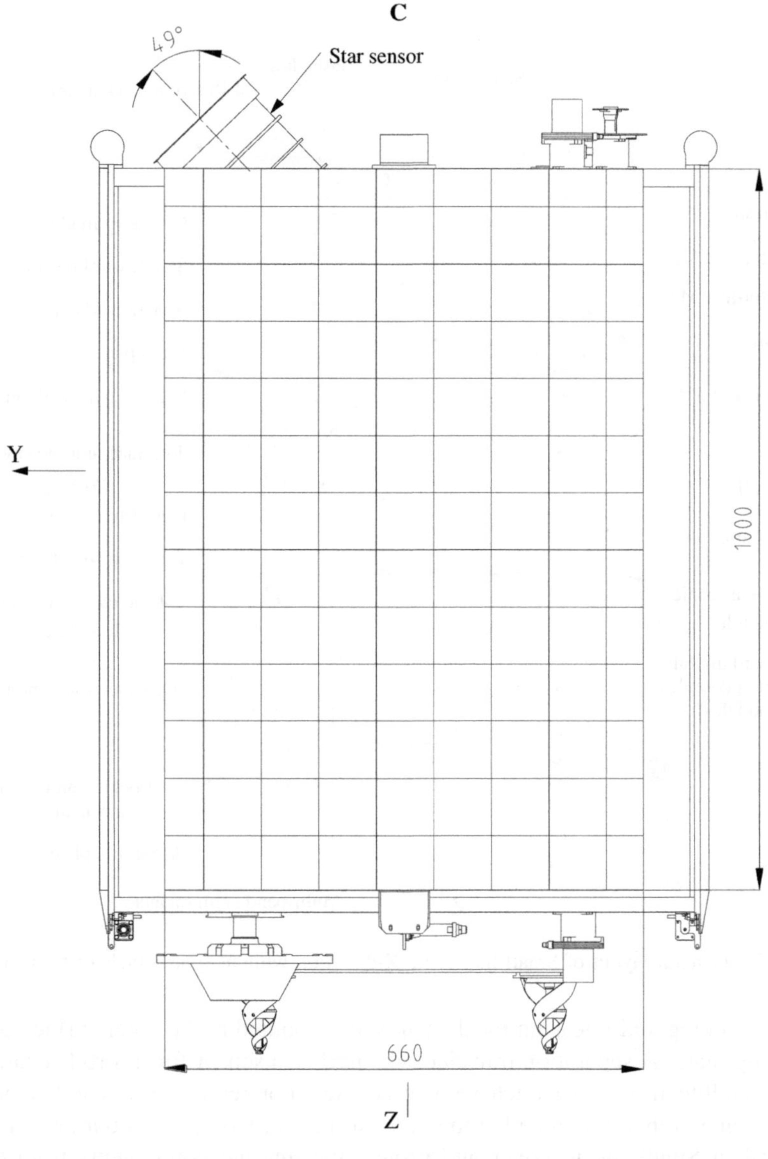

Fig. 2.16 General view of the satellite in stowed configuration in the opposite X-direction

940 × 1000 mm, while those of the second one are 660 × 1000 mm. The thickness of all solar panels is 12 mm. According to the different dimensions of solar panels, two different rotation mechanisms are used. They are mounted at the outer surface of base plate.

D

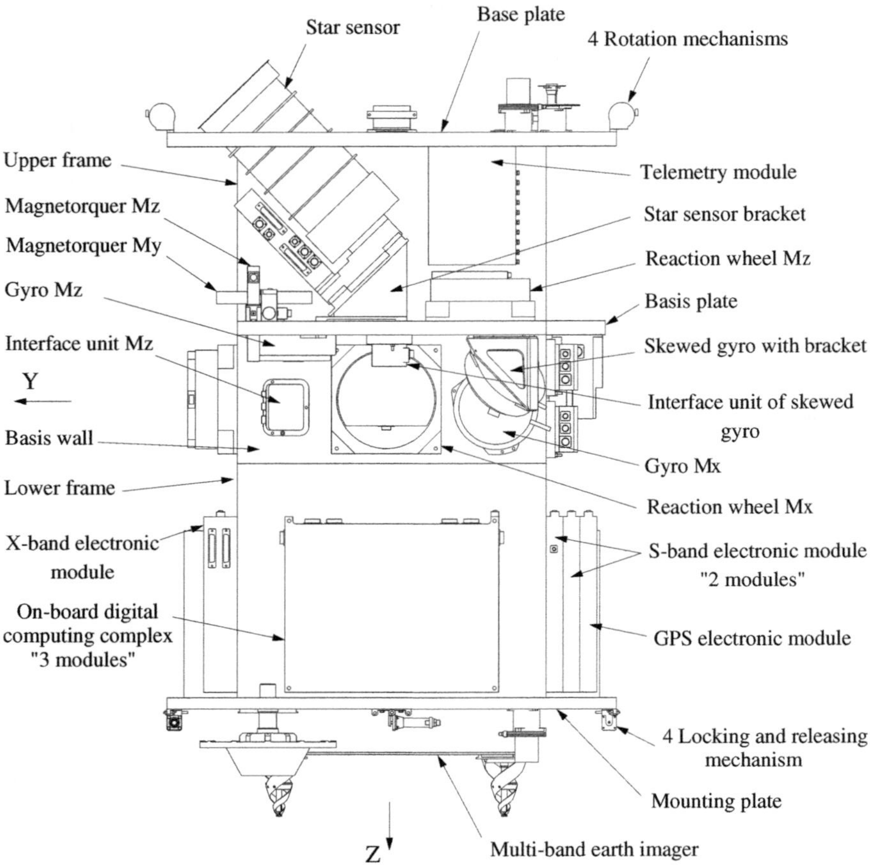

Fig. 2.17 General layout of Small Sat in the X-direction without solar panels or heat shields

Four locking and releasing mechanisms are mounted at the outer surface of the mounting plate. A separation transducer is used as a sensor for in-orbit separation of the satellite from the launch vehicle adapter. For redundancy, another one is installed and both are mounted at the outer surface of base plate. Four heat shields are used in Small Sat to cover and protect the internal components from environmental effects. They consist of two groups of panels to be suitable for covering the rectangular shape of the satellite.

Figure 2.12 shows the packaging arrangement of the conceptual configuration for Small Sat. It shows that a rectangular outer body with a square cross-section main load path can indeed hold the required equipment. The body size and packaging layout provide some room for growth and cable routing. Many of the packaging guidelines conflict, so compromises must be used. For example, all heavy components should be located aft to minimize bending moments at the

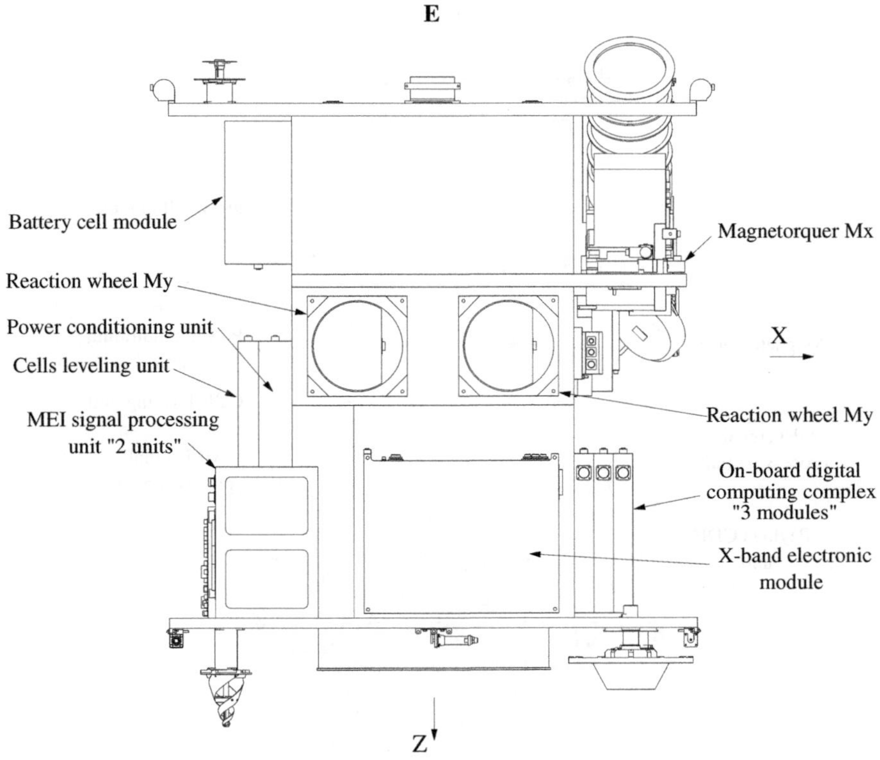

Fig. 2.18 General layout of Small Sat in the Y-direction without solar panels or heat shields

lunch vehicle interface, but the limited packaging volume prevents this. Although the on-board digital computing complex is heavy (11.1 kg), it is mounted at the forward end to balance mass distribution.

Finally, the results of our efforts in the packaging process for Small Sat configuration are presented in the form of configuration layouts. These layout drawings show as many details of the configuration. Moreover, these will become the basis for much effort on the part of subsystem and structural designers. Figures 2.13, 2.14, 2.15 and 2.16 show different views of Small Sat's preliminary stowed configuration. Figures 2.17, 2.18, 2.19, and 2.20 show different views of the general layout of Small Sat without solar panels and heat shields. Figure 2.21 shows different views of the preliminary deployed configuration for Small sat.

2.7.5 Mechanical Interfaces

Satellite configuration design should provide easy access between the satellite and both of the launch vehicle adapter and mechanical ground support equipment

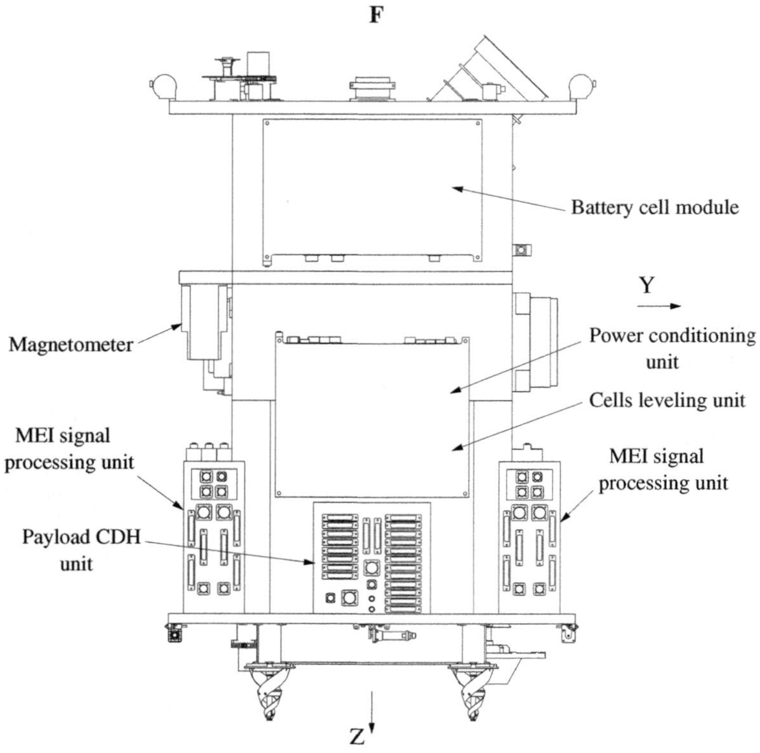

Fig. 2.19 General layout of Small Sat in the opposite side of X-direction without solar panels or heat shields

(MGSE). To provide this, the satellite configuration must enclose suitable mechanical interfaces to meet both launch vehicle adapter and mechanical ground support equipment. Mechanical interface of the satellite with the adapter of Dnepr launch vehicle is realized by pyro-locks screwed from the side of adapter in threaded holes located in the satellite base plate. Dimensions and location of these holes depend on the satellite body shape, main load path, and total weight of the satellite. As mentioned before, Small Sat conceptual configuration has a rectangular outer body with a square cross-section main load path, so four threaded holes are used to conduct mechanical interface between Small sat and launch vehicle adapter. Location of the threaded holes is given in general view of the satellite in stowed configuration in Fig. 2.14. Mechanical interfaces of the satellite with mechanical ground support equipments are provided by using the same four threaded holes located in the base plate. The mounting plate must contain another set of holes to provide mechanical interfaces with rigging devices of MGSE.

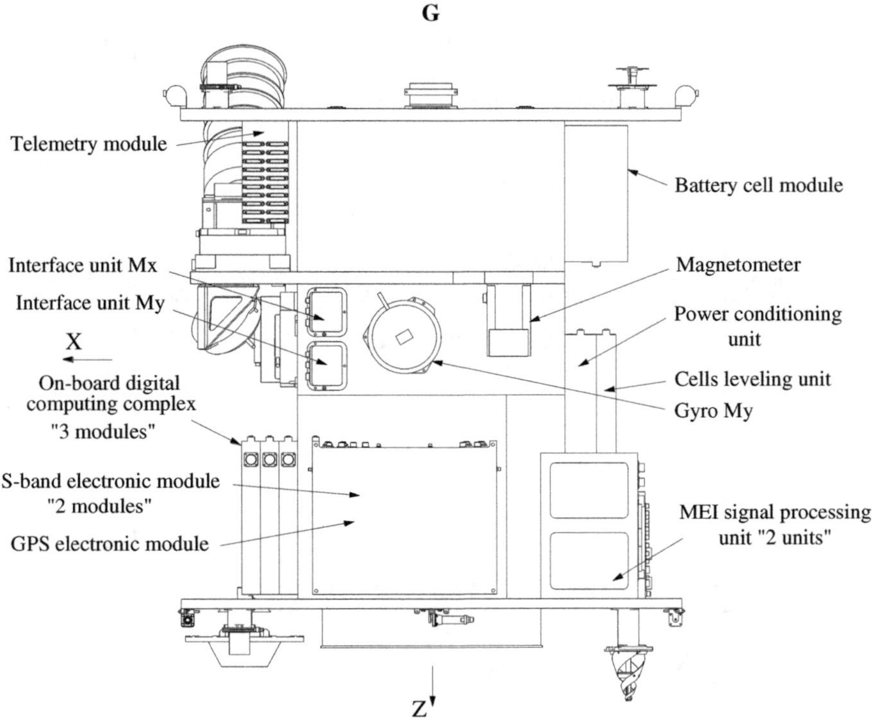

Fig. 2.20 General layout of Small Sat in the opposite side of Y-direction without solar panels or heat shields

2.7.6 Coordinate System

The following rectangular right-hand coordinates are applied.

- Design coordinate system "$O_dX_dY_dZ_d$"

Origin of coordinates is in LV/satellite interface plane on the center lines of the launch vehicle and satellite; it coincides with the geometrical center of the base plate. O_dZ_d axis is perpendicular to the LV/satellite interface plane with +Z toward the nose of the fairing that corresponds to nadir on orbit. O_dX_d axis lies in the interface plane and is pointed to direction of flight. O_dY_d axis lies in the interface plane and supplements the design coordinate system to make it a right-hand one (Fig. 2.22).

- Orbital coordinate system "O $X_oY_oZ_o$"

Origin of coordinates O coincides with the satellite's center of mass. OZ_o is directed along a radius vector which joins the satellite's center of mass and the Earth's center, where +Z in nadir direction. OX_o lies in the satellite's orbital plane and is orbital motion-directed. OY_o supplements the orbital coordinates to make them a right-hand system (Fig. 2.22).

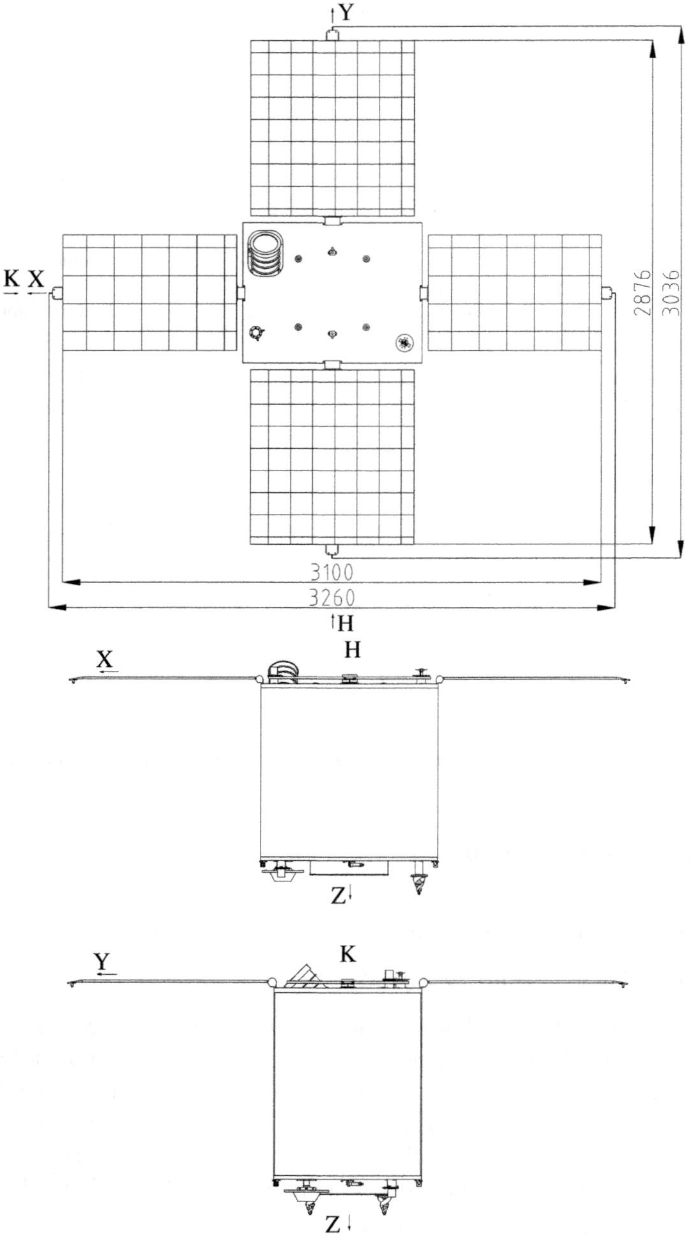

Fig. 2.21 General view of the satellite in deployed configuration

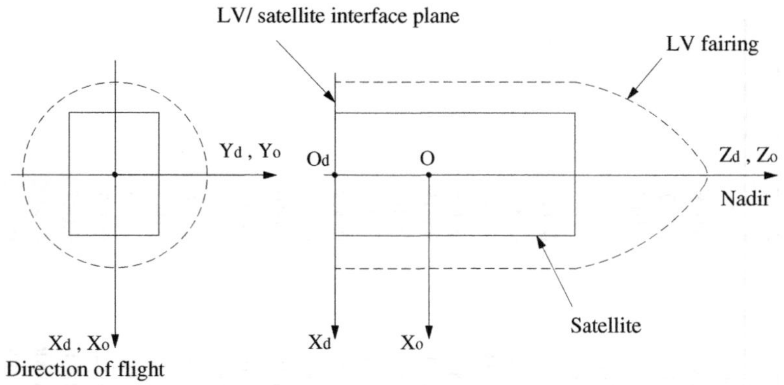

Fig. 2.22 Coordinate systems

2.8 Mass Properties

One of the products of configuration development is a summary of mass properties. This summary should show:

- Mass and centroidal mass moments of inertia of each component about axes aligned with a reference coordinate system
- Coordinates of each component's center of mass
- Mass, center of mass, and moments of inertia of the full satellite in both stowed and deployed configurations

This information is needed for various analyses. The satellite's stowed mass properties are needed to compare with launch vehicle limitations, design structures for launch loading, and predict natural frequencies. The deployed mass properties are needed to support structural and attitude control analyses. And mass properties for individual components are needed to generate math models, size mechanisms, and design secondary and tertiary structures. The calculation of mass properties of Small Sat is done with a computer aided design system using Mechanical Desktop package "MDT". Table 2.4 lists mass properties of each item in the equipments list (Table 2.3) in addition to other components identified during conceptual configuration design. The design coordinate system $O_dX_dY_dZ_d$ is used as a reference to calculate the center of mass coordinates for each component. On the other hand, centroidal mass moments of inertia of each item are calculated about axes aligned with an orbital coordinate system. For Small Sat, the 41 kg predicted for satellite structures are distributed to major structure modules in the configuration, based on rough estimates.

Table 2.5 lists mass properties for the full satellite in both stowed and deployed configurations. Center of mass coordinates are calculated relative to the design coordinate system $O_dX_dY_dZ_d$. Satellite moments of inertia in both stowed and

Table 2.4 Mass properties for each component in Small Sat

Component	Mass(kg)	Coordinates (mm)			Centroidal mass moment of inertia ($kg.m^2$)		
		x	y	z	I_{0x}	I_{0y}	I_{0z}
Multi-band earth imager	45	0	0	400	3.89	3.95	1.24
Payload CDH unit	7.2	−363	0	900	0.061	0.045	0.048
MEI signal processing unit-1	3.7	−328	−310	860	0.032	0.04	0.017
MEI signal processing unit-2	3.7	−328	310	860	0.032	0.04	0.017
Star sensor	4	386	136	140	0.056	0.038	0.033
Gyro M_x	1	297	−185	490	0.003	0.002	0.002
Gyro M_y	1	55	−297	470	0.002	0.003	0.002
Gyro M_z	1	375	175	382	0.002	0.002	0.003
Skewed gyro with bracket	1.2	421	−201	432	0.003	0.003	0.002
Interface unit M_x	0.92	216	−301	529	0.001	0.001	0.001
Interface unit M_y	0.92	216	−301	418	0.001	0.001	0.001
Interface unit M_z	0.92	301	189	489	0.001	0.001	0.001
Interface unit of skewed gyro	0.92	424	3	386	0.001	0.001	0.001
Magnetometer	1.5	−160	−330	431	0.004	0.004	0.002
Magnetorquers with bracket	1.22	420	232	309	0.003	0.003	0.003
Reaction wheel M_x	3.3	314	10	474	0.016	0.010	0.010
Reaction wheel M_y -1	3.3	146	314	475	0.010	0.016	0.010
Reaction wheel M_y -2	3.3	−149	314	475	0.010	0.016	0.010
Reaction wheel M_z	3.3	388	−156	296	0.010	0.010	0.016
X-band electronic module	3.8	56	305	846	0.030	0.075	0.047
X-band antenna	1.6	385	220	1089	0.005	0.005	0.006
S-band electronic module -1	2.2	55	−290	846	0.017	0.044	0.027
S-band electronic module -2	2.2	55	−320	846	0.017	0.044	0.027
S-band conical antenna w/b-1	0.3	−405	−200	1124	0.0006	0.0006	0.0002
S-band conical antenna w/b-2	0.3	−405	200	1124	0.0006	0.0006	0.0002

(continued)

Table 2.4 (continued)

Component	Mass(kg)	Coordinates (mm)			Centroidal mass moment of inertia (kg.m^2)			
		x	y	z	I_{0x}	I_{0y}	I_{0z}	
S-band dipole antenna w/b	0.15	−415	−285	−50	0.0001	0.0001	0.0001	
GPS electronic module	1.1	56	−350	846	0.009	0.022	0.014	
GPS antenna with bracket -1	0.17	439	−225	−45	0.0001	0.0001	0.0001	
GPS antenna with bracket -2	0.17	394	−240	1075	0.0001	0.0001	0.0001	
On-board computer -1	3.7	294	0	847	0.074	0.029	0.046	
On-board computer -2	3.7	332	0	847	0.074	0.029	0.046	
On-board computer -3	3.7	370	0	847	0.074	0.029	0.046	
Telemetry Module	2.8	340.5	144	130	0.016	0.012	0.008	
Battery cell module	16.5	−340	0	169	0.354	0.122	0.278	
Power-conditioning unit	3.6	−307	0	626	0.071	0.029	0.044	
Cells leveling unit	1.9	−360	0	628	0.038	0.015	0.023	
Cabling	15	–	–	–	–	–	–	
Solar panel-1 (+Y)	Stowed	2	0	414	512	0.172	0.313	0.141
	Deployed		0	950	−34	0.172	0.141	0.313
Solar panel-1 (−Y)	Stowed	2	0	−414	512	0.172	0.313	0.141
	Deployed		0	−950	−34	0.172	0.141	0.313
Solar panel-2 (+X)	Stowed	1.4	529	0	516	0.173	0.124	0.049
	Deployed		1066	0	−31	0.049	0.124	0.173
Solar panel-2 (−X)	Stowed	1.4	−529	0	516	0.173	0.124	0.049
	Deployed		−1066	0	−31	0.049	0.124	0.173
Heat shield-1 (+Y)	1	0	397	518	0.082	0.172	0.090	
Heat shield-1 (−Y)	1	0	−397	518	0.082	0.172	0.090	
Heat shield-2 (+X)	0.8	512	0	518	0.103	0.063	0.040	
Heat shield-2 (−X)	0.8	−512	0	518	0.103	0.063	0.040	
Insulation, coatings, sensors	1.5	–	–	–	–	–	–	

(continued)

Table 2.4 (continued)

Component	Mass(kg)	Coordinates (mm)			Centroidal mass moment of inertia (kg.m^2)		
		x	y	z	I_{0x}	I_{0y}	I_{0z}
Base plate	6	–	–	–	–	–	–
Upper frame	6.5	–	–	–	–	–	–
Basis plate	4	–	–	–	–	–	–
Basis walls	6.5	–	–	–	–	–	–
Lower frame	7	–	–	–	–	–	–
Mounting plate	5.5	–	–	–	–	–	–
Star sensor bracket	1.1	385	41	286	0.004	0.005	0.005
Rotation mechanism-1 (+Y)	0.5	0	408	−26	0.0002	0.0004	0.0004
Rotation mechanism-1 (-Y)	0.5	0	−408	−26	0.0002	0.0004	0.0004
Rotation mechanism-2 (+X)	0.35	523	0	−23	0.0002	0.0001	0.0002
Rotation mechanism-2 (-X)	0.35	−523	0	−23	0.0002	0.0001	0.0002
Locking mechanism (+Y)	0.1	20	388	1047	0.00001	0.00008	0.00007
Locking mechanism (-Y)	0.1	20	−388	1047	0.00001	0.00008	0.00007
Locking mechanism (+X)	0.1	503	−20	1047	0.00008	0.00001	0.00007
Locking mechanism (−X)	0.1	−503	−20	1047	0.00008	0.00001	0.00007
Separation transducer -1	0.05	0	232	−17	0.00001	0.00001	0.00001
Separation transducer -2	0.05	0	−232	−17	0.00001	0.00001	0.00001
Fastening	4	–	–	–	–	–	–
Satellite total mass	205 kg						

Table 2.5 Mass properties for the whole configuration of Small Sat in both stowed and deployed configurations

Small Sat mass properties	Satellite stowed configuration	Satellite operating configuration
Mass (kg)	205	
Center of mass (mm)		
X	-1.65 ± 10	-1.65 ± 10
Y	0.51 ± 10	0.51 ± 10
Z	487.26 ± 10	467.72 ± 10
Mass moments of inertia (kg.m^2)		
Ixx	26.69 ± 1	31.04 ± 1
Iyy	33.64 ± 1	37.41 ± 1
Izz	22.09 ± 1	27.93 ± 1
Ixy	−0.01	−0.01
Ixz	0.19	0.19
Iyz	0.06	0.06

deployed (operating) configuration are calculated relatively to axes aligned with an orbital coordinate system. The origin of both coordinates coincides with the satellite's center of mass.

References

1. Sarafin TP, Larson WJ (eds) (1995) Spacecraft structures and mechanisms—from concept to launch. Microcosm Press and Kluwer Academic Publishers, Torrance, CA
2. Dnepr SLS User's Guide, Issue 2, Nov. 2001. http://www.kosmotras.ru

Chapter 3
Satellite Structural Design

Abstract The process to develop the structural design of a "Small Sat" is discussed and applied on a practical design case. It shows how to derive Small Sat's structural requirements from the satellite configuration. Then, it addresses the iterative design procedure used to create the current design. It also discusses the worst load cases to which the satellite is exposed during ground, launch, and space environments. A simplified finite element analysis of the satellite structure is then performed to size the preliminary design of the structure. Finally, the structural module descriptions are presented. In this chapter, the process used to develop the structural design of "Small Sat" is described. It shows how to derive Small Sat's structural requirements from the satellite configuration. Then, it addresses the iterative design procedure used to create the current design. It also discusses the worst load cases to which the satellite is exposed during ground, launch, and space environments. A simplified finite element analysis of the satellite is then performed to size the preliminary design of the structure. Finally, the structural module descriptions are presented.

3.1 Definition and Function

The structure supports the payload and satellite subsystems with enough strength and stiffness to preclude any failure (rupture, collapse, buckling, or detrimental deformation) that may keep them from working successfully. To perform the mission requirements, the structure must perform the following functions:

- Structures support the satellite's key components in desirable locations, considering thermal control, fields of view for antennas and sensors, and lengths and weights of cables. The stowed configuration must fit within the launch vehicle's payload envelope, yet the design must provide access for installing and maintaining components.

G. F. Abdelal et al., *Finite Element Analysis for Satellite Structures*,
DOI: 10.1007/978-1-4471-4637-7_3, © Springer-Verlag London 2013

- Structures protect the satellite components from dynamic environments during ground operation, launch, deployment, and mission operations. They must deploy appendages and provide enough stiffness to keep them steady. But they must also stay light enough for the selected launch vehicle.
- Structural vibration must not interfere with the LV's control system, which must be able to differentiate between motion caused by a misaligned thrust vector and that resulting from booster vibration. Similarly, the satellite's vibration in its deployed configuration must not interfere with its own control system vibration.
- The materials used must survive ground, launch, and on-orbit environments (time-varying applied forces, pressure, humidity, radiation, contamination, thermal cycling, and atomic particles) without rupturing, collapsing, buckling, excessively distorting, or contaminating critical components.

3.2 Structural Requirements

Before spending much time designing any part of the satellite structure, key requirements for all the major substructures should be derived from the conceptual configuration [1]. As mentioned before in the literature review, structures are categorized into three categories: primary, secondary, and tertiary structures. Structures are categorized because different requirements typically drive their respective designs.

3.2.1 Primary Structure

The primary structure is the backbone, or the major load path, between the satellite's components and the launch vehicle. It carries shear, bending moments, axial loads, and torsion. For Small Sat, the primary structure consists of the base plate, upper frame, basis block case, lower frame, and the mounting plate. These structural modules are considered as the main body structure. The LV adapter is taken as a part of the primary structure. The primary structure takes its shape from the stowed packaging of satellite components, and dictates the configuration. Dnepr LV drives most requirements for the primary structure. The primary structure is usually designed for stiffness or natural frequency, and to survive steady-state accelerations and transient loading during ground operation, launch, and space operation.

General shape and purpose must provide load paths between supported components and LV. Loads are derived from LV environments and the satellite's mass and stiffness properties. The satellite natural frequencies must not interfere with the LV-imposed fundamental frequency. Requirements at the LV interface include flatness, bolt-hole pattern, and uniform load distribution. These requirements usually drive the need for an LV adapter. The primary structure may need to maintain a certain alignment among payload, sensors, and antennas to satisfy dynamic Positional stability. Mass is typically allocated to the structures subsystem, and is thus often treated as a target in preliminary design rather than a firm requirement. The primary structure must provide access to install and service components during ground operations and maintenance.

3.2.2 Secondary Structure

For Small Sat, the secondary structure includes the four solar panels. Most of the considerations for primary structures apply also to secondary structures. Acoustics, launch, and on-orbit thermal cyclic loading are often important as well. The load factors used to design the primary structure often are not suitable for designing secondary structures. During launch, solar panels (lightweight structures with large surface areas) will respond to acoustics at the same time they are excited by transients and steady-state accelerations. Secondary structures are often not well protected from thermal environments as is the primary structure. Wide variations in temperature can greatly stress solar panels and the mechanisms supporting them.

3.2.3 Tertiary Structure

For Small Sat, the tertiary structure includes mounting brackets, cable-support brackets, and connector panels. Base-driven vibration is the most severe environment for the most of these structures, and fatigue life is the driving requirement. For small satellites, most random vibrations are introduced at the launch vehicle interface, while for large satellites, acoustic and launch vibration response of solar panels, antenna dishes, and other lightweight structures with large surface areas generate most of the random vibration components. To reduce structural loads on these components, they can be mounted on isolators which add damping and flexibility, thus isolating the component from high-energy vibration sources. Spacers of a thermally nonconductive material are used to isolate devices from thermal stresses.

3.3 Design Procedure

The first step in the design procedure is to size the general satellite configuration taking into account the large number of equipment involved in the mission. The Small Sat's primary structure is configured as a rectangular box outer body with a square cross-section. This configuration provides the main load path and minimizes the inherent complications of integration. The dimensions of the preliminary configuration are represented in Chap. 2. The rectangular configuration with a square load path allows selection of a simple separation system between the satellite and LV adapter.

The next step requires a selection of the basic primary structural design. By reviewing the four basic designs discussed in the literature review and the preliminary configuration presented in Chap. 2, a skin-frame structure is found to be the appropriate type used to design the basic primary structure of Small Sat. This type is chosen because it has a heritage in small satellite structures. The design enables deployed-fixed solar panels and allows mounting the satellite's components to internal and external surfaces of its structural modules (plates and frames). The next step in the design process is to prepare a technological breakdown scheme for Small Sat. This scheme presents a total view for the satellite assembly and its subassemblies. Figure 3.1 shows the satellite technological breakdown scheme.

3.3.1 Used Materials

The next step in the design process is to select the optimum material for the primary structure. A large number of materials with space heritage are available for satellite structural design and are presented in the literature review. Solid metallic, honeycomb sandwich, and fiber composite are the three kinds of materials suitable for primary structure design. To select the best kind of material, each of these materials is evaluated in more detail based on several design criteria. The design criteria selected for the material selection are mass; design, development, and test cost; ease of fabrication and assembly; ease of inspection and repair; availability of material; and thermal performance. The evaluation clearly demonstrates that metallic materials are the most effective in structural design for Small Sat. There are several reasons for this decision. Fiber composite structures are expensive to manufacture and require costly structural testing. Honeycomb sandwich materials require potted inserts to attach fasteners, which create safety concerns and parasitic mass. The metallic materials are relatively simple to manufacture and minimize the safety and testing requirements.

The next step in the design process is to develop an optimum metallic material for structural design. Several metals are used in satellite fabrication, such as aluminum, titanium, beryllium, and stainless steel, which are discussed in detail in Chap. 2. These materials vary in cost, manufacturing time, mass, strength, and durability. Aluminum alloy AMg6 is selected as the main material to manufacture the primary structural modules for several reasons. The alloy is relatively abundant and economically feasible for a low budget program. It has a density of approximately 2,630 kg/m^3, which is roughly one-third the density of steel with slightly inferior strength. AMg6 alloy is used to manufacture and weld constructions working at temperatures from -196 to $+150$ °C. Finally, AMg6 is simple to

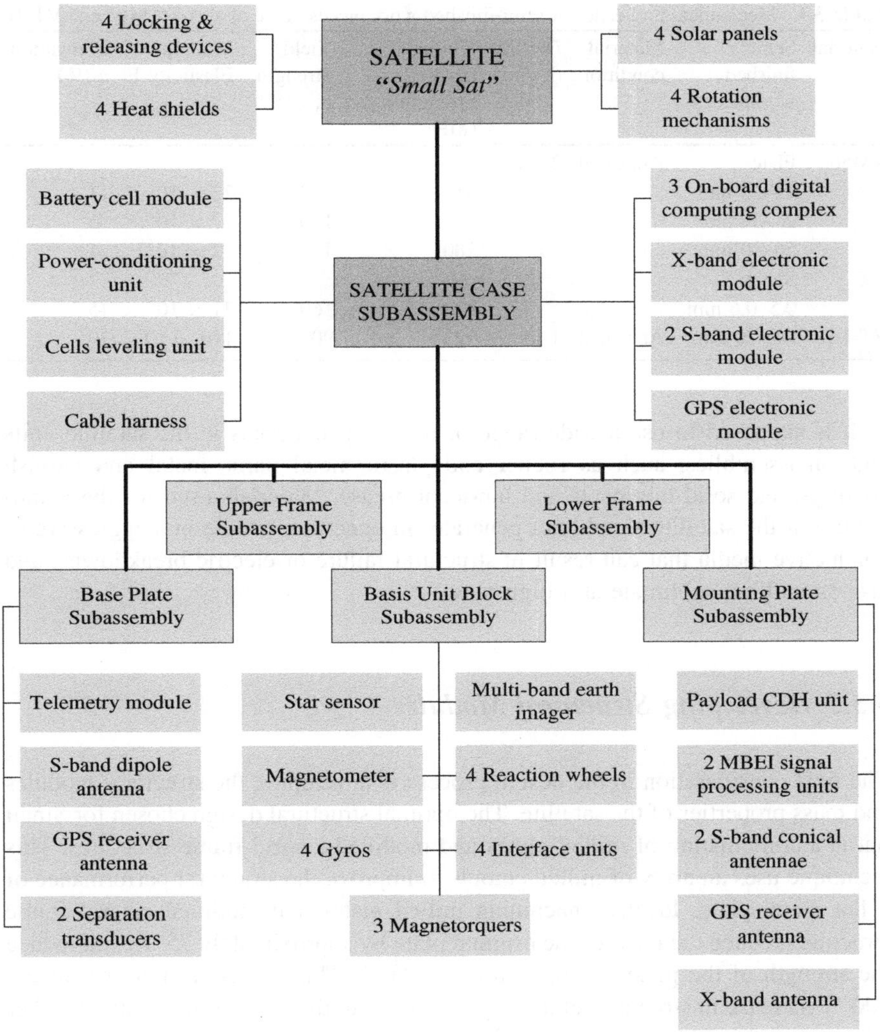

Fig. 3.1 Satellite technological breakdown scheme

manufacture and has relatively good workability. Mechanical properties of semi-finished work pieces made of alloy AMg6 according to the standards are given in Table 3.1.

Studs for fastening the structural modules together can be fabricated from bars made of titanium alloy VT-16. Titanium alloy VT-16 is used to manufacture fasteners operating within a temperature range from −120 to +300 °C. Titanium alloy is also selected because of its very low creep deformation, which minimizes loss of stud tightness during the satellite's service life. Mechanical properties of semi-finished work pieces made of titanium alloy VT-16 are given in Table 3.1.

Table 3.1 Mechanical properties of semi-finished work pieces made of alloys AMg6 and VT-16

Material	Semi finished workpieces	Material condition	Density (kg/m^3)	Ultimate tensile Strength σ_u (MPa)	Yield Strength σ_y	Modulus of Elasticity E	Elongation δ (%)
AMg6	Plate	Annealed	2,630				
	12–25 mm			310	150	72×10^3	11
	25–50 mm			300	140	72×10^3	6
	50–80 mm			280	130	72×10^3	4
	Sheet						
	0.5–0.6 mm			320	160	71×10^3	15
VT-16	Bar 20 mm	Annealed	4,430	923	903	110×10^3	10

It is supposed to use a wide range of nonmetal materials in the satellite units and subassemblies, such as rubber and plastic-metal parts, metal and varnish coatings, and solid lubricants and lubricant grease. Materials used for the manufacture of the satellite should not generate (in operation conditions) aggressive or conductive media that can result in structural failure or electric breakdowns, and also should not sublimate at a high rate.

3.3.2 Developing Structural Modules

The next consideration in the design process is developing the structural modules and mass properties of the satellite. The optimal structural design chosen for Small Sat is a combination of milled plates and modified isogrid frame structures. This technique uses an array of milled cutouts to improve the structural performance of a flat metal plate. Ideally, machining milled plates and modified isogrid frame structures reduce the mass of the original plate by approximately 75 %, and reduce the strength of the plate by approximately 25 %. This procedure may produce a 200 % increase in structural efficiency compared to the original flat plate. Another advantage of machining an array of milled cutouts is that the plate remains isotropic. Therefore, it exhibits similar strength in all directions, and minimizes locations of stress concentrations. Finally, milled integral constructions are usually easier and cheeper to manufacture than skin stringer structures.

3.3.3 Iterative Process of Design Development

The design development process consists of several steps. The first step ensures that all subsystems will fit into the satellite. The second step uses Mechanical Desktop package "MDT" and finite element analysis by ANSYS package to ensure acceptable structural through stiffness and mass properties. The next steps determine if the manufacturing cost and schedule of the structure are acceptable. The final steps in the design process examine the assembly, manufacturing, and operational aspects of the design. Many steps are taken to easy handling, assembly, and manufacturing issues throughout the process. Finite element analysis techniques are coupled with development considerations to arrive at the most efficient design. Upon successful completion of the design process, the final Small Sat structural design is obtained.

3.4 Preliminary Analysis

To size the preliminary design of the structure, a simplified static analysis is performed on the primary structure modules. An iterative procedure as described in Sect. 3.3.3 is used. The initial structural design for milled plates and modified isogrid frame structures is chosen according to the preliminary configuration described in Chap. 2. The initial structural design is established to provide appropriate mechanical interface between structural modules, and follow the arrangement of components installed on each structural module. Using the mechanical loading conditions described in Appendix A, finite element analyses for each of the structural modules are performed to arrive at the most efficient design.

3.4.1 Strength Analysis Cases

By reviewing the various mechanical loads described in Appendix A, the most critical quasi-static loads lead to two strength analysis cases:

Case 1: The satellite loading by operational g-loads during road transportation in the container. Table A.1 gives the operational g-load factors as:

$$\mathbf{n_x} = \pm 2, \quad \mathbf{n_y} = \pm 1.25, \quad \mathbf{n_z} = 1 \pm 2.$$

where: "±" is attached to the dynamic components, and "1" for n_z g-load is the loading caused by weight. The X-axis of vehicle is in the forward-motion direction, the Z-axis is vertically down, and the Y-axis makes the system right-handed.

Case 2: The satellite loading by maximum longitudinal and lateral g-loads during launch. From the operational g-loads given in Table A.8, the combinations of the maximum axial "$\boldsymbol{n_a}$" and lateral "$\boldsymbol{n_l}$" g-load factors during the first and second LV stages flight are found to be:

$$\boldsymbol{n_a} = 7.8 \pm 0.5, \quad \boldsymbol{n_l} = 0.1 \pm 0.5$$

The following values of safety factors are adapted from Appendix A:

- During road transportation in the container "case 1"
  - $f = 1.5$—for static g-load component
  - $f = 2.0$—for dynamic g-load component
- During the flight phase of launch "case 2"
  - $f = 1.3$

Hence, the critical limit loads for Small Sat are as follows:

Case 1:

$N_x = 2 \times 2 \times 9.81 = 39.24$ m/s^2
$N_y = 1.25 \times 2 \times 9.81 = 24.53$ m/s^2
$N_{z\ \max} = (1 \times 1.5 + 2 \times 2) \times 9.81 = 53.96$ m/s^2
$N_{z\ \min} = (1 \times 1.5 – 2 \times 2) \times 9.81 = –24.53$ m/s^2

where

N_x is the lateral "shear" acceleration in the forward-motion direction,
N_y is the lateral "shear" acceleration in the Y-axis,
$N_{z\ \max}$ is the maximum axial "longitudinal" acceleration in the Z-axis
$N_{z\ \min}$ is the minimum axial "longitudinal" acceleration in the Z-axis

The resultant (equivalent) shear acceleration is:

$$N_{\mathrm{I}} = \sqrt{(N_{\mathrm{x}})^2 + (N_{\mathrm{y}})^2} = 46.28 \quad m/s^2$$

Case 2:

$N_a = (7.8 + 0.5) \times 1.3 \times 9.81 = 105.85$ m/s^2
$N_l = (0.1 + 0.5) \times 1.3 \times 9.81 = 7.64$ m/s^2
where N_a is the axial "longitudinal" acceleration, and
N_l is the lateral "shear" acceleration.

3.4.2 Design of Fastening Studs and Small Sat/LV Interface Bolts

From the conceptual configuration of Small Sat described in Chap. 2, the satellite structure consists of a base plate, upper frame, basis unit, lower frame, and mounting plate. These structural modules must be fastened together to work as a load path to provide structural requirements. The most effective method to provide acceptable mounting accuracy and alignment is using fastening studs connecting structural modules in the satellite case subassembly. The connection between Small Sat and launch vehicle adapter or transportation container is done through interface bolts. The holes of these bolts are installed on specified locations in the base plate.

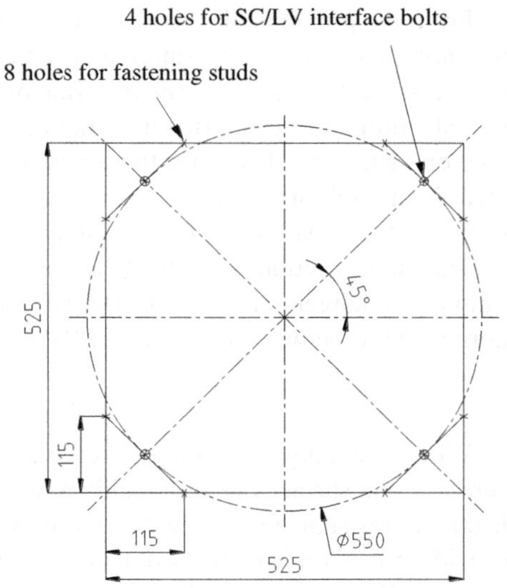

Fig. 3.2 Location of fastening studs and interface bolts

Selection of the number and location of fastening studs and interface bolts depends on the load path shape and structural modules configuration. For Small Sat, the main load path is a square cross-section with 525 mm side length, so the suitable number of fastening studs is four, six, or eight studs. To provide sufficient stability, using four or six studs is not acceptable. Therefore, eight studs are selected to connect the structural modules. Each side of the main load path contains two studs. The locations of the fastening studs are designed to provide symmetry and regularity in load carrying. The same requirements are required for designing the interface bolts. Four bolts are used to connect the satellite to LV adapter or transportation container. Therefore, four steel bushes are mounted on the base plate to work as an interface between the satellite body and launch vehicle adapter. Figure 3.2 shows the arrangement of eight fastening studs and the location of four interface bolts between the satellite base plate and launch vehicle adapter or transportation container.

To determine the final dimension and suitable material for the fastening studs, a close look should be made at the studs function. The major objective of the fastening studs is connecting structural modules in accurate alignment during all phases of the satellite's service life. To provide this objective, there are two essential conditions that must be met. The first is to provide accurate assembly of the structural module. This is started from the manufacturing process by assuring the accuracy of the relative locations of the fastening holes in each structural module, and ended by taking all precautions during the assembly process. The second condition is to prevent relative lateral sliding between structural modules during all phases of satellite operation. This condition can be met by applying a specific amount of longitudinal compression to the satellite during assembly process to provide sufficient lateral friction. This is done by applying a specific torque on each stud.

From the previous discussion, it is clear that fastening studs are mainly used in mounting and connecting satellite structural modules. They are pre-loaded with a compressive load to perform this function. This pre-load compression prevents lateral sliding of the structural modules. This function depends on the static friction criterion. Therefore, the amount of longitudinal compression load generated from applying torque to each stud must be sufficient to overcome the most critical lateral load affecting the satellite in all phases.

The static friction coefficient (μ_s) between two solid surfaces is defined as the ratio of the tangential (lateral) force (F) required to produce sliding divided by the normal force between the surfaces (N),

$$\mu_s = F/N$$

When the tangential force overcomes the frictional force between the two surfaces, the surfaces begin to slide relative to each other. In this case, the sliding frictional resistance is different from the static frictional resistance. The coefficient of sliding friction is expressed using the same formula as the static coefficient and is generally lower than the static coefficient of friction. For dry surfaces, the coefficient of friction is independent of the surface area. In Small Sat case, structural modules are separated by thin steel washers which work as thermal isolators besides helping in overcoming surfaces irregularity between structural modules. The friction coefficient in this case is $\mu = 0.3$ between aluminum and steel in static dry case. To prevent relative sliding, the friction force (f) must be greater than the maximum lateral force ($F_{l\ \max}$) during all satellite loading phases. The friction force is equal to the friction coefficient (μ) multiplied by the net normal force (N_{net}), which is the total longitudinal compressive force (F_c) generated by applying a torque to the studs plus the longitudinal force (F_a) generated during the maximum lateral force case.

$$\left.\begin{aligned} \mathrm{f} &> \mathrm{F}_{\mathrm{l\ max}} \\ \mathrm{f} &= \mu \times \mathrm{N}_{\mathrm{net}} \\ \mathrm{N}_{\mathrm{net}} &= \mathrm{F}_{\mathrm{c}} + \mathrm{F}_{\mathrm{a}} \end{aligned}\right\} \ldots \text{Friction Equations}$$

By reviewing Sect. 3.4.1, the most critical lateral acceleration occurs in case 1 during road transportation in the container. This acceleration is the resultant shear acceleration (N_l), which can be considered uniform along the satellite body. So the maximum lateral force due to the total gross mass of the satellite is equal to:

$$F_{l\ \max} = 256 \times 46.28 = 11847.68N$$

where (256 kg) is the total predicted gross mass of the satellite plus a 25 % growth allowance, which is a historical average. In case 1, there are two situations for axial loading one downward and the other upward. From friction equations, the most critical situation is the upward axial loading. So the longitudinal (axial) force is equal to:

$$F_a = 256 \times (-24.53) = -6279.68N$$

Hence:
$N_{\text{net}} = F_c - 6279.68$
$f = 0.3 \times (F_c - 6279.68)$
So,
$0.3 \times (F_c - 6279.68) > 11847.68$
Leading to: $F_c > 45771.95\ N$

This value gives the minimum total longitudinal compressive force to be generated by the eight studs. Therefore, applying a preload tension force in each stud equal to 6,000 N can be considered a suitable amount to prevent lateral sliding in all cases. Fastening studs are made of titanium alloy VT-16, which is the suitable material to manufacture a precise long rod with a relatively small diameter and minimum creep requirements. The manufacturing process applies several constraints on stud fabrication, and the selection of stud diameter is affected by these restrictions. The eight studs are identically 6 mm diameter with external thread M6 at both ends with 20 mm depth. The design stress (σ) is calculated using the formula:

$$\sigma = \frac{F}{A}$$

where A is the stud cross-section area. This gives:

$$\sigma = 212.2\,\text{MPa}$$

The yield margin of safety (MS_y) is calculated from the formula:

$$MS_y = \frac{\text{Allowable yield stress } (\sigma_Y)}{\text{Design yield stress } (\sigma)} - 1$$

For titanium alloy, VT-16 the allowable yield stress (σ_y) is 903 MPa. Therefore,

$$\text{MSy} = 3.25$$

The stud elongation (δ) due to preload (F) can be computed using the formula:

$$\delta = \frac{Fl}{AE}$$

where: (l) is the length of stud, and (E) is the modulus of elasticity of titanium alloy VT-16. The length of the active part of the stud is equal to 1,030 mm from the preliminary configuration of Small Sat. Therefore,

$$\delta = 2\ \text{mm}$$

The torque (T) applied to each stud to achieve the required pretension force is calculated using the formula:

$$T = KFd$$

where: (K) is the torque coefficient, and (d) is the stud diameter. When the stud condition is not stated, the torque coefficient (K) is taken as 0.2. Therefore,

$$T = 3.6 \text{ N.m}$$

3.4.3 Preliminary Sizing of Structural Modules

This section describes the preliminary process used to size each module of the primary structure of Small Sat. An iterative procedure is performed to achieve a suitable initial design for each structural module. The loading condition used to compute the maximum stresses and deformations for each module is the maximum longitudinal g-load during launch. The margins of safety, MS, are calculated using the formula given before, where the allowable yield stress (σ_y) for aluminum alloy AMg6 is equal to 150 MPa. The margins must be positive for all strength cases. The analysis is done by ANSYS package for each module separately to calculate the maximum deformations and stresses. Each structural module is first modeled using arrays of beam and shell elements. The preliminary dimensions of the structure are defined according to the arrangement of equipment. An element size and its arrangement must provide a stiff load path. Appropriate mechanical interfaces between structural modules must be provided. A combination of intersecting lines and in-plane areas is used as a graphical model for each structural Module. These lines represent internal and external stiffeners of the Modules. They connect the main points to support bending, torsion, and axial forces. The main points are the locations of the eight fastening studs, equipment seats, and four interface bolts in the base plate. The in-plane areas represent the skin, which reinforces the structure by supporting the shear forces introduced by the interior member connections.

For simplicity during analyses, the Small Sat primary structure is divided into the following modules: base plate, upper frame, basis plate, basis unit walls, lower frame, and mounting plate. The graphical models created on ANSYS package are presented in Figs. 3.3, 3.4, 3.5, 3.6, 3.7, and 3.8. All primary structural modules have skin except the upper and lower frames. For the base and mounting plates, the skin is used as a kind of environmental shield besides supporting the shear forces. The main function of the skin in the basis unit block is reinforcing the structure and decreasing the deformation to keep mounting accuracy of high precise equipments. Skin is removed from the upper and lower frames to reduce structural weight with little effect on shear deformation.

The material used during preliminary analysis is aluminum alloy AMg6 with the following properties: modulus of elasticity (E) = 72×10^3 MPa, Poisson's ratio (ν) = 0.33, density (ρ) = 2630 kg/m^3, yield stress (σ_y) = 150 MPa, and ultimate stress (σ_u) = 310 MPa. The ANSYS element types used for the meshing process are:

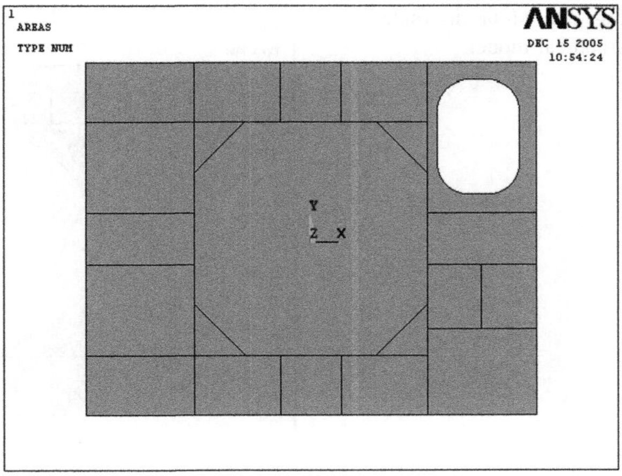

Fig. 3.3 Base plate graphical model

- BEAM189: used for lines to create stiffeners
- SHELL91: used for areas to create skin
- MASS21: used for mounting points to represent equipment masses.

Stress analyses are performed for the structural modules by using different dimensions of beam element sections and shell element thicknesses. An iterative analysis is performed by calculating the stiffeners stresses and strain deformations at different section dimensions. The main objective of this iterative process is to arrive at appropriate values for the maximum stresses and deformations for each structural module. The values of the maximum stresses must be less than the allowable yield stress for aluminum alloy AMg6 (150 MPa), so the calculated values of the margins of safety must be larger than one. This way, the preliminary design for each module can be considered as a safe one and taken as a good start for the detailed analysis.

In reality, the connection points between structural modules are not fixed at all degree of freedoms. They have limited deformations depending on the kinetics of the whole primary structure. For simplicity, the boundary conditions for all structural modules are fixing all degree of freedoms at the eight connection points except the base plate, which has boundary conditions at the four points intended of fixation with the launch vehicle adapter. This condition is not the real situation for the actual primary structure, but it can be taken as a severe condition. The external loading is a longitudinal acceleration g-load during launch as presented in Table A.8.

The process is started by performing analysis for a finite element model of the mounting plate. The final finite element model created on ANSYS for the mounting plate is shown in Fig. 3.9. It consists of the following beam cross-sections: rectangular 6×20 mm c.s. for the main octagonal stiffeners, rectangular 2×20 mm c.s. for the boundary stiffeners, and rectangular 4×15 and 2×15 mm c.s. for internal stiffeners. The skin is 1.5 mm thick. Reaction forces

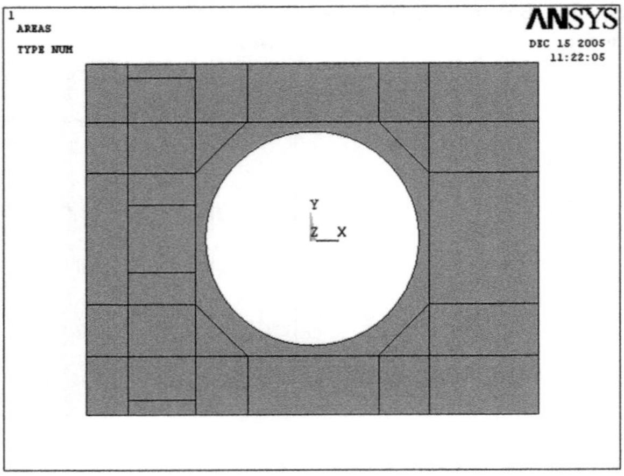

Fig. 3.4 Mounting plate graphical model

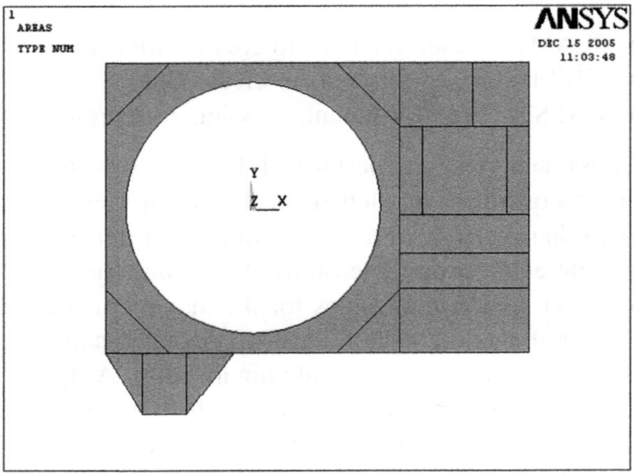

Fig. 3.5 Basis plate graphical model

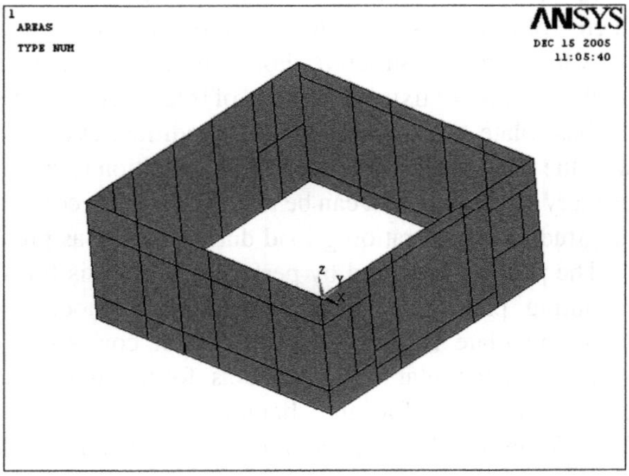

Fig. 3.6 Basis walls graphical model

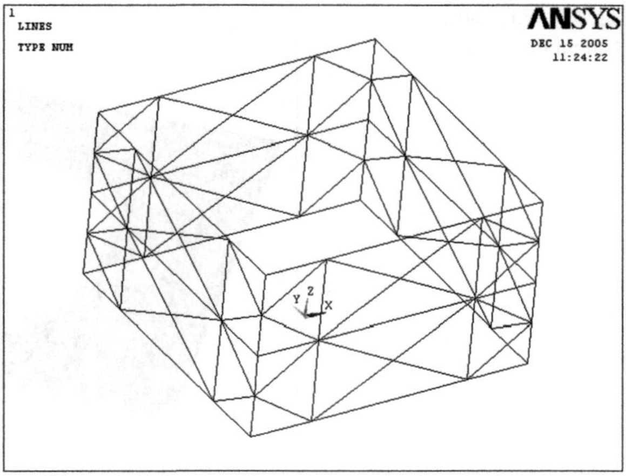

Fig. 3.7 Upper frame graphical model

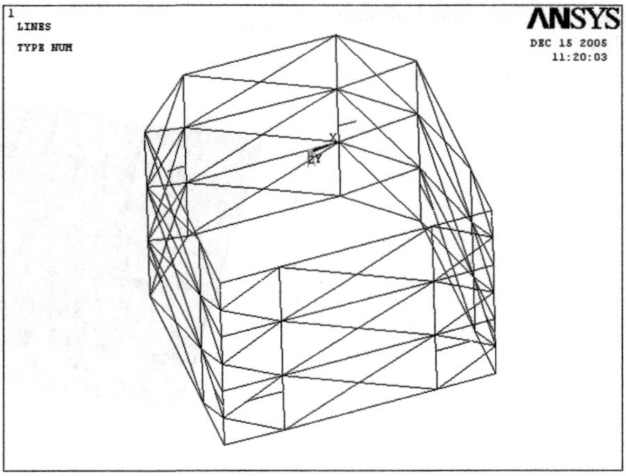

Fig. 3.8 Lower frame graphical model

are calculated at fixation points of the mounting plate, and are considered as external forces for the lower frame.

F.E. analysis is made for the rest of the primary structural modules. The same analysis process used for the mounting plate is applied to the lower frame, basis unit walls, basis plate, upper frame, and base plate, respectively. Final F.E. models created on ANSYS are shown in Figs. 3.10, 3.11, 3.12, 3.13, and 3.14. The final F.E.M. for the base plate consists of the following beam cross-sections: rectangular 6 × 25 mm c.s. and 25 × 25 × 8 × 5 L c.s. for the main octagonal stiffeners, rectangular 2 × 25 mm c.s. for the boundary stiffeners, and rectangular 4 × 25 and 2 × 15 × mm c.s. for internal stiffeners. The skin is 1.5 mm thick. The final basis plate F.E.M. consists of the following beam cross-sections: 25 × 22 × 3 × 2, 25 × 16 × 3 × 2, and 25 × 11 × 3 × 2 L c.s. for the main octagonal

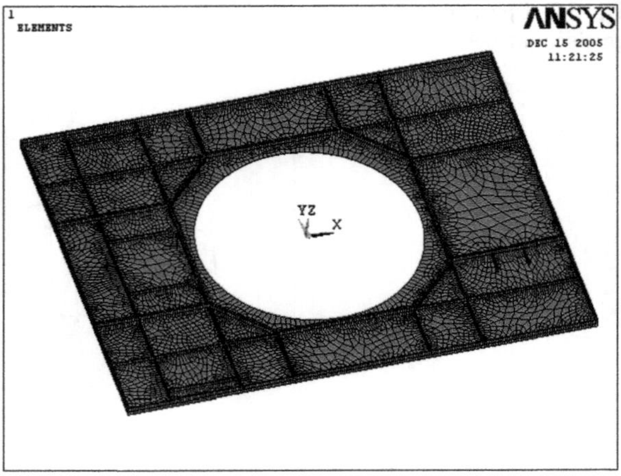

Fig. 3.9 F.E. model of mounting plate

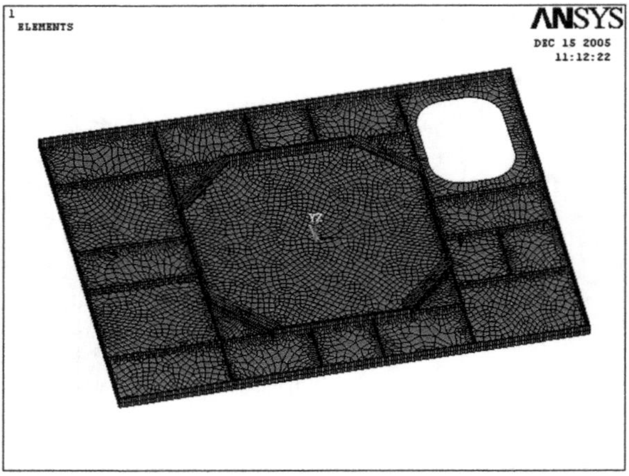

Fig. 3.10 F.E. model of base plate

stiffeners, rectangular 2 × 25 mm c.s. for the boundary stiffeners, and rectangular 2 × 25 and 2 × 15 mm c.s. for internal stiffeners. The skin is 1.5 mm thick.

Figures 3.20 and 3.21 show the stress and deformation distributions in the basis plate. As shown in the figures, the maximum stress and deformation in the basis plate are not acceptable, so two diagonal struts are added to connect the basis plate with the basis unit walls. The selected cross-section of the diagonal struts is a 16 × 16 × 18 × 2 × 2 × 3 channel c.s.

For the basis unit walls, the final F.E.M. consists of the following beam cross-sections: channel 30 × 30 × 22 × 4 × 4 × 4 c.s. for the main eight stringers intended for structure fastening and load carrying path, L 22 × 22 × 6 × 6 c.s. for the four wall connection corners, and rectangular 1.5 × 16.5 mm c.s. for internal and boundary stiffeners. The skin is 1.5 mm thick. The final F.E.M. for the upper frame has no skin and consists of the following beam cross-sections: channel

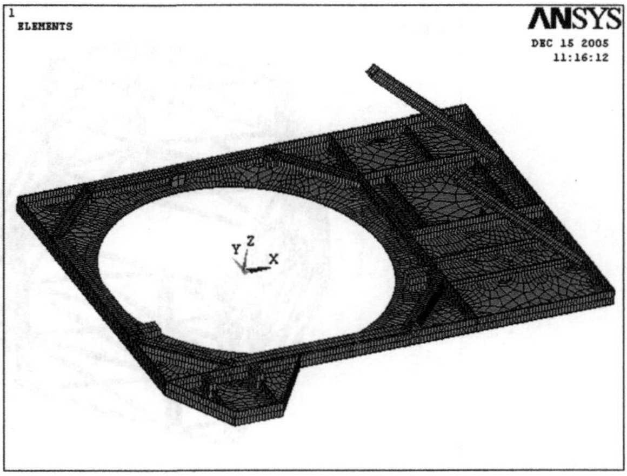

Fig. 3.11 F.E. model of basis plate

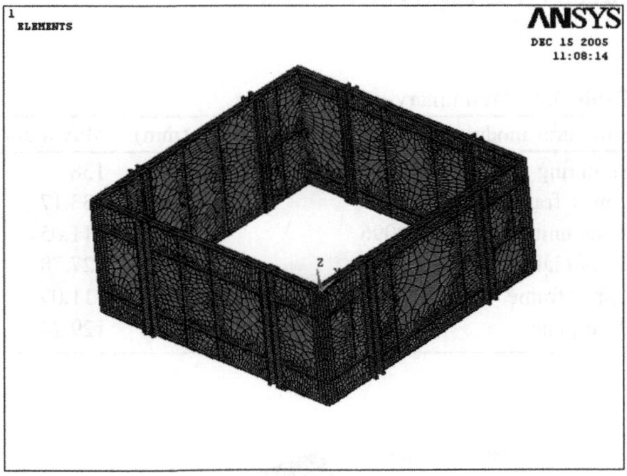

Fig. 3.12 F.E. model of basis unit walls

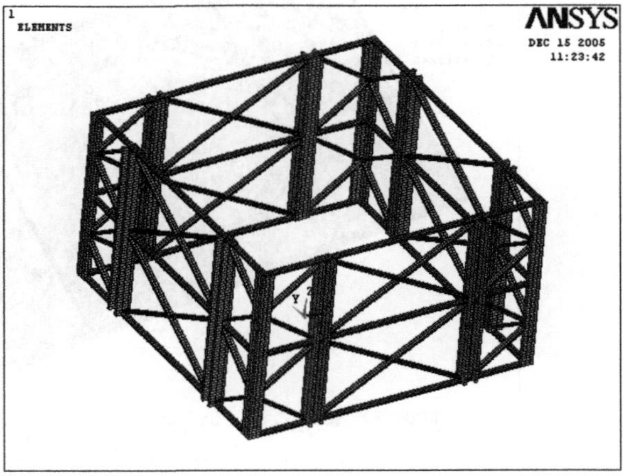

Fig. 3.13 F.E. model of upper frame

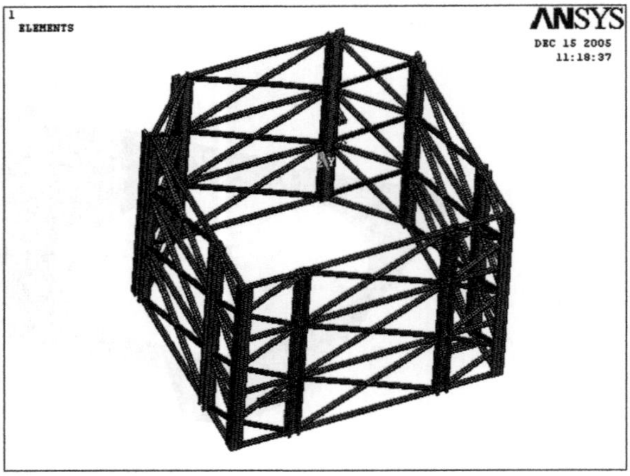

Fig. 3.14 F.E. model of lower frame

Table 3.2 Preliminary analysis results

Structural module	Maximum deformation (mm)	Maximum stress (MPa)	Margin of safety
Mounting plate	3.69	138	0.087
Lower frame	0.13	33.17	3.522
Basis unit walls	0.0096	11.05	12.57
Basis plate	1.15	127.78	0.174
Upper frame	0.097	11.09	12.53
Base plate	5.93	129.24	0.161

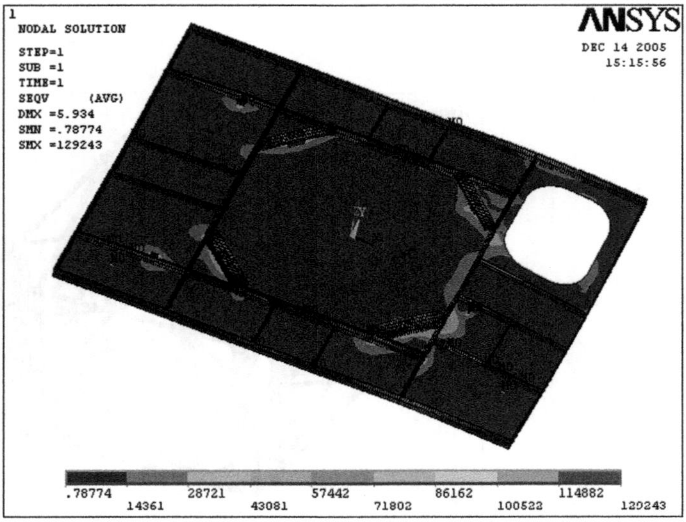

Fig. 3.15 Stress distribution of preliminary base plate

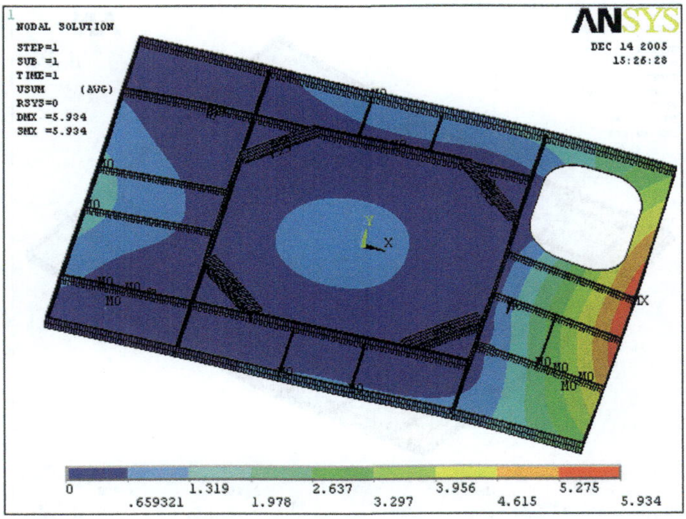

Fig. 3.16 Deformation of preliminary base plate

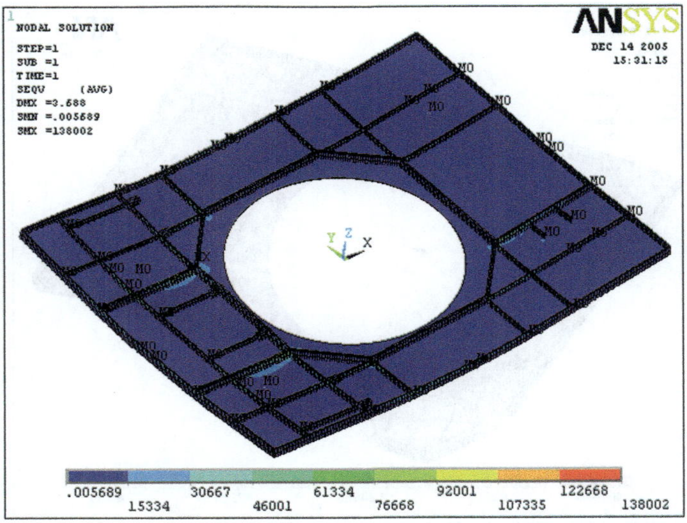

Fig. 3.17 Stress distribution of preliminary mounting plate

26 × 26 × 26 × 5 × 5 × 5 c.s. for the main eight stringers intended for structure fastening and load carrying path, L 22 × 22 × 6 × 6 c.s. for the four wall connection corners, and rectangular 4 × 10 mm c.s. for internal and boundary stiffeners. The final F.E.M. for lower frame has no skin and consists of the following beam cross-sections: channel 22 × 22 × 22 × 4 × 4 × 4 c.s. for the main eight stringers intended for structure fastening and load carrying path, L 22 × 22 × 6 × 6 c.s. for

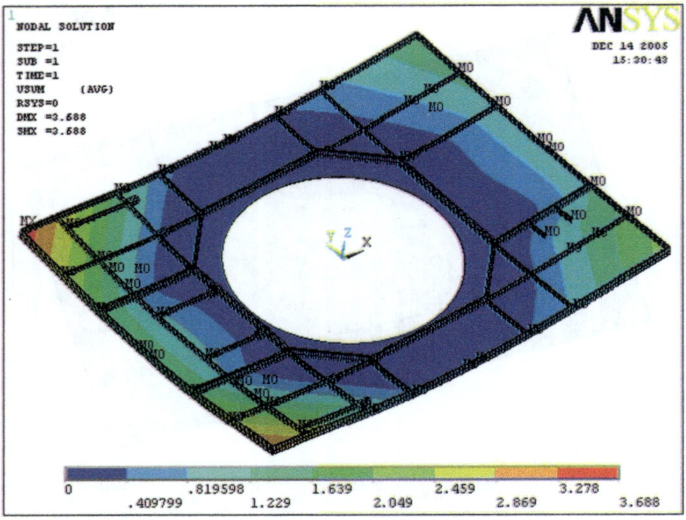

Fig. 3.18 Deformation of preliminary mounting plate

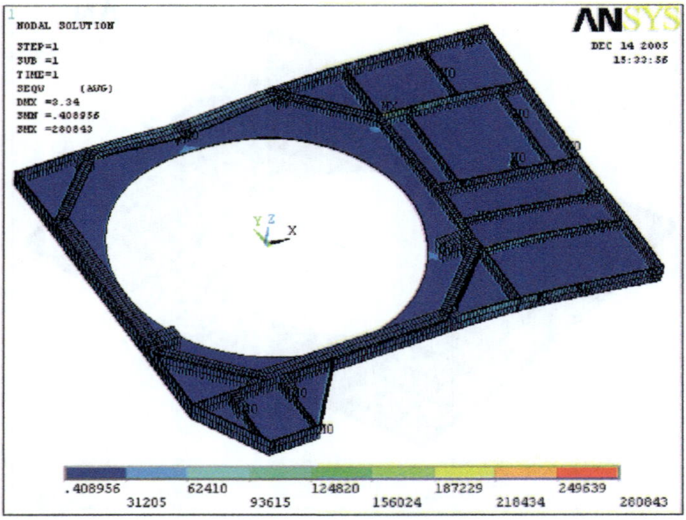

Fig. 3.19 Stress distribution of preliminary basis plate without struts

the wall connection corners, and rectangular 3 × 10 mm c.s. for internal and boundary stiffeners. Table 3.2 and Figs. 3.14, 3.15, 3.16, 3.17, 3.18, 3.19, 3.20, 3.21, 3.22, 3.23, 3.24, 3.25, 3.26, 3.27, and 3.28 give the final stress and deformation distributions in the various structural models. It is seen that all structural modules of Small Sat have acceptable strength and stiffness.

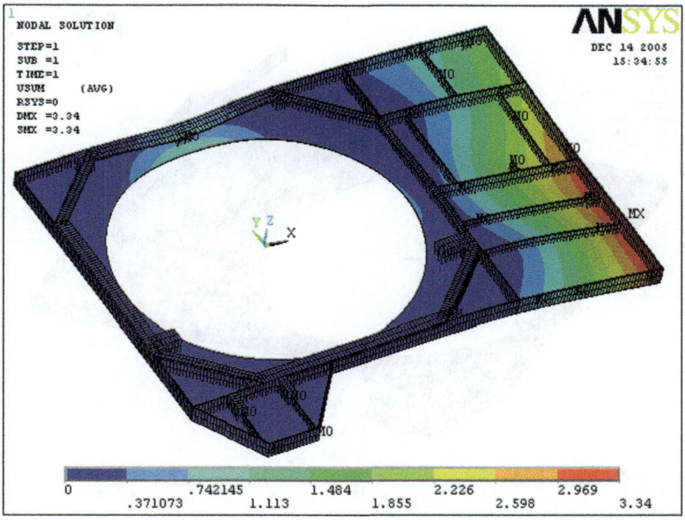

Fig. 3.20 Deformation of preliminary basis plate without struts

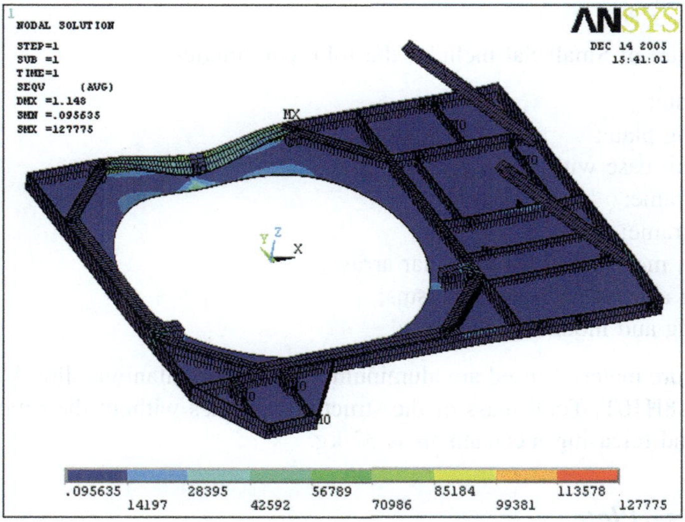

Fig. 3.21 Stress distribution of preliminary basis plate with struts

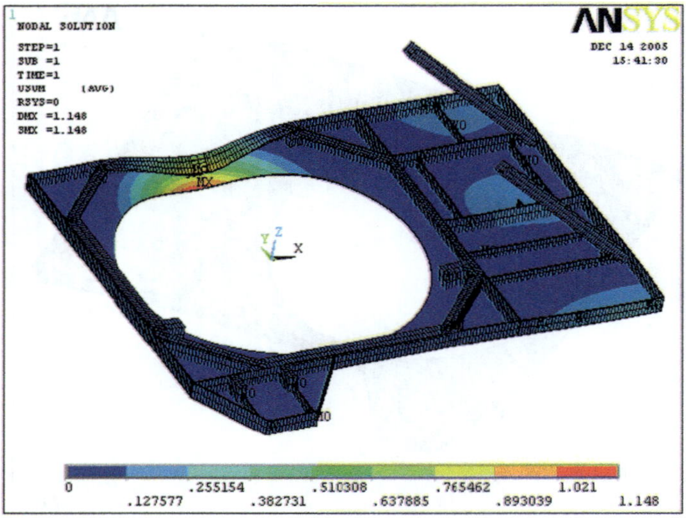

Fig. 3.22 Deformation of preliminary basis plate with struts

3.5 Detailed Structure Description

The structure of Small Sat includes the following modules:

- Base plate;
- Mounting plate;
- Basis unit case with star sensor bracket;
- Upper frame;
- Lower frame;
- Rotation mechanisms of the solar arrays;
- Locking and releasing mechanisms;
- Fastening and mounting elements.

The structure materials used are aluminum alloy AMg6, titanium alloy VT16, and steel 12X18H10T. Total mass of the structure modules without the rotation, and locking, and releasing mechanisms is 32 kg.

3.5.1 Base Plate

The base plate is a 1,030 × 800 × 26 mm milled plate made of aluminum alloy AMg6. The built-in milled stiffeners are oriented as shown in Fig. 3.29 and are milled in the following shapes:

- L-shaped cross-section stiffeners—along lines between holes for eight studs of the satellite case fastening and holes for four bushes intended for fastening of the satellite to the launch vehicle adapter;

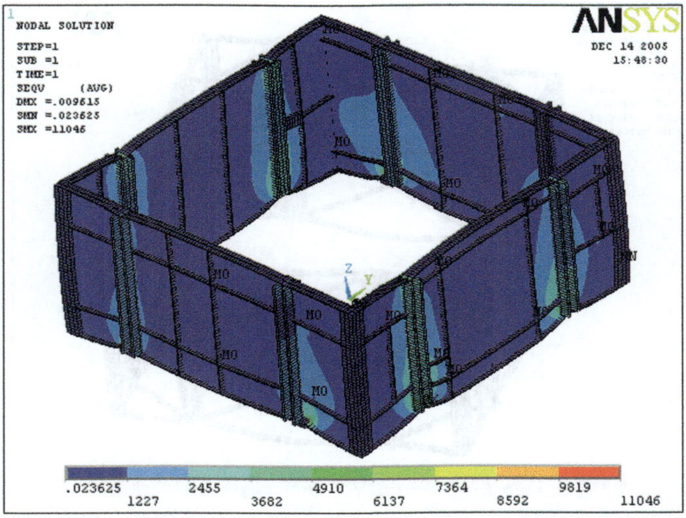

Fig. 3.23 Stress distribution of preliminary basis unit walls

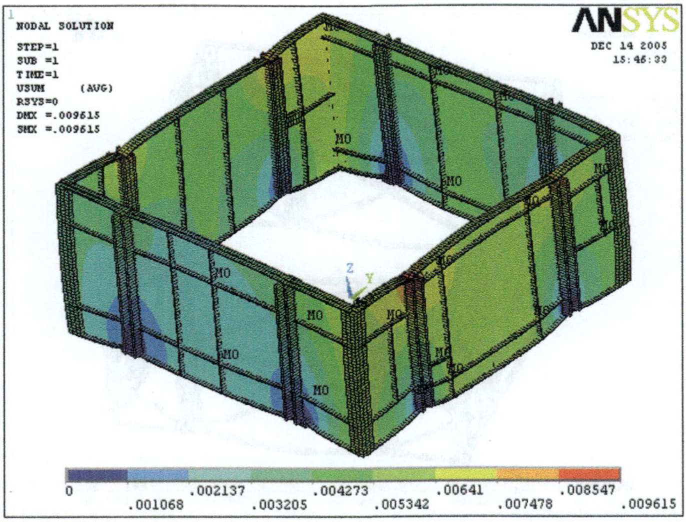

Fig. 3.24 Deformation of preliminary basis unit walls

- Rectangular cross-section stiffeners—along the perimeter of the plate, and between installation seats of the satellite components.

Thickness of the skin is 1.5 mm, while the outer and internal stiffeners of the plate have 2 mm thickness. Thickness of the rectangular cross-section stiffeners connecting holes for eight studs is 6 mm. Outer stiffeners and stiffeners connecting

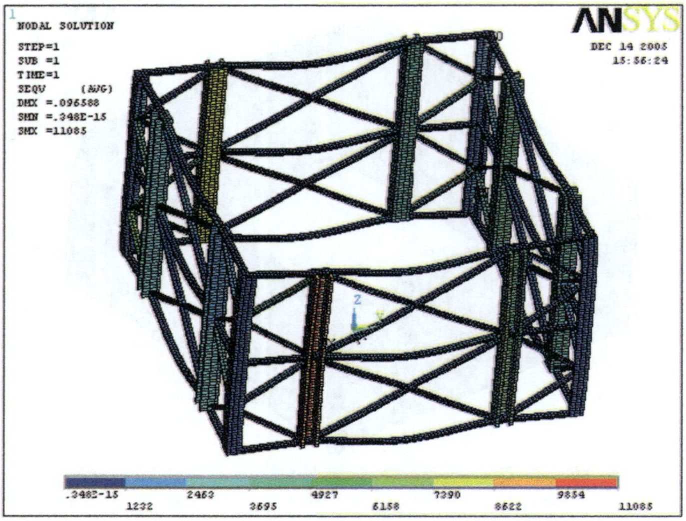

Fig. 3.25 Stress distribution of preliminary upper frame

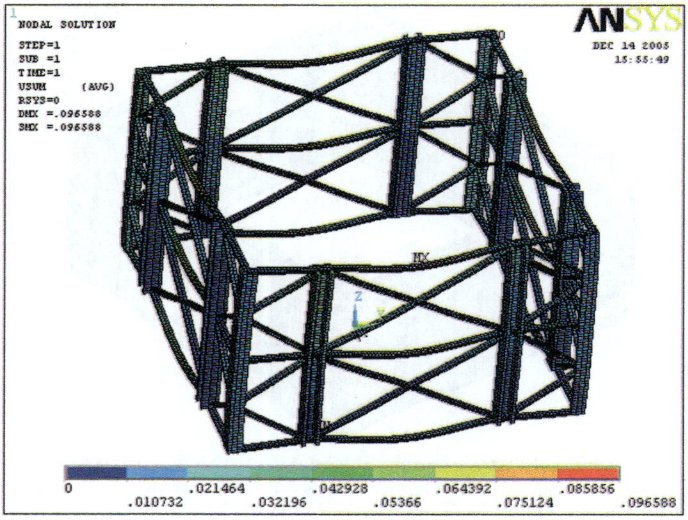

Fig. 3.26 Deformation of preliminary upper frame

eight studs have 23.5 mm height, while the height of internal stiffeners is 13.5 mm. Four bushes with internal M10 thread hole are screwed to the base plate and intended for cantilever fastening of the satellite to launch vehicle adapter and transportation container. Bushes are made of steel 12X18H10T. The base plate contains eight 6 mm diameter holes to provide installation of the eight studs

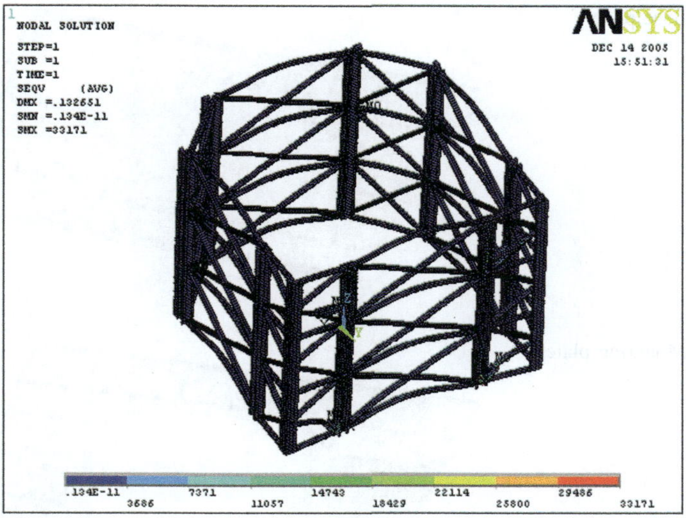

Fig. 3.27 Stress distribution of preliminary lower frame

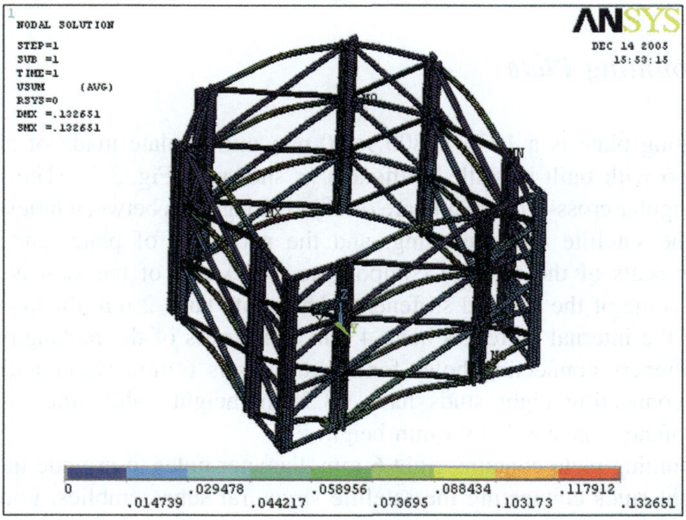

Fig. 3.28 Deformation of preliminary lower frame

connecting the satellite structural subassemblies. Base plate provides mounting seats for the telemetry module, bracket of GPS antenna, bracket of dipole S-band antenna, four rotation mechanisms, and two separation transducers. It has also two additional seats for mounting battery bracket. The base plate contains elliptical hole working as a window for the star sensor. Mass of the base plate is 5.3 kg.

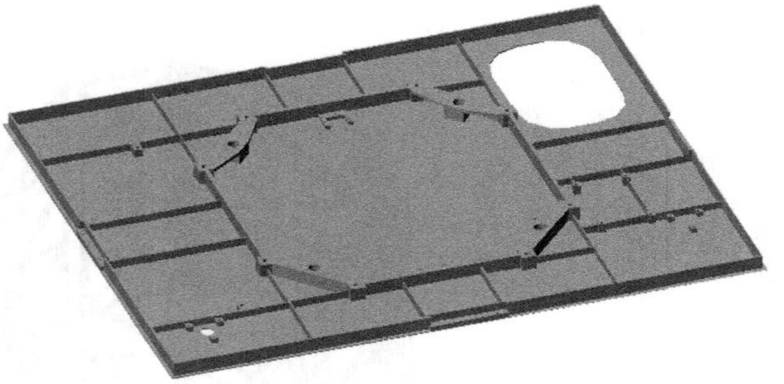

Fig. 3.29 Base plate

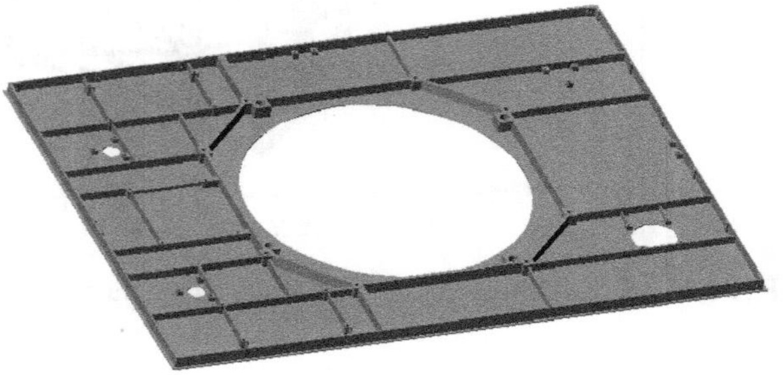

Fig. 3.30 Mounting plate

3.5.2 Mounting Plate

The mounting plate is a 1030 × 800 × 20 mm milled plate made of aluminum alloy AMg6 with built-in milled stiffeners as shown in Fig. 3.30. The stiffeners have rectangular cross-section, and are oriented along lines between holes for eight studs of the satellite case fastening, and the perimeter of plate, and between installation seats of the satellite components. Thickness of the skin is 1.5 mm. Outer and some of the internal stiffeners of the plate have 2 mm thickness, while the rest of the internal stiffeners have 4 mm. Thickness of the rectangular cross-section stiffeners connecting holes for eight studs is 6 mm. Outer and internal stiffeners connecting eight studs have 18.5 mm height, while the rest of the internal stiffeners have 8.5–13.5 mm height.

The mounting plate contains eight 6 mm diameter holes to provide installation for the eight studs connecting the satellite structural subassemblies. Four bushes with internal M8 thread hole are screwed to the mounting plate. These holes are intended for handling operations. Bushes are made of steel 12X18H10T. The mounting plate provides mounting seats for the payload CDH unit, two signal processing units, bracket of GPS antenna, two brackets of conical S-band antennas, X-band antenna, four locking and releasing devices, and four heat shields. It contains a circular hole working as a window for the MBEI. Mass of the mounting plate is 4.2 kg.

Fig. 3.31 Basis unit case

3.5.3 Basis Unit Case

The basis unit case consists of a basis plate and four wall frames connected together and reinforced with the help of two struts. It contains also the star sensor bracket. The plate, frame, bracket, and struts are made of aluminum alloy AMg6. General view of the basis unit case is shown in Fig. 3.31.

Basis plate is a 25 mm height milled plate made of aluminum alloy AMg6. It has L-shaped and rectangular cross-section stiffeners. Selection of the modules cross-sections is made to provide design rigidity for the plate to satisfy the requirement of high-accuracy relative position of the MBEI and ADCS sensors. Stiffeners are oriented as shown in Fig. 3.32. These stiffeners are milled as follows:

- L-shaped cross-section stiffeners—along lines between holes for eight studs of the satellite case fastening and four holes intended for fastening of the plate with the basis frame;
- Rectangular cross-section stiffeners—along the perimeter of plate, and between installation seats of the satellite components.

Thickness of the skin is 1.5 mm. Outer and internal stiffeners of the basis plate have 2 mm thickness and 13.5–23.5 mm height. The basis plate provides mounting seats for MBEI, bracket of star sensor, bracket of magnetorquers, Z-direction reaction wheel, Z-direction gyro, bracket of skewed gyro, interface unit of skewed gyro, and magnetometer. It contains a circular hole working as a window for the MBEI. Mass of the basis plate is 2.66 kg.

Basis frame (230 mm height) consists of four milled walls connected together by rivets as shown in Fig. 3.33. The basis frame has a square cross-section with outer and inner dimensions equal to 565 and 503 mm, respectively. The basis frame walls are made of aluminum alloy AMg6. Stiffeners of the basis walls have rectangular cross-section, and are oriented along the perimeter of each wall and between installation seats of the satellite components. Stiffeners height of the basis frame is 10–16.5 mm. And thickness of the skin is 1.5 mm.

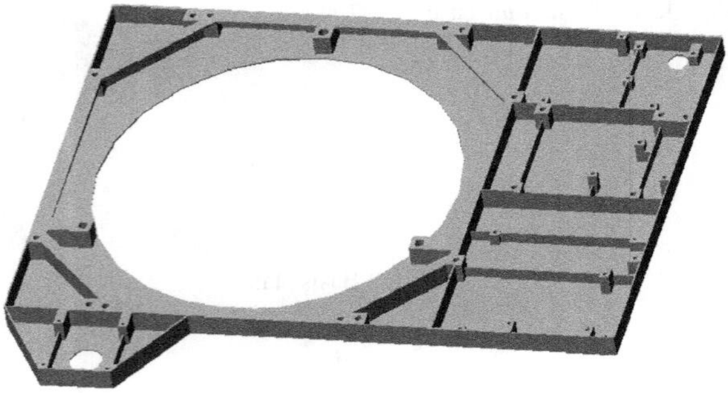

Fig. 3.32 Basis plate

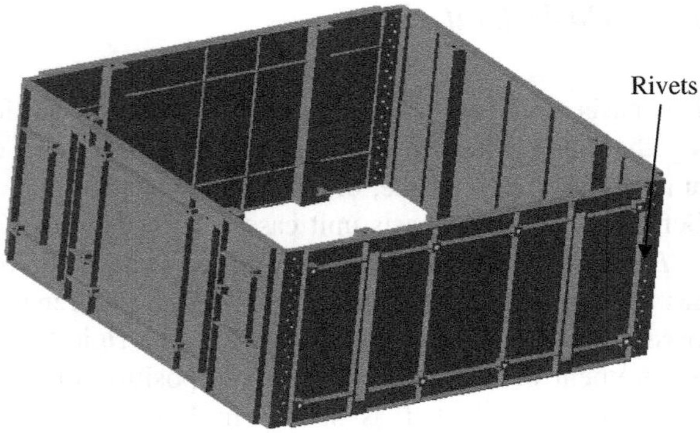

Fig. 3.33 Basis frame

Mounting seats and precise holes are provided on the outer surface of the frame to install the ADCS devices. The basis frame provides mounting seats for two reaction wheels in the Y, and X-directions, both X and Y-direction gyros, and three interface units of X, Y, and Z-direction gyros. One of the frame walls contains two seats for mounting PCU and CLU. The mass of the basis frame is 4.88 kg.

The basis plate and basis frame are fastened together using four M8 screws and reinforced by two diagonal struts as shown in Fig. 3.31. Each of the two diagonal struts is connected to the basis plate and frame using two M6 screws. The diagonal struts are channel bars made of aluminum alloy AMg6. The mass of each strut is 0.08 kg. 86 mm diameter holes in the basis plate and frame are provided for installation of the eight studs connecting the satellite case subassembly. A bracket for installation of the inclined star sensor on the basis unit plate is a welded structure made of aluminum alloy AMg6. The star sensor bracket is installed on the basis plate using four M8 bolts. The total mass of the basis unit case without the star sensor bracket is 7.7 kg.

3.5.4 Upper Frame

The upper frame consists of four milled walls connected together by rivets as shown in Fig. 3.34. The upper frame walls are made of aluminum alloy AMg6.

Fig. 3.34 Upper frame

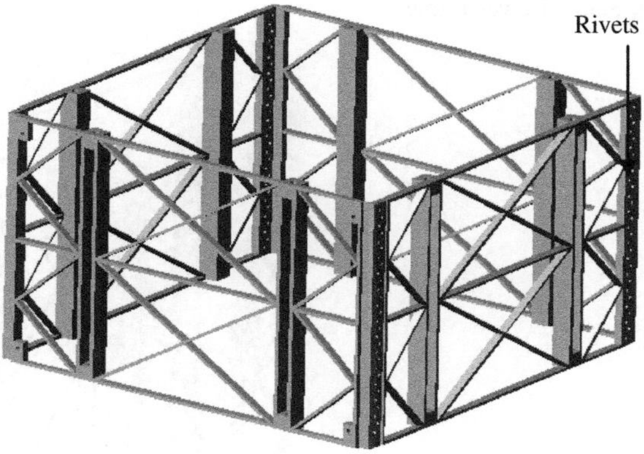

The height of the upper frame is 310 mm, and it has a square cross-section with outer and inner dimensions equal to 551 and 499 mm, respectively. The upper frame milled walls have rectangular cross-section stiffeners of 10 mm width and 4 mm thickness. 86 mm diameter holes are provided in the upper frame for installation of the eight studs connecting structural subassemblies to the satellite case. One wall contains four seats for mounting the bracket of the battery. The mass of the upper frame is 5 kg.

3.5.5 Lower Frame

The lower frame consists of six milled walls connected together by rivets as shown in Fig. 3.35. The lower frame structural walls are made of aluminum alloy AMg6. The height of upper frame is 420 mm, and it has a modified square cross-section with outer dimensions equal to 547 and 556 mm and inner dimension equal to 503 and 503 mm. The lower frame walls have rectangular cross-section stiffeners of 10 mm width and 3 mm thickness. Eight 6 mm diameter holes are provided in the lower frame for installation of the eight studs connecting structural subassemblies at the satellite case. The lower frame provides mounting seats for the X-band electronic module, two S-band electronic modules, GPS electronic module, and three modules of the on-board digital computing complex. One wall contains two seats for mounting PCU and CLU. The mass of the lower frame is 5.3 kg.

3.5.6 Rotation Mechanisms

Rotation mechanisms provide connection of the solar arrays with the satellite body, smooth rotation of the solar arrays into ready-to-work position, and their fixation in the ready-to-work position during satellite lifetime. A rotation

Fig. 3.35 Lower frame

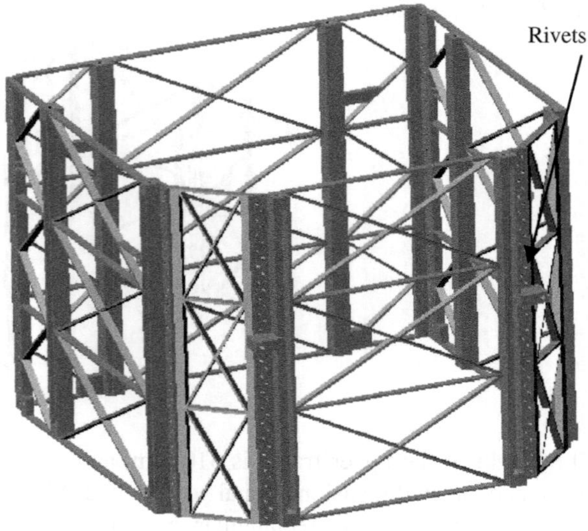

mechanism consists of a fixed part mounted to the satellite body and a rotary part connected to the solar panel as shown in Fig. 3.36. There are two different sizes of rotation mechanisms for two different sets of solar arrays mounted on the base plate. The connection between each rotation mechanism and the base plate is made by two M4 bolts. The masses of the two rotation mechanisms are 0.5 and 0.35 kg.

3.5.7 Locking and Releasing Mechanisms

Locking of the four solar arrays is carried out independently with the help of four pyrotechnic pin-pullers. The structure of the locking and releasing mechanisms of the solar arrays provides:

- Fixation of the solar array in stowed position during storage, transportation, and orbital injection of the satellite;
- Releasing of the solar arrays in orbit.

The locking and releasing mechanisms are mounted on the mounting plate by two M4 bolts for each one. The mass of each locking and releasing mechanism is 0.1 kg. General view of the locking and releasing mechanism is shown in Fig. 3.37.

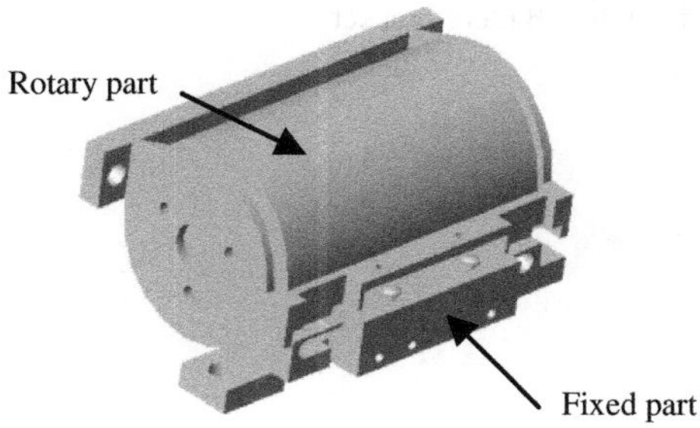

Fig. 3.36 Rotation mechanism

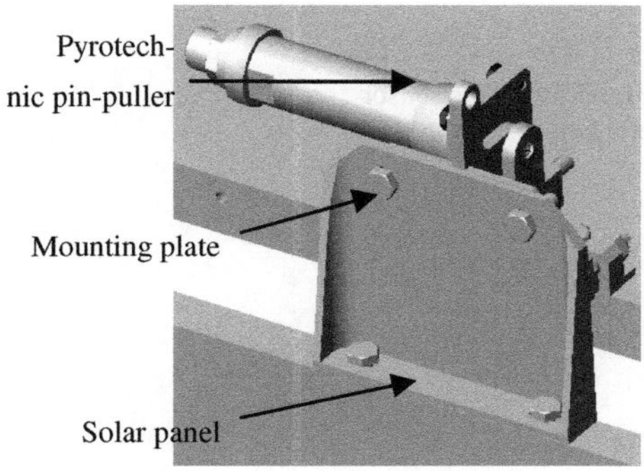

Fig. 3.37 Locking and releasing mechanism

3.5.8 Fastening and Mounting Elements

The fastening and mounting elements include:

- Eight studs connecting the base plate, upper frame, basis unit, lower frame, and mounting plate together. They are 6 mm diameter, 1,070 mm long with external M6 thread at both ends with 20 mm depth. During assembly of the satellite case, these studs are tightened using nuts on the mounting plate and the base plate. Studs are made of titanium alloy VT-16.
- Standard and specially developed elements, and fastening and mounting elements (brackets, bolts, … etc.) for fastening and mounting of the sensors, devices, antennae, socket connectors, heat shields, and other mechanical and electrical units.

The brackets used in Small Sat include the battery bracket, star sensor bracket, GPS antenna bracket, S-band dipole antenna bracket, S-band conical antenna bracket, and skewed gyro bracket. Brackets are made of aluminum alloy AMg6. The total mass of the fastening elements is 1.3 kg, and that of the mounting elements is 3.2 kg. General views of Small Sat structural brackets are shown in Figs. 3.38, 3.39, 3.40, 3.41, 3.42, and 3.43.

Fig. 3.38 Battery bracket

Fig. 3.39 Star sensor bracket

Fig. 3.40 Skewed gyro bracket

3.6 Launch Vehicle Adapter

When a satellite is to be launched by an expendable booster, the satellite developer must provide a launch vehicle adapter, which structurally links the satellite to the launch vehicle. The purpose of the launch vehicle adapter is to provide a compatible interface with the launch vehicle booster, and distribute spacecraft loads uniformly so the booster structure would not be locally overloaded. A tall adapter may be able to distribute concentrated loads introduced at the corners, but length appears to be a scarce resource. The design of a launch vehicle adapter is a very sensitive, because it has a strong effect on the satellite frequencies and mode shapes. The adapter should be quite heavy to successfully transfer loads to the launch vehicle. It often stays behind with the launch vehicle when the satellite separates Fig. 3.44.

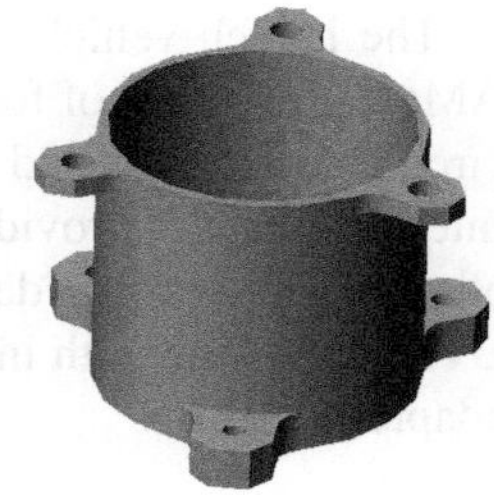

Fig. 3.41 GPS antenna bracket

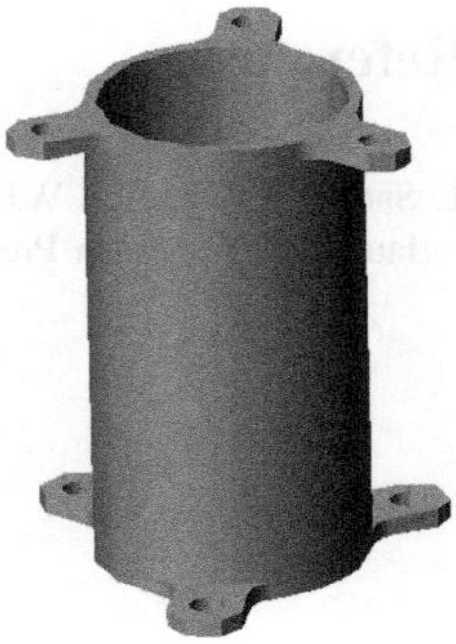

Fig. 3.42 S-band conical antenna bracket

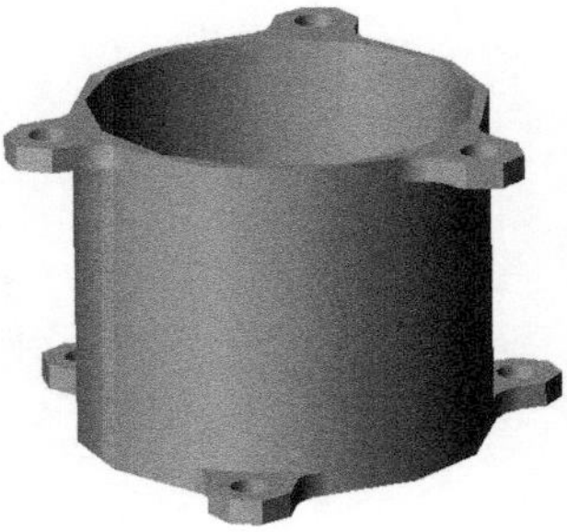

Fig. 3.43 S-band dipole antenna bracket

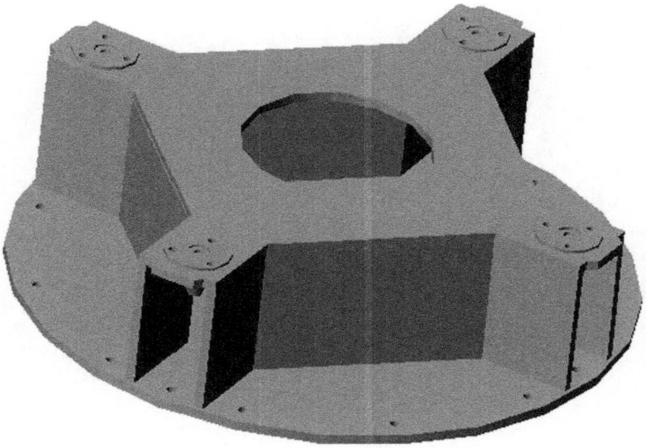

Fig. 3.44 Launch vehicle adapter

The launch vehicle adapter is a welded construction made of aluminum alloy AMg6. It consists of four bended plates working as connection walls between the circular base plate and the interface plate. Four steel bushes are mounted on the interface plate to provide suitable interface between the satellite base plate and LV adapter. Pyrotechnic devices intended for satellite separation are installed on the LV adapter side with interface bushes. The total mass of Small Sat launch vehicle adapter is 19 kg.

Reference

1. Sarafin TP, Larson WJ (eds) (1995) Spacecraft structures and mechanisms—from concept to launch. Microcosm Press and Kluwer Academic Publishers, Torrance

Chapter 4
Strength Analysis

Abstract The satellite structure must be designed to withstand all static and dynamic loads encountered during manufacturing, transportation, launch, and operational life of the satellite. A detailed analysis is made for the satellite structure for all strength analysis cases of loading specified for it. Strength analysis of Small Sat is divided into two categories, one is static analysis and the other is vibration (dynamic) analysis. Strength calculations under static and dynamic loads are conducted according to special calculation models by finite-element method with the help of ANSYS package. Modal calculations are conducted to determine frequencies and forms of natural oscillations for the satellite structure. Dynamic strength calculations are carried out taking into account damage accumulation on the basis of fatigue characteristics of the material.

The satellite structure must be designed to withstand all static and dynamic loads encountered during manufacturing, transportation, launch, and operational life of the satellite. The satellite must be able to withstand the highest loads encountered during its lifespan, and it is also required that the satellite structure be durable during its service life. This is known as designing for the worst case. By ensuring that the satellite will not fail under the worst loading conditions it can be shown that the satellite will not fail under any static or dynamic loads during its lifecycle.

Verification means providing confidence through disciplined steps that a product will do what it is supposed to do. For a high quality product, mechanical requirements are verified by combining analysis, inspection, and testing. With inspection and testing, a lot of time and money is spent before finding out whether the product is any good. Analysis, process development, and development testing are acceptable ways of ensuring quality. The analysis is the first step in the verification process. It provides confidence that the satellite structure will meet requirements for strength, stiffness, natural frequency, and dynamic envelope.

In this chapter, the analysis methods used to verify the integrity of the Small Sat structure are described. The preliminary static analysis that guided the mechanical

G. F. Abdelal et al., *Finite Element Analysis for Satellite Structures*,
DOI: 10.1007/978-1-4471-4637-7_4, © Springer-Verlag London 2013

design explained in the last chapter is considered as the first step in the verification process. To refine the preliminary analysis, a finite-element model is created by ANSYS package to conduct a detailed analysis for Small Sat structure to ensure survivability of the stack configuration during its lifecycle.

A detailed analysis must be made for the satellite structure for all strength-analysis cases of loading specified for it. Strength analysis of Small Sat is divided into two categories, one is static analysis and the other is vibration (dynamic) analysis. Strength calculations under static and dynamic loads are conducted according to special calculation models by finite-element method with the help of ANSYS package. Modal calculations are conducted to determine frequencies and forms of natural oscillations for the satellite structure. Dynamic strength calculations are carried out taking into account damage accumulation on the basis of fatigue characteristics of the material.

4.1 Satellite Model

The first step before starting strength analysis of Small Sat structure is to build the finite-element model on ANSYS package. One of the most difficult problems in structural analysis is idealizing the structure, or representing it in a simple model easy to analyze. Assumptions have to be made to idealize a structure; these assumptions can be the strength or weakness of the analysis. They may simplify a problem, so that its solution may be estimated in a couple of hours while it would otherwise take days, or they may give wrong answers.

Finite-element modeling should provide the following objectives:

- Provide a model that will allow obtaining the necessary information at the desired accuracy (e.g., strains, stresses, mode shapes, displacements, and stiffness).
- Simplify the problem to the greatest extent while still satisfying the objective.

The effective use of any analysis requires insight regarding the problem, awareness of the sensitivity of assumptions, and disciplined documentation and checking. The following procedure describes the process that may lead to reliable, cost-effective finite-element analysis:

1. Understand the problem. Define the objectives of the analysis and identify any constraints such as cost, schedule, and available hardware and software.
2. Define the desired output from the finite-element analysis.
3. Decide on the modeling strategy, including class of model, types of elements, level of detail, boundary conditions, manner of loading, and method of generating the model.
4. Estimate cost and time for the analysis to make sure the plan is acceptable.
5. Generate the model.
6. Check the model.

7. Run analysis.
8. Check the analysis results.
9. Check how sensitive the results are to modeling assumptions.
10. Process the finite-element analysis results.
11. Document the analysis.

The approach that will be taken in finite-element modeling depends on the analysis purposes. Therefore, the satellite model should be appropriate to determine the level of detail required by the analysis to get the needed results. In general, finite-element models fall into three classes:

- One-dimensional beam models
- Two- and three-dimensional models
- Detailed models.

The effective selection of any class depends on Small Sat model objectives, which are listed as following:

- Predict stresses and deformations to assess strength and fatigue life.
- Predict natural frequencies and modes of vibration.
- Predict structural response (displacement, accelerations, and member loads) to time-varying forces.

Hence, the most efficient class for Small Sat structure is the detailed model, which includes stiffeners and skin in each structural module. Skin is represented by shell finite elements, while stiffeners are represented by solid finite elements. A detailed model is developed to predict strength or stiffness in a more accurate manner than a simple method can provide. It identifies peak stresses and provides stress-contour plots. The detailed model of Small Sat structure is a complex system, so minor details are neglected to simplify the model. The analysis is carried out using ANSYS package by means of a substructuring technique.

4.2 Substructuring Technique

Substructuring (superelement method) is a procedure that condenses a group of finite elements into one element represented as a matrix [1]. This single matrix element is called a superelement. A superelement can be used in the analysis as any other element type. The only difference is that the superelement is first created by performing a substructure generation analysis. Substructuring reduces computer time and allows solution of very large problems with limited computer resources.

Substructuring includes the following operations:

- Hypothetically, divide the structure into a set of substructures
- Separate model portions of a structure (substructures)
- Simplify and remove the fine details in each substructure and use the reduction method to reduce the size of its mass and stiffness matrices

- Assemble the simplified substructure mass and stiffness matrices to form a coupled model
- Calculate the response of the assembled structure model to the applied forces
- Back-solve to obtain displacements, strains, and stresses in each substructure.

4.2.1 Substructuring Analysis

The substructure analysis uses the technique of matrix reduction to reduce the system matrices to a smaller set of degrees of freedom DOFs. Matrix reduction is also used by the reduced modal, harmonic, and transient analyses. A superelement (substructure) may be used in any analysis type. It simply represents a collection of elements that are reduced to act as one element. This one (super) element may then be used in the actual analysis (use pass) or be used to generate more superelements (generation or use pass). To reconstruct the detailed solutions (e.g., displacements and stresses) within the superelement, an expansion pass may be done. The static analysis solution method is valid for all DOFs). Inertial and damping effects are ignored, except for static acceleration fields. The overall finite-element equilibrium equations for linear structural static analysis are:

$$[K]\{u\} = \{F\} \tag{4.1}$$

Equation (4.1) may be partitioned into two groups; the master (retained) DOFs, here denoted by the subscript "m", and the slave (removed) DOFs, here denoted by the subscript "s".

$$\begin{bmatrix} [K_{mm}] & [K_{ms}] \\ [K_{sm}] & [K_{ss}] \end{bmatrix} \begin{Bmatrix} \{u_m\} \\ \{u_s\} \end{Bmatrix} = \begin{Bmatrix} \{F_m\} \\ \{F_s\} \end{Bmatrix} \tag{4.2}$$

Expanding:

$$[K_{mm}]\{u_m\} + [K_{ms}]\{u_s\} = \{F_m\} \tag{4.3}$$

$$[K_{sm}]\{u_m\} + [K_{ss}]\{u_s\} = \{F_s\} \tag{4.4}$$

The master DOFs should include all DOFs of all nodes on surfaces that connect to other parts of the structure. If accelerations are to be used in the use pass or if the use pass is a transient analysis, master DOFs throughout the rest of the structure should also be used to characterize the distributed mass. Solving Eq. (4.4) for $\{u_s\}$,

$$\{u_s\} = [K_{ss}]^{-1}\{F_s\} - [K_{ss}]^{-1}[K_{sm}]\{u_m\} \tag{4.5}$$

Substitution $\{u_s\}$ into Eq. (4.3)

$$\Big[[K_{mm}] - [K_{ms}][K_{ss}]^{-1}[K_{sm}]\Big]\{u_m\} = \{F_m\} - [K_{ms}][K_{ss}]^{-1}\{F_s\} \tag{4.6}$$

Or alternatively:

$$[\hat{K}]\{\hat{u}\} = \{\hat{F}\} \tag{4.7}$$

where:

$$[\hat{K}] = [K_{mm}] - [K_{ms}][K_{ss}]^{-1}[K_{sm}] \tag{4.8}$$

$$\{\hat{F}\} = \{F_m\} - [K_{ms}][K_{ss}]^{-1}\{F_s\} \tag{4.9}$$

$$\{\hat{u}\} = \{u_m\} \tag{4.10}$$

$[\hat{K}]$ and $\{\hat{F}\}$are the superelement reduced stiffness matrix and load vector, respectively.

In the preceding development, the load vector for the superelement has been treated as a total load vector. The same derivation may be applied to any number of independent load vectors, which in turn may be individually scaled in the superelement use pass. For example, the analyst may wish to apply thermal, pressure, gravity, and other loading conditions in varying proportions. Expanding the right-hand sides of Eqs. (4.3) and (4.4):

$$\{F_m\} = \sum_{i=1}^{N}\{F_{mi}\} \tag{4.11}$$

$$\{F_s\} = \sum_{i=1}^{N}\{F_{si}\} \tag{4.12}$$

where:

N = number of independent load vectors.

Substitution into Eq. (4.9)

$$\{\hat{F}\} = \sum_{i=1}^{N}\{F_{mi}\} - [K_{ms}][K_{ss}]^{-1}\sum_{i=1}^{N}\{F_{si}\} \tag{4.13}$$

To have independently scaled load vectors in the use pass, expand the left-hand side of Eq. (4.13) as:

$$\{\hat{F}\} = \sum_{i=1}^{N}\{\hat{F}_i\} \tag{4.14}$$

Substituting Eq. (4.14) into Eq. (4.13):

$$\{\hat{F}_i\} = \{F_{mi}\} - [K_{ms}][K_{ss}]^{-1}\{F_{si}\} \tag{4.15}$$

If the load vectors are scaled in the use pass such that:

$$\{\hat{F}\} = \sum_{i=1}^{N} b_i\{\hat{F}_i\} \tag{4.16}$$

where b_i is the scaling factor, then Eq. (4.5) becomes:

$$\{u_s\} = [K_{ss}]^{-1}\sum_{i=1}^{N} b_i\{F_{si}\} - [K_{ss}]^{-1}[K_{sm}]\{u_m\} \tag{4.17}$$

Equation (4.17) is used in the expansion pass to obtain the DOF values at the slave DOFs if the back-substitution method is chosen. If the resolve method is chosen for expansion pass, then the program will use Eq. (4.2) to resolve for $\{u_s\}$. The program makes $\{u_m\}$ as the internally prescribed displacement boundary conditions since $\{u_m\}$ are known in the expansion pass. As the program treats DOFs associated with $\{u_m\}$ as displacement boundary conditions, the reaction forces by the resolve method will be different from those computed at those master DOFs by the back-substitution method. However, they are all in self-equilibrium satisfying Eq. (4.2).

4.2.2 Employing Substructuring

A substructure analysis [1] involves three distinct steps, called passes:

1. Generation pass
2. Use pass
3. Expansion pass

Figure 4.1 shows the data flow for a complete substructure analysis and some of the files involved. The three passes are explained next in detail.

4.2.2.1 Generation Pass: Creating the Superelement

The *generation pass* is where a group of "regular" finite elements are condensed into a single superelement. The condensation is done by identifying a set of master degrees of freedom, used mainly to define the interface between the superelement and other elements and to capture dynamic characteristics for dynamic analyses. The procedure to generate a superelement consists of two main steps:

A. *Build the model.* In this step, the jobname and analysis title are specified, and then the element types, element real constants, material properties, and the model geometry are defined. These tasks are common to most analyses.

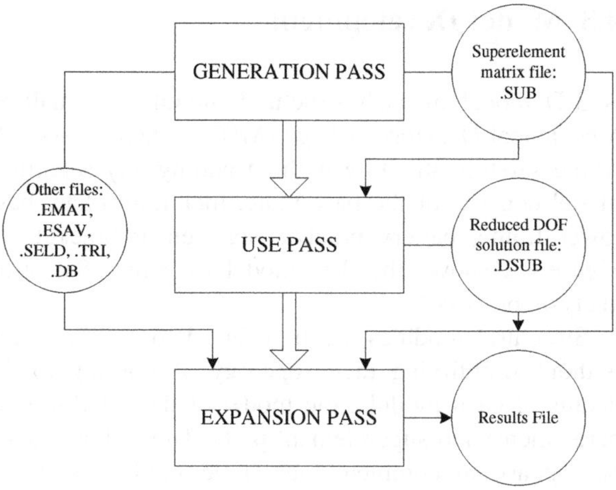

Fig. 4.1 Data flow for a complete substructure analysis [1]

B. *Apply loads and create the superelement matrices.* The "solution" from a substructure generation pass consists of the superelement matrix (or matrices). As with any other analysis, define the analysis type and options, apply loads, specify load step options, and initiate the solution.

In a substructure, the master DOFs serve three purposes:

- They serve as interface between the superelement and other elements.
- If the superelement is to be used in a dynamic analysis, master DOFs characterize the dynamic behavior of the structure.
- Master DOFs are used to calculate the deflection at any points in the superelement.

4.2.2.2 Use Pass: Using the Superelement

The *use pass* is where the superelement is used in an analysis by making it part of the model. The entire model may be a superelement or the superelement may be connected to other nonsuperelements. The solution from the use pass consists only of the reduced solution for the superelement (that is, the DOF solution only at the master DOF), and the complete solution for nonsuperelements. The use pass can involve any analysis type. The only difference is that one or more of the elements in the model are a superelement that has been previously generated.

4.2.2.3 Expansion Pass: Expanding Results Within the Superelement

The *expansion pass* is where the analysis starts with the reduced solution, and calculates the results at all DOF in the superelement. If multiple superelements are used in the use pass, a separate expansion pass will be required for each superelement.

4.3 Model Development

A 3-D model of each structural module of Small Sat is created by means of Mechanical Desktop package (MDT). Then an assembly model is formed for the whole satellite structure without adding any satellite equipment. The assembled model consists of the base plate, mounting plate, basis unit block, upper frame, lower frame, battery bracket, star sensor bracket, and launch vehicle adapter. Figure 4.2 shows the 3-D model of Small Sat structure used in the strength analysis process.

Structural modules are transferred to ANSYS package using “SAT” format without simplifying their topology. According to the complexity of structural details of each module, the model of the satellite is divided into virtual (nonsuperelement) and superelement parts. The virtual parts are created through the use pass phase to complete the whole satellite structure model in ANSYS. The superelement parts of the model are divided into seven modules:

1. Base plate with four connection bushes
2. Mounting plate
3. Basis plate with two diagonal struts and star sensor bracket
4. Basis walls
5. Upper frame with battery bracket
6. Lower frame
7. Launch vehicle adapter

4.4 Material Model Behavior

The materials of the satellite structural modules are assumed to behave in the elastic region. Therefore, linear isotropic behavior is used. Structural modules are made of aluminum alloy AMg6 ($E = 7.2 \times 10^3$ MPa, $v = 0.33$, $\rho = 2630$ kg/m^3, $\sigma_y = 150$ MPa, $\sigma_u = 310$ MPa), while the connection elements (bushes, bolts, studs … etc.) are made of steel ($E = 20.7 \times 10^3$ MPa, $v = 0.33$, $\rho = 7650$ kg/m^3, $\sigma_y = 620$ MPa, $\sigma_u = 862$ MPa).

4.5 Modeling of Equipments

Each satellite’s equipment is fixed in its specific location with connation elements (bolts or studs). Most equipment are connected at three or four fixation points. Only rotation mechanisms and locking and releasing devices are fixed at two points each. For modeling simplicity, rotation mechanisms, locking and releasing devices, and solar panels are modeled as localized masses attached to their

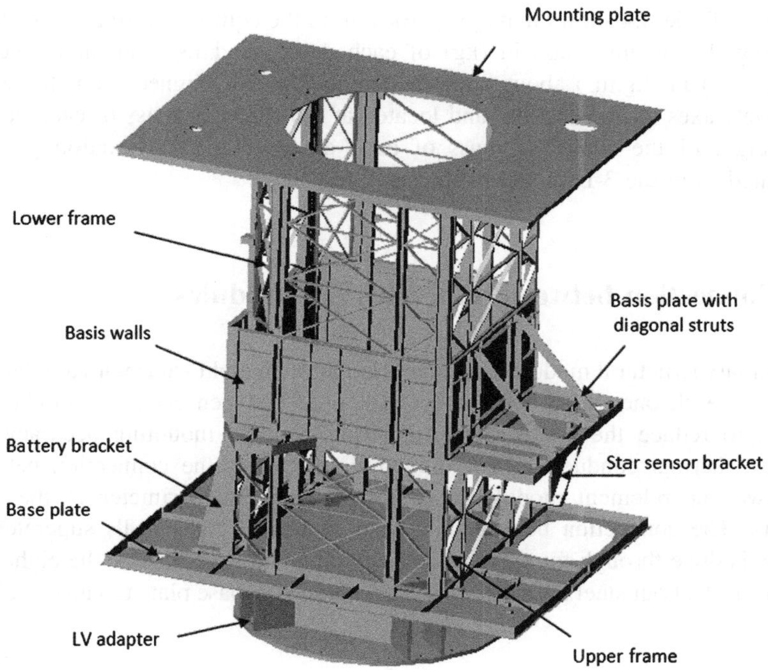

Fig. 4.2 3-D model of Small Sat structure used in the strength analysis process [6]

fastening points, because the rotation hinge do not transfer rotation inertia loads to the satellite structure. For other equipment, one of two methods can be used for equipment modeling:

1. The equipment is modeled as a pyramid with zero-density elements. The head of the pyramid is located at the equipment's center of mass and its base points are at the center of the fixation locations with its structural module. A mass element is attached at the head of the pyramid to represent the mass of the equipment with its moments of inertia. Zero-density elements used in the pyramid modeling have the same properties of aluminum alloy AMg6.
2. The equipment is modeled by connecting its center of fixation locations and then this closed contour is extruded to double the center of mass of the equipment. The volume of such a shape is calculated, and the suitable density that will make it equal the equipment in mass is evaluated. Equivalent density elements are used to model the equipment with the same properties of aluminum alloy AMg6.

The first method is more commonly used for equipment modeling because it provides more accurate modeling. This method represents the closest model for the actual equipment, because it models the mass of the equipment with its moments of inertia. Table 4.4 shows mass properties of all the equipment used in Small Sat modeling. It contains mass in (kg) of each device and its moment of inertias (I_{ox}, I_{oy}, I_{oz}) in (kg.m^2) about three perpendicular axes aligned with the design coordinate axes of the satellite, and located at the center of mass of each device. The height of the center of mass of each device above its fixation plane is calculated from the 3-D model of the satellite.

4.6 Connection between Superelement Modules

The primary structural modules are connected through eight studs with a tightening torque on each one. Steel thin washers are used between each two modules as spacers to reduce the contact area, and thus provide mounting accuracy and decrease thermal conduction. For modeling simplicity, the connection between each two superelement modules is assumed along the perimeter of the eight washers. The connection between the basis plate and basis walls superelement module is done through four screws and two diagonal struts beside the eight connection struts. Four steel bushes are used to connect the base plate to launch vehicle adapter.

4.7 Finite-Element Models of Structural Modules

All details that are unnecessary to strengthen the structural modules are removed from the 3-D model. All small-radius fillets are deleted and unnecessary small holes are filled. Each superelement module is exported separately from MDT as "*.SAT" file, then it is imported at ANSYS package. Exportation process must be done for all superelement modules from the same source file, which is the 3-D assembly model of the whole satellite structure. This is done to ensure accurate integrity during model generation in the use pass process. The modeling technique of all superelement modules can be summarized in the following steps:

1. Import the file "*.SAT" from MTD package. Remove volumes and details that are unnecessary to strength analysis of the structural modules.
2. Select the type of elements. In the present analysis, SOLID92 and MASS21 are used. The first type models the structural modules and the satellite equipment's body. The second type models the equipment's mass with its moments of inertia.
3. Define real constants of the mass elements (MASS21) for each equipment. They consist of the equipment's mass and its moments of inertia about three perpendicular axes (I_{ox}, I_{oy}, and I_{oz}) aligned with the design coordinate axes of the satellite and located at the center of mass of each device.

4. Define material properties (modulus of elasticity, Poisson's ratio, and density) of the elements. There are three materials used in the modeling process, aluminum alloy AMg6, steel, and zero-density material.
5. Assign the location of the interface between each superelement and other elements (master DOF location).
6. Fill holes that are used for equipments fixation with volumes (cylinders). These cylinders represent the fastening elements (bolts and studs), which are modeled by SOLID92 and steel material. Locate the center of the circular face of each cylinder, which represents one of the base points of the equipment models.
7. Build the equipment models as described in Sect. 4.5 according to the first method.
8. Meshing process is started by applying some criterion to control element seizing. This will result in a smaller number of elements with little effect on the analysis accuracy. Moreover, the computer run time for each superelement module is reduced. Before meshing the volume of each structural module the following steps are suggested:

   a. Identify lines with length (0–3 mm), and choose the number of element divisions = 1.
   b. Identify lines with length (3–30 mm), and choose the number of element divisions = 3.
   c. Identify lines with length (>30 mm), and choose the element edge length = 15 mm.

9. Mesh the volumes of the structural modules with SOLID92 elements and AMg6 material. And mesh the volumes of the attachment elements (bushes, bolts, and studs) with SOLID92 elements and steel material.
10. Mesh the volumes of the equipment models with SOLID92 elements and zero-density material. These volumes are meshed with very coarse meshing to ease the analysis process without large effect on the result accuracy, as only their weight is needed in the analysis.
11. Generate the mass elements (MASS21) at the heads of pyramids, which represent the equivalent equipments mass and their moments of inertia.
12. Apply rotational boundary conditions at all mass elements by preventing free rotation around the three perpendicular axes.
13. Pick master DOF at nodes attached to locations specified in step 4.
14. Solve substructures to generate the superelement files. Thus, the generation pass phase is ended.

Figures 4.3, 4.4, 4.5, 4.6, 4.7, 4.8, and 4.9 show finite-element models of structural modules. They illustrate the location of master DOF at connection nodes with other superelement modules. Different satellite's equipment are presented by pyramid shapes and mass elements with rotational boundary conditions. Master DOFs are chosen as follows:

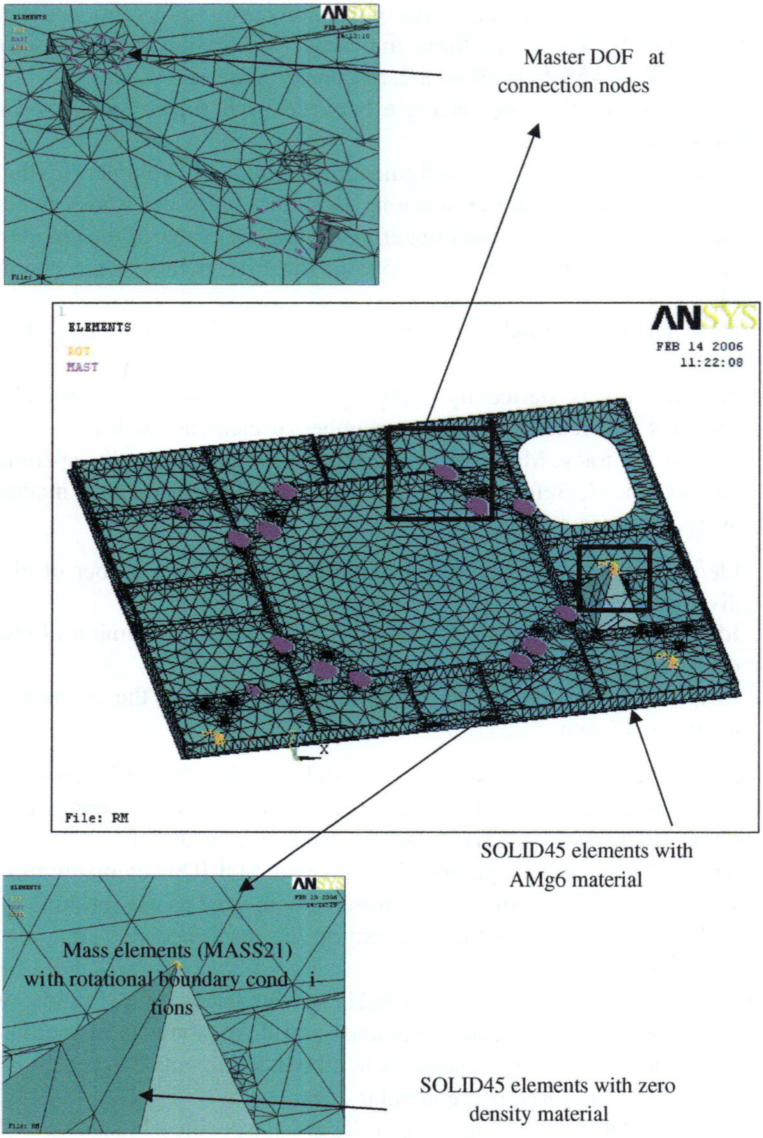

Fig. 4.3 Detailed F.E. model of base plate

- *LV adapter*: 48 nodes divided into four circles (12 nodes each) representing the connection with the base plate; and 117 nodes located at the interface plane with the LV platform, which are the fixation nodes for the whole satellite model.
- *Base plate module*: 48 nodes representing the connection with the LV adapter, 96 nodes divided into eight circles representing the connection with the upper

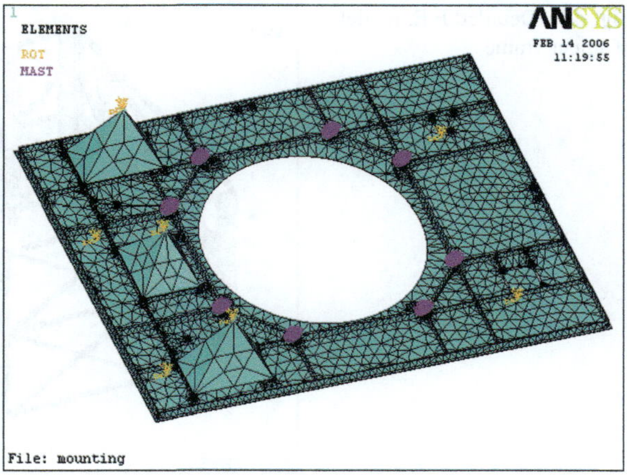

Fig. 4.4 Detailed F.E. model of mounting plate [6]

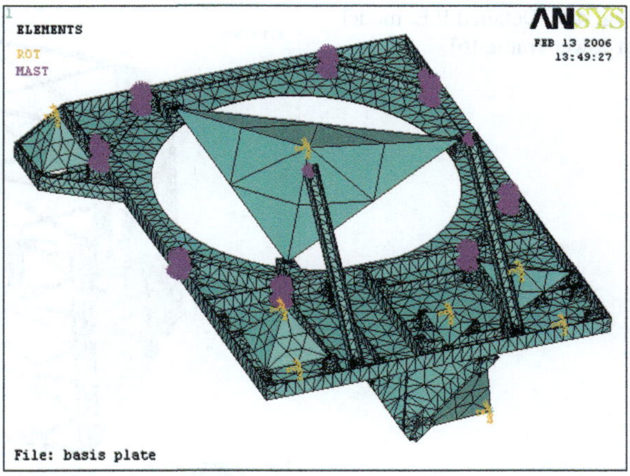

Fig. 4.5 Detailed F.E. model of basis plate [6]

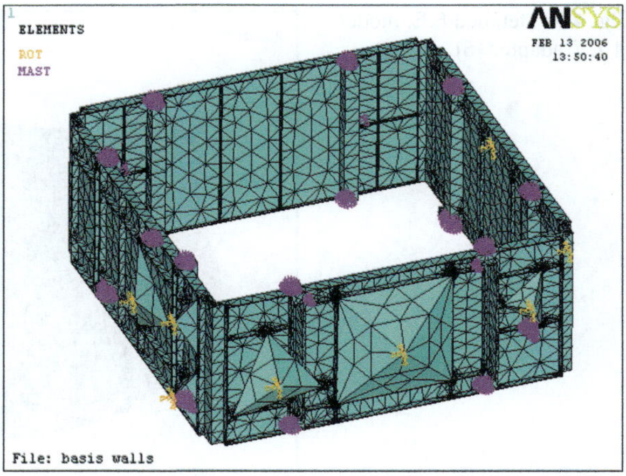

Fig. 4.6 Detailed F.E. model of basis walls

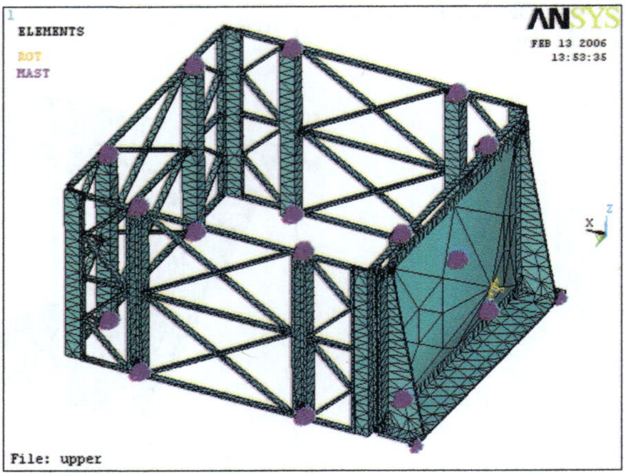

Fig. 4.7 Detailed F.E. model of upper frame

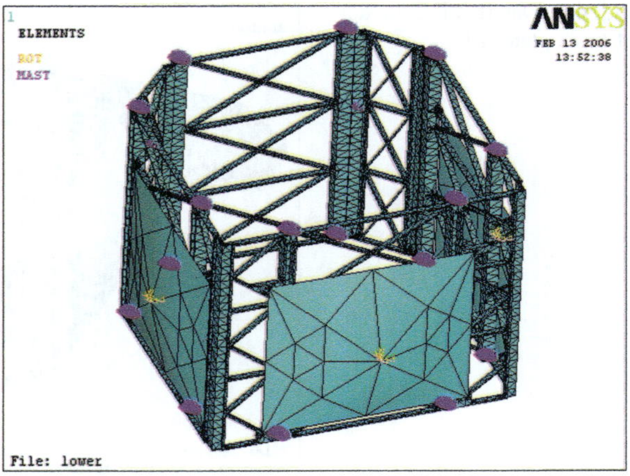

Fig. 4.8 Detailed F.E. model of lower frame [6]

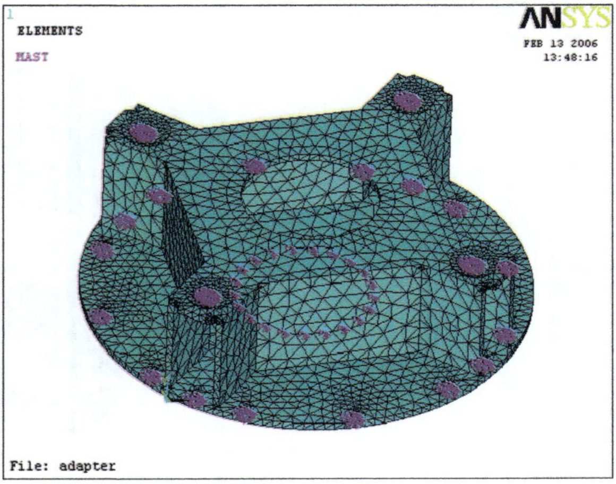

Fig. 4.9 Detailed F.E. model of LV adapter [6]

frame, and 12 nodes divided into two circles representing the connection with the battery bracket.

- *Upper frame module*: 96 nodes representing the connection of the frame with the base plate, 12 nodes representing the connection of the battery bracket with the base plate, and 96 nodes divided into eight circles representing the connection with the basis plate module.
- *Basis plate module*: 96 nodes representing the connection with the upper frame, 96 nodes divided into eight circles representing the connection with the basis walls module, 24 nodes divided into four circles representing the connection of the plate with the basis walls through four bolts, and 12 nodes divided into two circles representing the connection of the diagonal struts with the basis walls through two bolts.
- *Basis walls module*: 132 nodes representing the connection with the basis plate module, 96 nodes divided into eight circles representing the connection with the lower frame module, and two nodes located at the same wall representing two points of the base connections for the equipment model of Power Conditioning Unit (PCU) and Cell Leveling Unit (CLU). In the use pass phase, this equipment model is created as a virtual element during building the satellite model.
- *Lower frame module*: 96 nodes representing the connection with the basis walls module, 96 nodes divided into eight circles representing the connection with the mounting plate module, and two nodes located at the same wall representing the other two points of the base connections for the equipment model of PCU and CLU.
- *Mounting plate module*: 96 nodes divided into eight circles representing the connection of the plate with the lower frame module.

Table 4.1 lists the total number of elements and nodes used to mesh each structural module, in addition to the total number of selected master DOF. As such, the total number of master DOF is 3,819 associated with 1,273 nodes. Each selected node has only three DOF (U_x, U_y, and U_z) because they belong to the ANSYS solid element "SOLID92". It has a quadratic displacement behavior and is well suited to model irregular meshes. SOLID92 is defined by 10 nodes having three translational DOFs at each node.

4.8 Global Finite-Element Model of Small Sat

The entire finite-element model of Small Sat is generated during use the pass phase. It is used in static, modal, and dynamic analyses. It consists of the seven superelement modules of the primary structure with the launch vehicle adapter and a virtual element (nonsuperelement). The virtual part is the equipment model of the PCU and the CLU. They are modeled together as a single part. This virtual part is modeled as a pyramid with zero-density elements like all other satellite

Table 4.1 Number of elements and nodes used to mesh each structural module and the number of selected master DOF

Structure module	Number of elements	Number of nodes	Number of master DOF
LV adapter	20,411	38,823	495
Base plate	29,881	64,071	468
Upper frame	32,425	72,269	612
Basis plate	35,015	65,014	684
Basis walls	49,096	93,063	690
Lower frame	38,155	83,357	582
Mounting plate	35,727	68,642	288

equipment. The modeling procedure of the entire finite-element model of Small Sat can be summarized in the following steps:

1. Select the type of elements. MATRIX50, SOLID92, and MASS21 types are used. The first type models the seven superelement modules of the primary structure with the launch vehicle adapter. The second type models the satellite equipment's body (PCU and CLU). The third type models the equipment's mass and moments of inertia.
2. Define real constants of MASS21 elements for the PCU and CLU model. It consists of their total masses and moments of inertia about three perpendicular axes.
3. Define material properties (modulus of elasticity, Poisson's ratio, and density). Only one material is used in the modeling process, which is zero-density material.
4. Create a modified superelement from an existing superelement and write the new one to a new file (*.SUB). A new superelement is created from the original one by offsetting its node numbers without transferring its geometry from the active coordinate system into another coordinate system, or offsetting its geometry in the global Cartesian coordinate system. This step is done to overcome node number overlap during assembly of the entire model from different superelement modules having the same node numbers. Be sure that offsetting increments are suitable and sufficient to prevent nodal number repetition. New superelement files are created for all the superelement modules except the LV adapter.
5. Define the superelement by pointing to the proper element type (MATRIX50) reference number, and reading in the superelement matrix of the LV adapter and the new six matrixes of structural modules. As such, there are seven superelements containing 1,273 nodes with 3,819 master DOF.
6. Verify the location of each superelement using graphical displays and listings. Superelements are represented by an edge outline display, the data for which are written to the matrix file in the generation pass. The interface master node locations in each superelement must exactly match the locations of the corresponding master nodes on the other superelements.

7. Couple coincident nodes with a small tolerance for all DOF. Check each connection separately, and manually couple the necessary nodes if not already coupled. Never share any node to more than one set of coupling. The total number of coupled sets until this step is 576.
8. Create four keypoints at the existing node locations; two nodes located in the basis walls module, and the others in the lower frame module. These keypoints are the base points of the PCU and CLU equipment models.
9. Build the equipment model for PCU and CLU as described in Sect. 4.2 according to the first method.
10. Mesh the volume of the equipment model (48 solid elements including 119 nodes), generate the mass element and apply rotational boundary condition as described before.
11. Manually couple the coincident nodes at the four base points (four coupled sets) of the equipment model for all DOF.
12. Apply displacement boundary conditions at the nodes (117 nodes) attached to the launch vehicle adapter and located at the interface plane with the launch vehicle platform. Fix the 96 outer nodes at all DOF, but fix the 21 inner nodes in the Z-direction only.
13. Apply loads and obtain the solution. This step is performed during the solution phase of the analysis. The procedure to obtain the use-pass solution depends on the analysis type, since each type has its specific loads.

Solutions include of the complete solution for nonsuperelements, and the reduced solution (DOF solution at master nodes) for the superelements. The complete solution for nonsuperelements is written to the results file (*.RST), and the reduced solution is written to the file (*.DSUB). Figure 4.10 shows the entire finite-element model of Small Sat.

4.9 Static Strength Analysis

In this section, a detailed steady-state static strength analysis is performed for Small Sat structure. The entire finite-element model of the satellite beside the substructural superelement modules is used during the static analysis process. The main purpose of this analysis is to ensure that the satellite structure can withstand all static loads encountered during the manufacturing, transportation, and launch without excessive deformation.

4.9.1 Definition of Static Analysis

A static analysis calculates the effects of steady loading conditions on a structure, while ignoring inertia and damping effects, caused by time-varying loads. A static analysis can include steady inertia loads (such as gravity and rotational velocity),

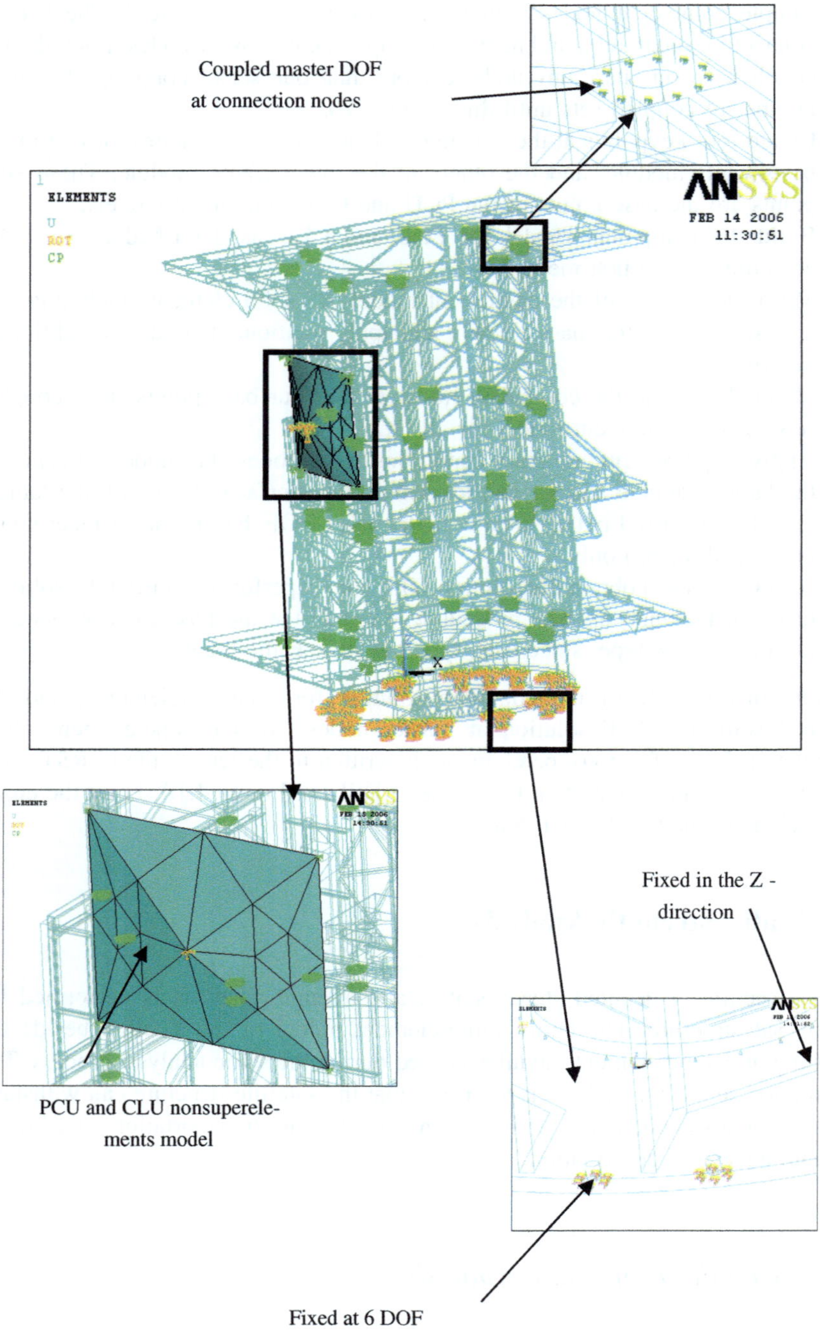

Fig. 4.10 The entire finite-element model of Small Sat [6]

and time-varying loads that can be approximated as equivalent static loads (such as static equivalent wind). A static analysis can be either linear or nonlinear. All types of nonlinearities are allowed: large deformations, plasticity, creep, stress stiffening, contact elements, and so on. This section focuses on linear static analyses only. Static analysis is used to determine the displacements, stresses, strains, and internal forces in structures or components caused by loads that do not induce significant inertia or damping effects. Steady loading and response conditions are assumed; that is, the loads and the structure's response are assumed to vary slowly with respect to time. The types of loading that can be applied in a static analysis include:

- Externally applied forces and pressures
- Steady-state inertial forces (such as gravity or rotational velocity)
- Imposed (nonzero) displacements
- Temperatures (for thermal strain)

4.9.2 Small Sat Quasi-Static Loads

For Small Sat, the most critical quasi-static loads are the two strength analysis cases described in Sect. 3.4.1. The first case is the satellite loading by operational g-loads during road transportation in the container. The second is the maximum longitudinal and lateral g-loads during flight of the launch vehicle. Table 4.2 shows the critical limit loads during the worst cases of inertia loading.

4.9.3 Performing a Static Analysis

The procedure for a static analysis of Small Sat structure consists of these steps:

1. Build the model: The entire finite-element model of the satellite beside the substructural superelement modules is used during the static analysis. These models are built as shown in Sect. 4.1. The following points are taken into consideration during building these models:
2. Linear structural elements are used:
   a. Linear, isotropic, and constant temperature-independent material properties are used.
   b. Young's modulus, Poisson's ratio, and density are defined for stiffness and mass calculations.
   c. Regions with large variation in stresses or strains require a relatively finer mesh than regions where stresses or strains are nearly constant.
3. Apply the loads and define boundary conditions: All load types are applicable in a static analysis. For Small Sat, the two strength analysis cases listed in

Table 4.2 Critical limit loads during the worst cases of inertia loading

Loading case	The critical limit loads (m/s^2)	
Case 1	N_x	39.24
	N_y	24.53
	N_z	53.96
Case 2	N_a	105.85
	N_l	7.64

where
N_x lateral acceleration in the forward-motion direction during road transportation
N_y lateral acceleration in the Y-axis during road transportation
N_z axial acceleration in the Z-axis during road transportation
N_a axial "longitudinal" acceleration during launch
N_l lateral "shear" acceleration during launch

Table 4.2 are applied individually for the entire finite-element model of the satellite. Displacement boundary conditions are DOF constraints usually specified at model boundaries to define rigid support points. They are defined in Sect. 4.1.

4. Set solution controls and solve the Analysis: A structural static analysis type is defined during performing the analysis on ANSYS package. The default settings that will work well for many structural static analyses are selected. The entire model of the satellite is first solved for static analysis for each load case. Then each superelement module is resolved in each case of loading to calculate the results at all DOF in the superelement for each load case.
5. Reviewing of results: Results from static analysis of the entire model and the expansion pass for each superelement module are written to the structural results files. They contain the following data:
6. Primary data: Nodal displacements and rotations (U_X, U_Y, U_Z, ROT_X, ROT_Y, and ROT_Z).
7. Derived data: Nodal and element stresses, Nodal and element strains, Element forces, Nodal reaction forces, … etc.

4.9.4 Static Analysis Results

The stress–strain state of the structure modules is determined as a result of the static analysis. The diagrams of equivalent Von Mises stress distribution (10^3 MPa) and displacement distribution (mm) of each structural module are shown in Figs. 4.11, 4.12, 4.13, 4.14, 4.15, 4.16, 4.17, 4.18, 4.19, 4.20, 4.21, 4.22, 4.23, 4.24, 4.25, 4.26, 4.27, 4.28, 4.29, 4.30, 4.31, 4.32, 4.33, 4.34, 4.35, 4.36, 4.37, and 4.38. Displacements in these figures are relative to the points of the LV adapter attachment to the LV platform, which are the fixation points for the entire model of Small Sat structure. The stress values are determined according to Von Mises criterion represented by the equation:

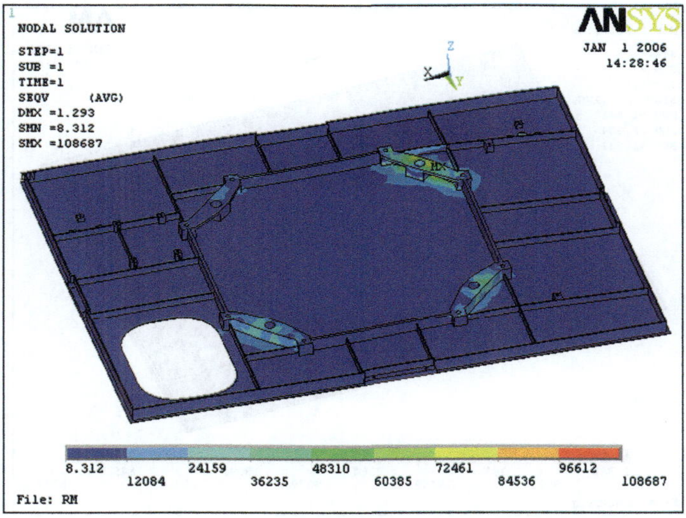

Fig. 4.11 Von Mises stress distribution in base plate in case 1

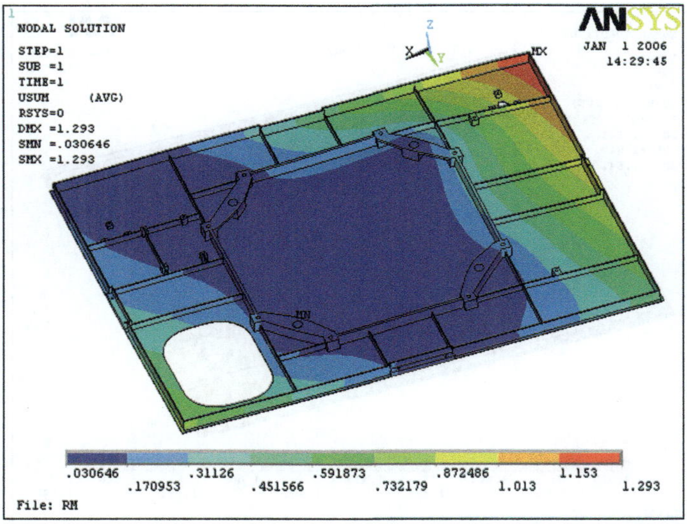

Fig. 4.12 Displacement distribution in base plate in case 1 Mounting plate

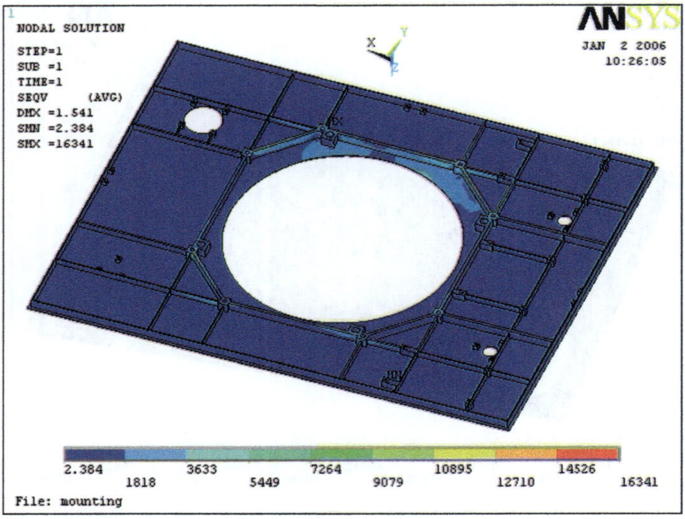

Fig. 4.13 Von Mises stress distribution in mounting plate in case 1

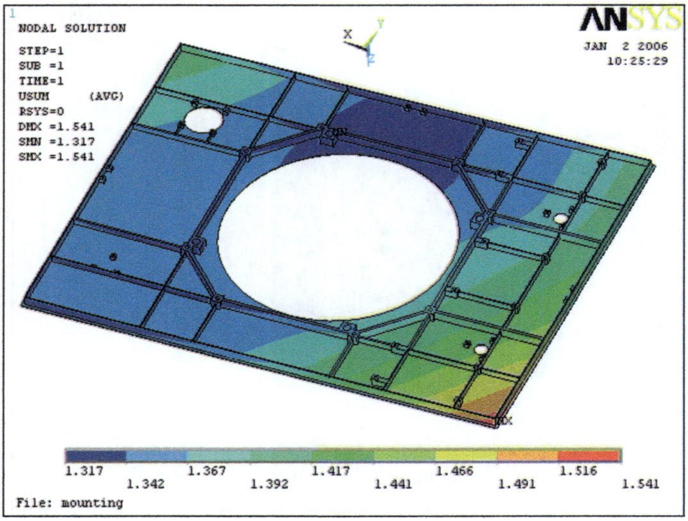

Fig. 4.14 Displacement distribution in mounting plate in case 1 Basis plate with star sensor bracket

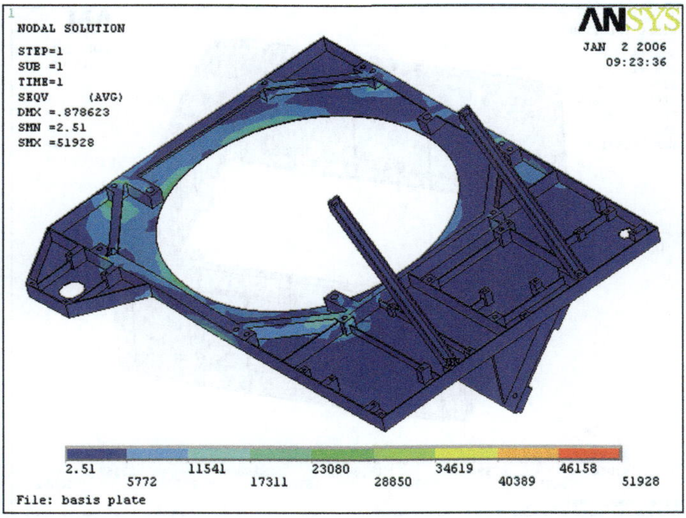

Fig. 4.15 Von Mises stress distribution in basis plate in case 1

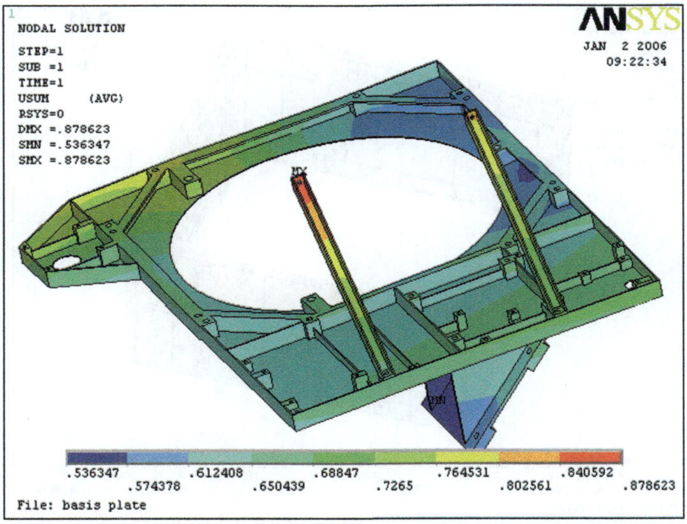

Fig. 4.16 Displacement distribution in basis plate in case 1 Basis walls

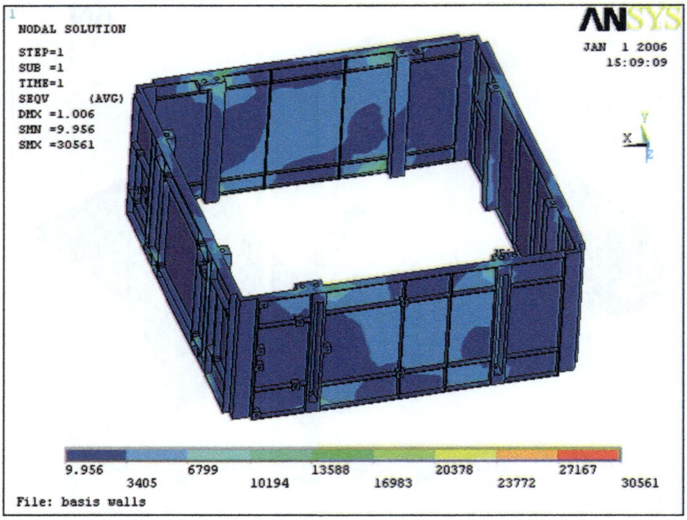

Fig. 4.17 Von Mises stress distribution in basis walls in case 1

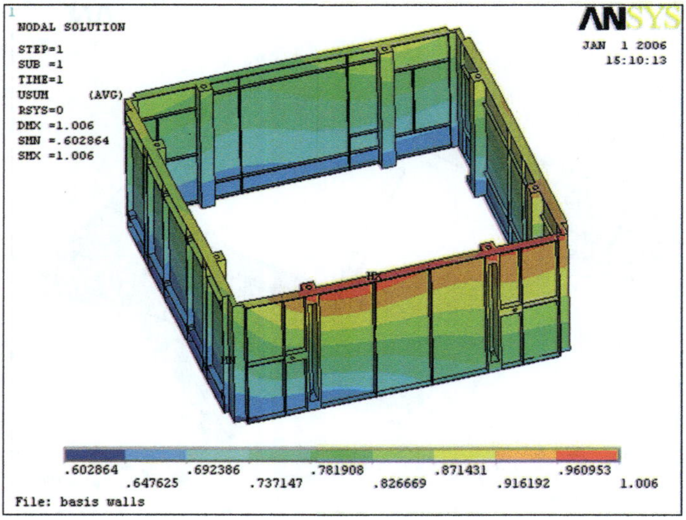

Fig. 4.18 Displacement distribution in basis walls in case 1 Upper frame with battery bracket

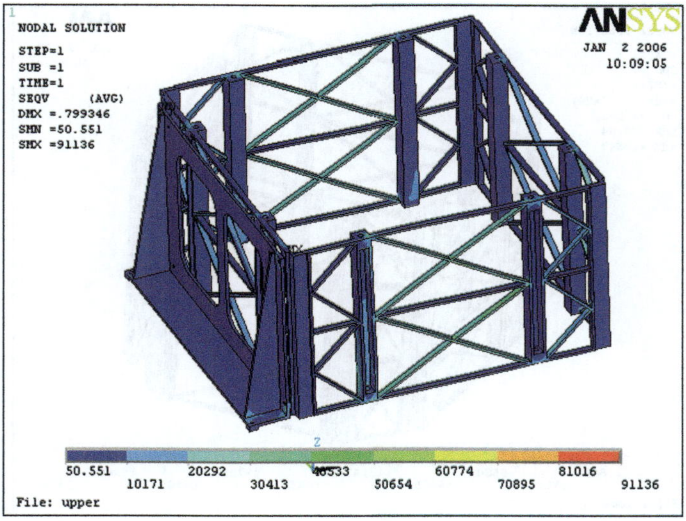

Fig. 4.19 Von Mises stress distribution in upper frame in case 1

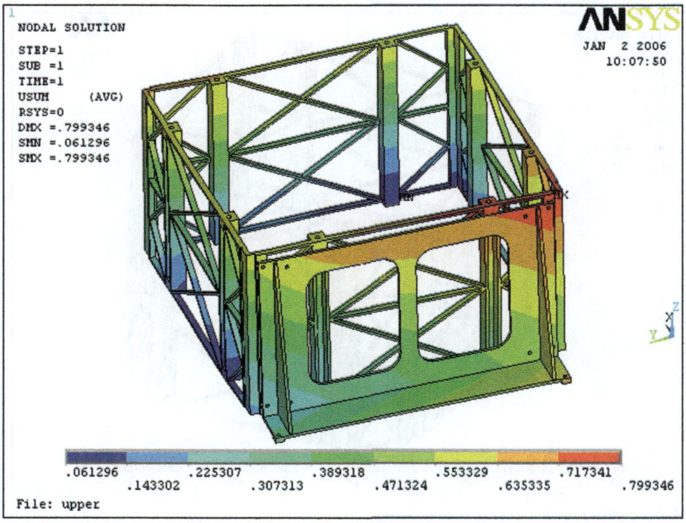

Fig. 4.20 Displacement distribution in upper frame in case 1 Lower frame

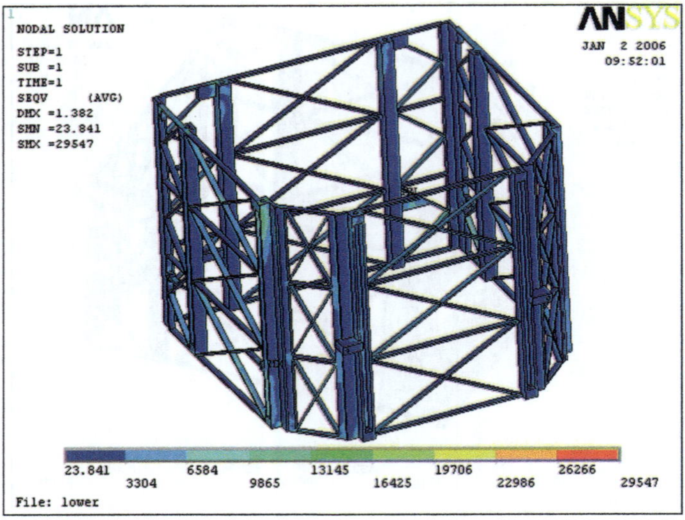

Fig. 4.21 Von Mises stress distribution in lower frame in case 1

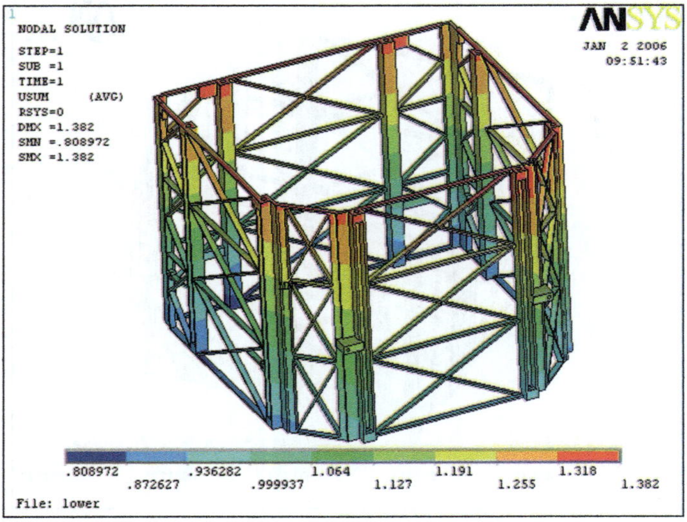

Fig. 4.22 Displacement distribution in lower frame in case 1 Launch vehicle adapter

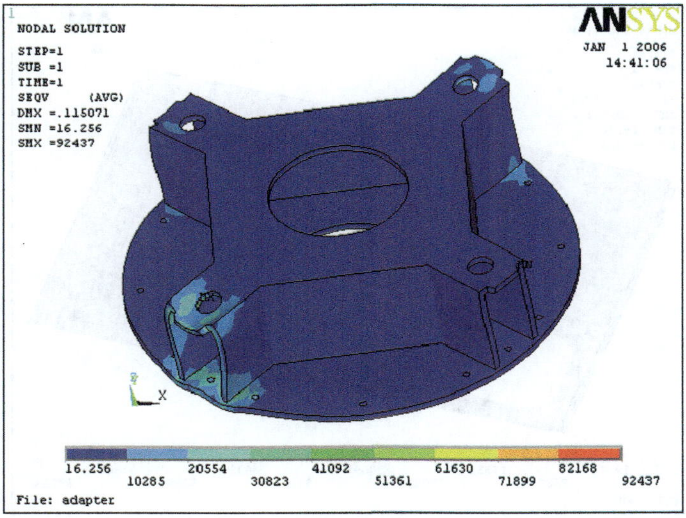

Fig. 4.23 Von Mises stress distribution in LV adapter in case 1

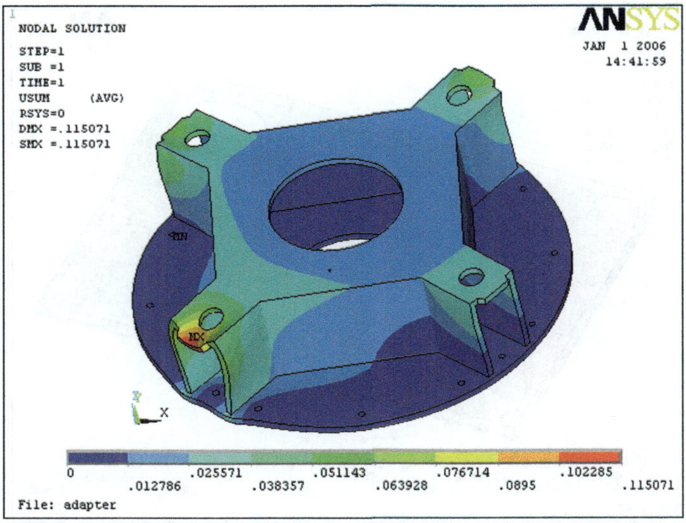

Fig. 4.24 Displacement distribution in LV adapter in case 1

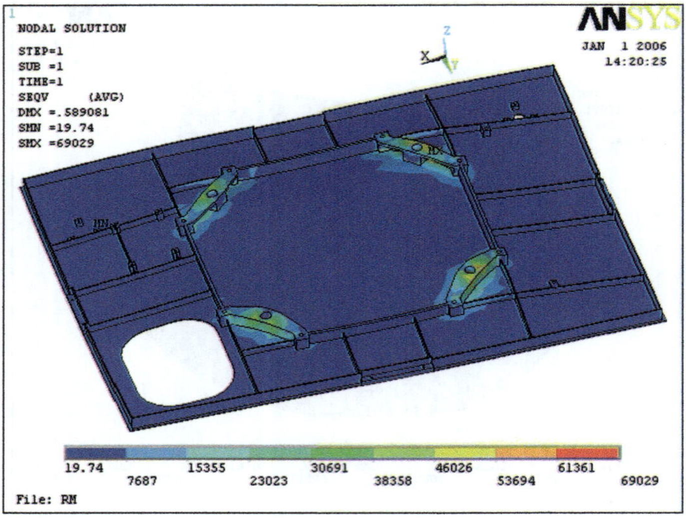

Fig. 4.25 Von Mises stress distribution in base plate in case 2

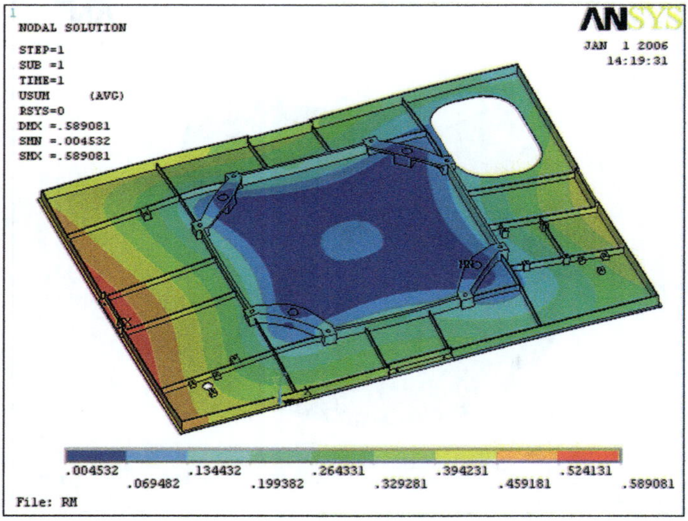

Fig. 4.26 Displacement distribution in base plate in case 2

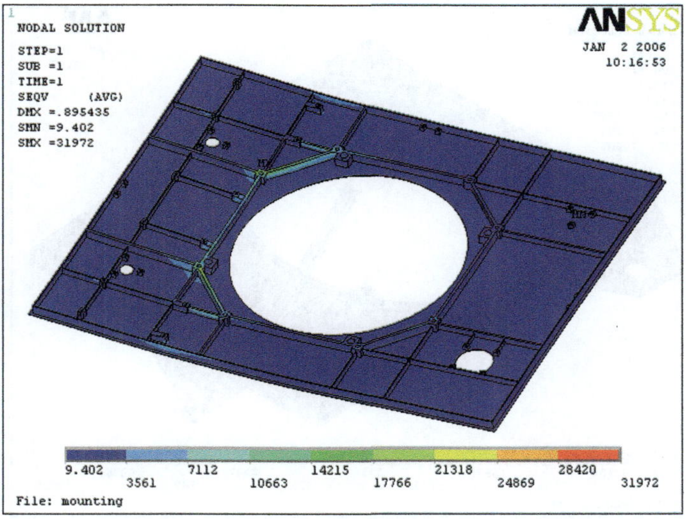

Fig. 4.27 Von Mises stress distribution in mounting plate in case 2

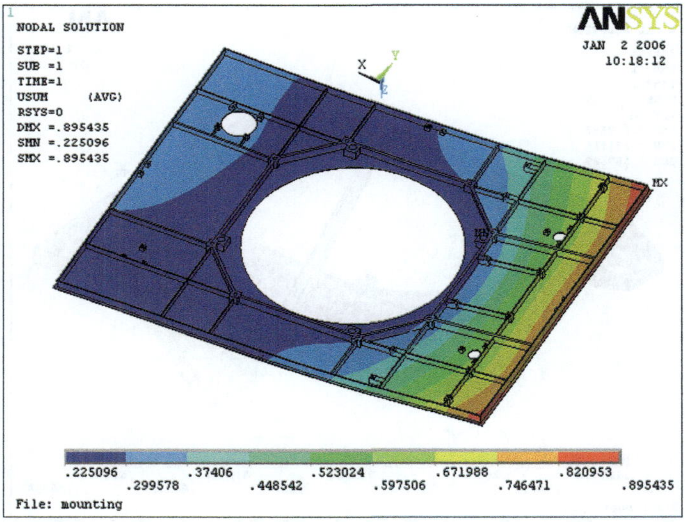

Fig. 4.28 Displacement distribution in mounting plate in case 2

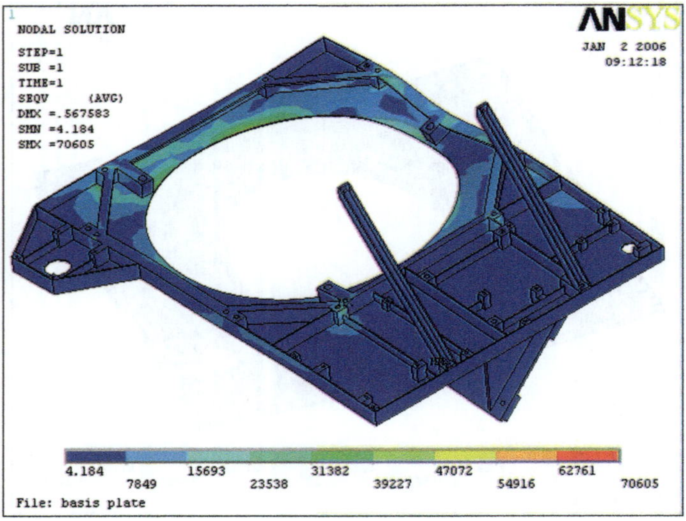

Fig. 4.29 Von Mises stress distribution in basis plate in case 2

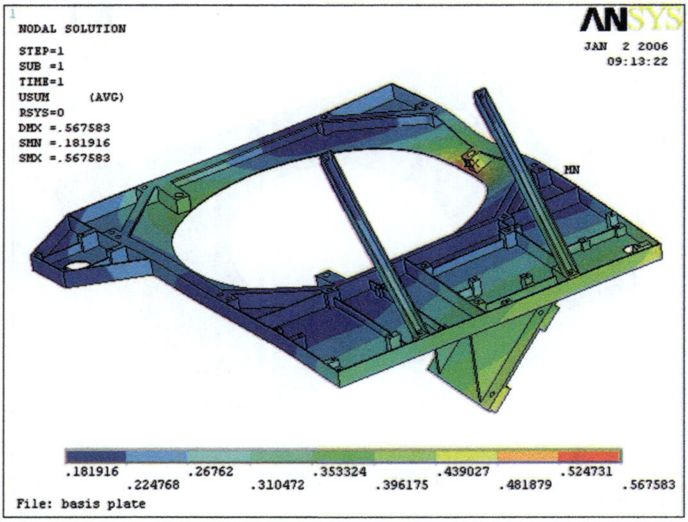

Fig. 4.30 Displacement distribution in basis plate in case 2

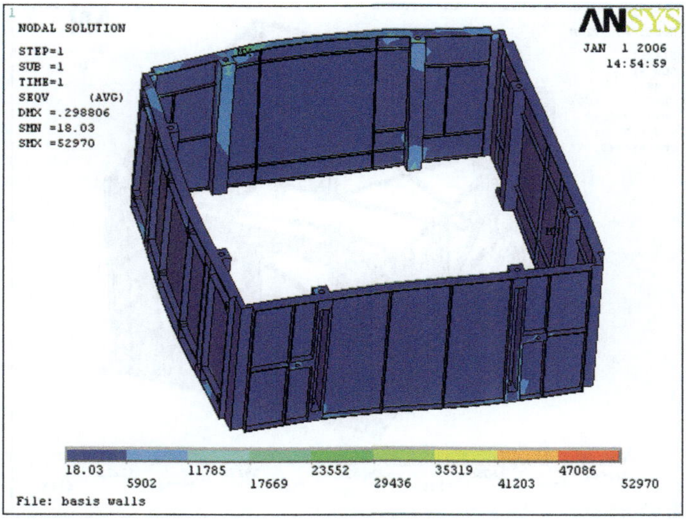

Fig. 4.31 Von Mises stress distribution in basis walls in case 2

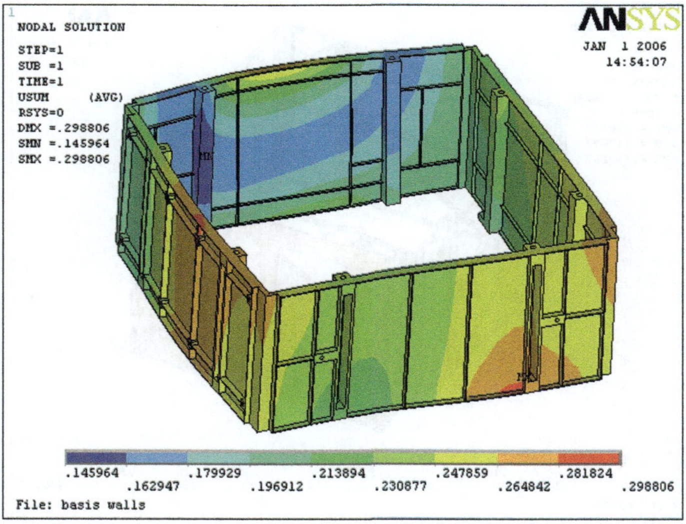

Fig. 4.32 Displacement distribution in basis walls in case 2

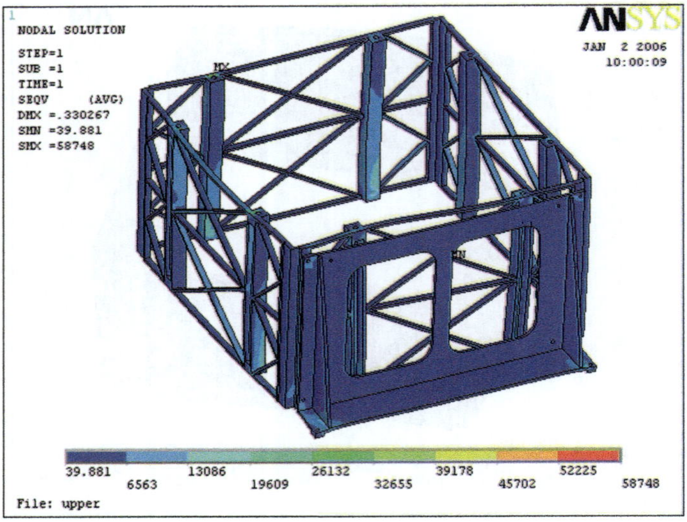

Fig. 4.33 Von Mises stress distribution in upper frame in case 2

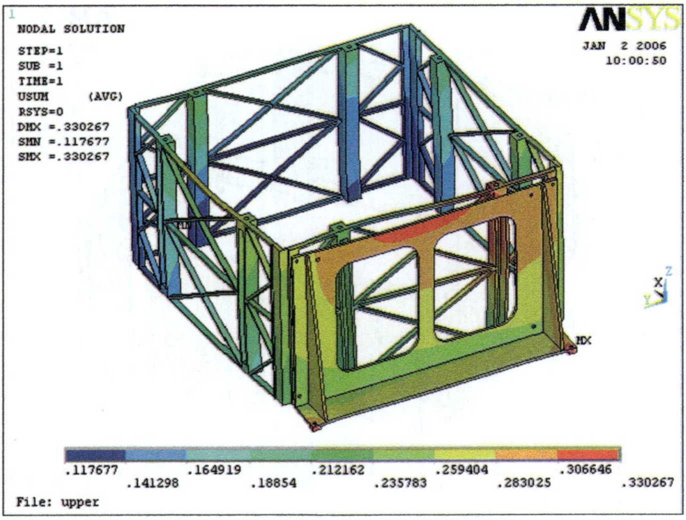

Fig. 4.34 Displacement distribution in upper frame in case 2

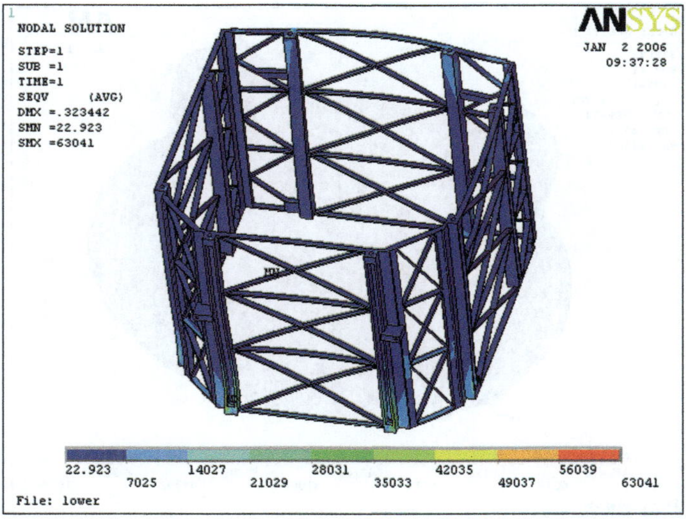

Fig. 4.35 Von Mises stress distribution in lower frame in case 2

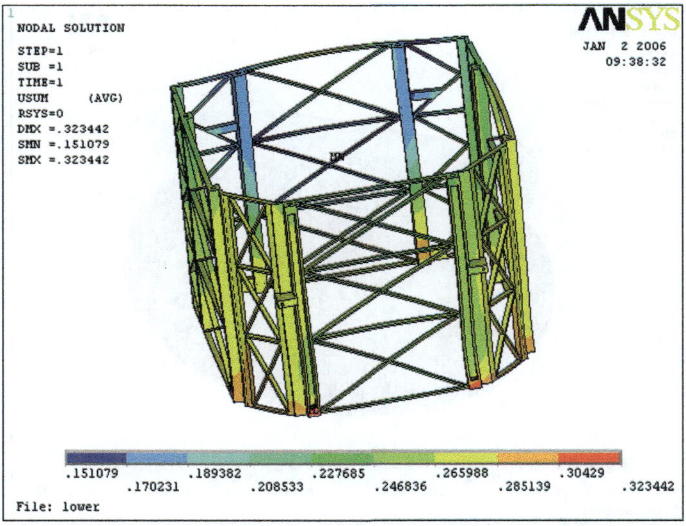

Fig. 4.36 Displacement distribution in lower frame in case 2

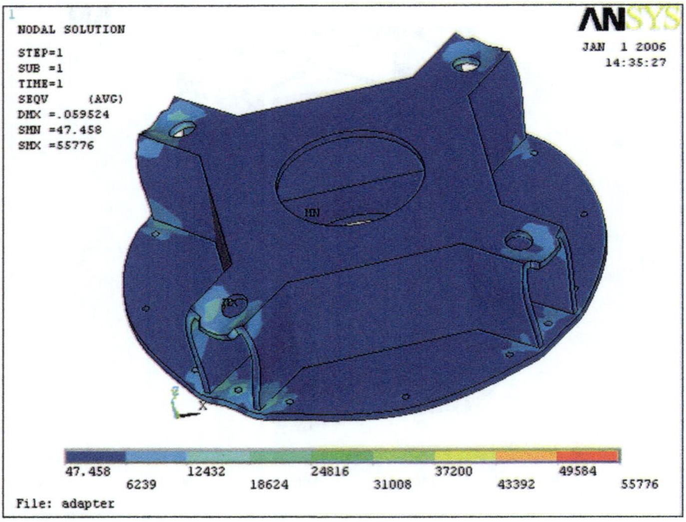

Fig. 4.37 Von Mises stress distribution in LV adapter in case 2

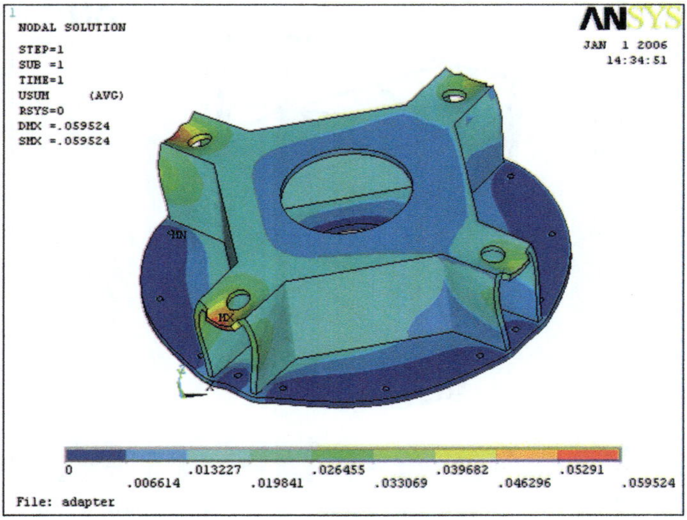

Fig. 4.38 Displacement distribution in LV adapter in case 2

Table 4.3 Maximum Von Mises equivalent stresses and yield margins of safety for structural modules in each design load case

Structural modules	σ_e (MPa)		MS_y	
	Design case		Design case	
	«case 1»	«case 2»	«case 1»	«case 2»
Base plate	108.69	69.03	0.38	1.17
Mounting plate	16.34	31.97	8.18	3.69
Basis plate	51.93	70.4	1.89	1.13
Basis walls	30.56	52.97	3.91	1.83
Upper frame	91.14	58.75	0.65	1.55
Lower frame	29.55	63.04	4.08	1.38
Launch vehicle adapter	92.44	55.78	0.62	1.69

$$\sigma_e = \frac{1}{\sqrt{2}}\sqrt{(\sigma_x - \sigma_y)^2 + (\sigma_y - \sigma_z)^2 + (\sigma_z - \sigma_x)^2 + 6(\tau_{xy}^2 + \tau_{yz}^2 + \tau_{zx}^2)}$$

where:

$$\{\sigma\} = \text{stress vector} = \begin{bmatrix} \sigma_x & \sigma_y & \sigma_z & \sigma_{xy} & \sigma_{yz} & \sigma_{zx} \end{bmatrix}^{\mathrm{T}}$$

The maximum stress values of σ_e in the structural modules in each design load case are given in Table 4.2. This table also shows the yield margins of safety, MS_y. The margins must be positive for all strength cases to satisfy safety criteria. The values of the maximum displacements under operational loads are given in Table 4.3. The maximum value of displacement in any structural module must not exceed 3 mm to avoid component collapse.

By reviewing the results of the yield margin of safety listed in Table 4.3, it is found that the first load case has the most critical influence on the base plate, upper frame, and launch vehicle adapter. On the other hand, the mounting plate, basis plate, basis walls, and lower frame are more influenced by the second load case. The minimum values of margin of safety for each structural module satisfy the safety criteria. The mounting plate and basis walls module are statically over safe. However, the size of these modules will not be diminished to reduce structural weight because of many technical reasons. Manufacturing process techniques have limitations and constraints on the sizes of structural modules. Satisfactory performance of the satellite requires verifying pointing and alignment requirements for sensors and highly precise equipments (Mounting Accuracy), which are installed on both the basis plate and basis walls modules. Therefore, these modules require high stiffness to provide the desired mounting accuracy. The natural frequency constraint imposed on Small Sat structure is satisfied with a small margin as shown in next section. Reduction of the dimension of these two structural modules or any other one will lower the first natural frequency leading to unacceptable structural design. Finally, fatigue and damage accumulation checks may also prevent any size reduction in structural modules (Table 4.4).

Table 4.4 Maximum displacements for structural modules at each design load case

Structural modules	Displacements under operational loads, (mm)	
	Design case	
	«case 1»	«case 2»
Base plate	1.29	0.59
Mounting plate	1.54	0.9
Basis plate	0.88	0.57
Basis walls	1.01	0.3
Upper frame	0.8	0.33
Lower frame	1.38	0.32
Launch vehicle adapter	0.12	0.06

4.9.4.1 Results of Loading Case 1 (Road Transportation)

Base plate
Mounting plate
Basis plate with star sensor bracket
Basis walls
Upper frame with battery bracket
Lower frame
Launch vehicle adapter

4.9.4.2 Results of Loading Case 2 (Launch)

Base plate
Mounting plate
Basis plate with star sensor bracket
Basis walls
Upper frame with battery bracket
Lower frame
Launch vehicle adapter

4.10 Modal Analysis

In this section, a modal analysis is performed for Small Sat structure. The entire finite-element model of the satellite based on the sub-structural superelement modules are used during the modal analysis process. The main purpose of this analysis is to define the natural frequencies of the satellite structure when it is attached to the launch vehicle adapter during transportation and launch. The results of modal analysis are used in the next sections to perform harmonic response analysis and stress fatigue analysis for Small Sat structure. Therefore, modal analysis can be considered as intermediate calculations to specify the natural frequencies and mode shapes of the satellite structure, and then sort them to define which ones have the greatest contribution to forced vibration response.

4.10.1 Definition of Modal Analysis

Modal analysis [1] is used to determine the vibration characteristics (natural frequencies and mode shapes) of a structure while it is being designed. The natural frequencies and mode shapes are important parameters in the design of a structure for dynamic loading conditions. Modal analysis is also a starting point for more detailed dynamic analyses, such as transient dynamic analysis, harmonic response analysis, or spectral analysis. When frequencies of loading vibrations match one of these natural frequencies, then resonance takes place. For any dynamic structure, resonance is one of the most critical problems that control its design.

4.10.2 Performing a Modal Analysis

The procedure of modal analysis of Small Sat structure consists of the following steps:

1. Build the model: The entire finite-element model of the satellite based on the substructural superelement modules is used during modal analysis. This model is built in Sect. 4.1.
2. Apply the loads and define boundary conditions: The only "loads" valid in a typical modal analysis are zero displacement constraints. If a nonzero displacement constraint is specified, the solution will assign a zero constraint to some DOF instead. Other loads—forces, pressures, temperatures, accelerations, … etc.—can be specified in a modal analysis, but they are ignored in the mode-extraction procedure. For directions in which no constraints are specified, the program calculates rigid body (zero frequency) as well as higher (nonzero frequency) free body modes.
3. Set solution controls and solve the analysis: A modal analysis type is defined during performing the analysis on ANSYS. The default settings that will work well for many structural modal analyses are selected. Modal analysis in the ANSYS is a linear analysis. There are several mode-extraction methods. The default method on ANSYS which is used during analysis is Block Lanczos. The number of modes to be extracted during modal analysis is selected to cover all frequency bands affecting the satellite during transportation and launch. From the satellite mechanical loads listed in Appendix A, launch has the largest band of random vibrations in the three mutually perpendicular directions which

extend from 20 up to 2,000 Hz. These values were gathered from the historical experience in carrying out dynamic tests of similar small satellites. Therefore, the entire model of the satellite is solved for modal analysis to calculate the satellite natural frequencies up to 2,000 Hz.

4. Review the results: Results from modal analysis of the entire finite-element model of the satellite structure are written to the structural results files. They contain the following data:

   - Natural frequencies.
   - Expanded mode shapes.
   - Participation factor and effective mass calculations for the expanded mode shapes in each excitation direction (U_X, U_Y, U_Z, ROT_X, ROT_Y, and ROT_Z).

4.10.3 Modal Analysis Results

Natural frequencies of Small Sat structure are determined as a result of the modal analysis performed by ANSYS. The total numbers of natural frequencies of Small Sat up to 2,000 Hz are found to be 121 modes. The user's guide of Dnepr Launch Vehicle cautions that the payload of LV (Small Sat satellite) should be designed with a structural stiffness which ensures that the values of fundamental frequencies of the satellite, hard mounted (rigidly constrained) at the launch vehicle interface, are not less than:

- 20 Hz in the longitudinal axis; and
- 10 Hz in the lateral axis.

This constraint must be satisfied to ensure that the satellite's dynamic characteristics do not adversely affect the LV's control system. For Small Sat structure, the first natural frequency is found to be 33 Hz and the 121th is 1963.5 Hz. Thus, the Small Sat first natural frequency (33 Hz) satisfies the minimum fundamental frequency constraint of Dnepr LV.

The modal analysis is intermediate calculations to define the natural frequencies of the satellite structure. The results of modal analysis are used to perform harmonic response analysis and stress fatigue analysis for Small Sat structure, which will be a very complicated process if the first 121 modes (up to 2,000 Hz) are used. For simplicity, a reduction method should be applied to reduce the size of modal frequencies used during subsequent analyses. The problem to be solved in this section is: Determine the smallest number of natural frequencies which accurately construct the frequency response characteristics through a given frequency range (up to 2,000 Hz). The first step is to define the eigenvector elements for all modes for only the input and output DOF which have a contribution on the satellite response. The second step is to analyze the model contributions of all modes, and sort them to define which ones have the greatest contribution.

One method for reducing the size of a modal model is to simply truncate the higher frequency modes. If this truncation is performed without understanding the contribution of each mode to the response, several problems could arise. One problem is that a high frequency mode with a significant response contribution may be eliminated, adversely affecting the model. Typically, contributions of the modes decrease as their frequencies increase. However, this is not always the case, especially for complicated models. Excluding a specific higher frequency mode can result in a model with less than desired accuracy. Therefore, the satellite's natural frequencies should be sorted, so that only the most important modes are kept.

For Small Sat, the applied mechanical loads during transportation and launch are mainly axial and lateral components, while the rotational components are relatively neglected. Hence, the rotational eigenvector elements (ROT_X, ROT_Y, and ROT_Z) can be eliminated during sorting the natural frequencies. Eigenvectors for U_X, U_Y, and U_Z components of the entire finite-element model of the satellite are used to determine the most important modes. One of the applicable methods of sorting natural frequencies is the *effective mass concept*, which can be useful in ranking the relative importance of modes, and determine the number of modes to be included in the modal analysis.

4.10.3.1 Sorting of Modes of Free Vibration

The transient dynamic equilibrium equation of interest for a linear multi degree of freedom (MDOF) structure is [2]:

$$[M]\{\ddot{u}\} + [C]\{\dot{u}\} + [K]\{u\} = \{F^a\} \tag{4.18}$$

where:

- $[M]$ = structural mass matrix
- $[K]$ = structural stiffness matrix
- $[C]$ = structural damping matrix
- $\{\ddot{u}\}$ = nodal acceleration vector
- $\{\dot{u}\}$ = nodal velocity vector
- $\{u\}$ = nodal displacement vector
- $\{F^a\}$ = applied load vector

The participation factor for the given excitation direction is defined as [2]:

$$\gamma_i = \frac{\{\phi\}_i^T[M]\{D\}}{\{\phi\}_i^T[M]\{\phi\}_i} = \frac{\{\phi\}_i^T[M]\{D\}}{M_r} \tag{4.19}$$

where:

- γ_i = participation factor for the ith mode
- $\{\phi\}_i$ = eigenvector for the ith mode
- $\{D\}$ = vector describing the excitation direction
- M_r = generalized modal mass

If each eigenvector $\{\phi\}_i$ is normalized to the mass matrix:

$$\{\phi\}_i^T[M]\{\phi\}_i = M_r = 1 \tag{4.20}$$

Hence:

$$\gamma_i = \{\phi\}_i^T[M]\{D\} \tag{4.21}$$

The vector describing the excitation direction has the form:

$$\{D\} = [T]\{e\} \tag{4.22}$$

where:

$$\{D\} = \left[D_1^a D_2^a D_3^a \ldots\right]^T$$

D_j^a = excitation at DOF j in direction a; a may be either X, Y, Z, or rotations about one of these axes.

$$[T] = \begin{bmatrix} 1 & 0 & 0 & 0 & (Z-Z_0) & -(Y-Y_0) \\ 0 & 1 & 0 & -(Z-Z_0) & 0 & (X-X_0) \\ 0 & 0 & 1 & (Y-Y_0) & -(X-X_0) & 0 \\ 0 & 0 & 0 & 1 & 0 & 0 \\ 0 & 0 & 0 & 0 & 1 & 0 \\ 0 & 0 & 0 & 0 & 0 & 1 \end{bmatrix}$$

X, Y, Z = global Cartesian coordinates of a point

X_o, Y_o, Z_o = global Cartesian coordinates of point about which rotations are done (reference point)

$\{e\}$ = six possible unit vectors

The effective mass for the ith mode (which is a function of excitation direction) is:

$$M_{ei} = \{\phi\}_i^T[M]\{D\}\gamma_i \tag{4.23}$$

Note from Eq. (4.20) that:

$$M_{ei} = \gamma_i^2 \tag{4.24}$$

The sum of the effective masses of all modes ($i = 1, 2, 3, \ldots, n$) is equal to the total mass of the structure. This results in a means of determining the number of individual modal responses necessary to accurately represent the structural response. If the total response of the system is to be represented in terms of a small number of modes "p" (where $p \ll n$), and if the sum of the "p" effective masses is greater than a predefined percentage of the total mass of the structure, then the number of modes "p" considered in the analysis is adequate. However, if this is not the case, then additional modes must be considered. Dynamic analysis procedures of large MDOF systems specify that for the "p" modes considered in the analysis, at least 90 % of the participating mass of the structure must be included in the response calculations for each principal horizontal direction.

For Small Sat, the participation factor and effective mass are calculated for the expanded mode shapes in each excitation direction from the results of modal analysis of the entire finite-element model of the satellite structure. The natural frequencies of the satellite are sorted according to the values of the effective masses from largest to smallest for all expanded mode shapes (121 modes) in each important excitation direction (*X*, *Y*, and *Z* direction) separately. *X* is the axial direction, while *Y* and *Z* are the lateral directions. The effective mass percentage values of the total mass of the satellite are calculated for all expanded mode shapes. The total mass of the entire model of Small Sat is found to be 200 kg.

For each excitation direction, certain sorted modes are selected to provide at least 90 % of the participating mass of the satellite. Performing this criterion leads to several problems, the first one is that each excitation direction has different ranking order, which represents difficulty in selecting the most important modes for the whole satellite system. Another problem is the large number of modes which should be selected for each direction. The first 19 sorted modes are selected in the X direction, 17 in the Y direction, and 15 in the Z direction. Therefore, this criterion should be modified to determine a reduced ranking method for the whole satellite in all directions. This can be done by sorting the natural frequencies by the root square values of the effective mass percentage in each mode for all directions. Table 4.5 lists the participation factors and effective masses for only the first 15 modes in the *X*, *Y*, and *Z* directions.

Tables 4.6, 4.7 4.8 list the sorted natural frequencies and mode number of Small Sat structure in the *X*, *Y*, and *Z* directions, respectively. They also represent the effective mass percentage values of the total mass of the satellite for each sorted mode. Table 4.9 lists the first 15 sorted natural frequencies for the whole satellite structure and their damping ratio, which are used in the harmonic analysis and fatigue damage calculations. From the single DOF dynamic analysis theory, the damping ratio can be approximated by the following formula:

$$\xi = \frac{1}{10 + 0.05 f_i}$$

where: ξ is the damping ratio, and f_i is the satellite natural frequency in (Hz).

4.11 Harmonic Response Analysis

In this section, a harmonic response analysis is performed for Small Sat structure. The entire finite-element model of the satellite beside the substructural superelement modules is used during the harmonic analysis process. The main purpose of this analysis is to investigate whether the satellite structure can withstand all dynamic loads encountered during transportation and launch. The results of harmonic analysis are used in the next section to perform stress fatigue damage analysis for Small Sat structure due to mechanical dynamic vibration.

Table 4.5 Participation factors and effective masses for the first 15 modes in the X, Y, and Z directions

Mode	Frequency (Hz)	X direction		Y direction		Z direction	
		Participation factor	Effective mass	Participation factor	Effective mass	Participation factor	Effective mass
1	33	11.801	127.423	−0.046	0.002	−0.322	0.095
2	35	0.026	0.001	12.237	137.014	−0.150	0.021
3	62.6	−2.738	6.859	0.032	0.001	−4.654	19.819
4	68.4	−0.097	0.009	0.196	0.035	0.241	0.053
5	73.6	1.650	2.492	3.936	14.175	0.579	0.307
6	96	−2.171	4.311	0.196	0.035	−9.470	82.055
7	98.4	0.123	0.014	−0.216	0.043	−0.143	0.019
8	104.6	3.241	9.613	0.004	0.000011	−6.163	34.756
9	119.8	2.990	8.178	−2.899	7.691	−1.571	2.258
10	128.3	0.021	0.0004	0.068	0.004	3.214	9.452
11	137	0.239	0.052	0.122	0.014	−2.311	4.889
12	152.5	−0.016	0.0002	−0.005	0.000023	−1.419	1.843
13	153.8	−0.009	0.00007	−1.040	0.989	1.503	2.066
14	157.3	−0.163	0.024	0.425	0.165	1.017	0.947
15	165.2	0.192	0.034	0.824	0.622	−0.534	0.261

Table 4.6 Sorted natural frequencies in the *X* direction

No.	Mode	Frequency (Hz)	Effective mass	Effective mass percentage (%)
1	1	33	127.423	63.711
2	8	104.7	9.613	4.807
3	9	119.8	8.178	4.089
4	3	62.6	6.859	3.429
5	40	471.7	4.787	2.394
6	6	96	4.311	2.156
7	28	311.3	3.019	1.510
8	5	73.6	2.492	1.246
9	39	446.7	2.357	1.178
10	25	279.7	2.137	1.069
11	29	332.4	1.550	0.775
12	17	187.9	1.265	0.632
13	65	887.4	1.157	0.578
14	66	906.1	1.091	0.545
15	52	627	0.921	0.461
16	57	726	0.774	0.387
17	33	367.4	0.747	0.373
18	68	929.4	0.729	0.365
19	72	985.7	0.724	0.362
Total effective mass percentage				90.067

Table 4.7 Sorted natural frequencies in the *Y* direction

No.	Mode	Frequency (Hz)	Effective mass	Effective mass percentage (%)
1	2	35	137.014	68.507
2	5	73.6	14.175	7.088
3	9	119.8	7.691	3.846
4	22	247.58	2.954	1.477
5	16	173.46	2.471	1.236
6	27	304.852	2.156	1.078
7	30	341.853	1.974	0.987
8	17	187.9	1.692	0.846
9	65	887.4	1.624	0.812
10	33	367.4	1.519	0.759
11	57	726.1	1.457	0.729
12	52	627	1.143	0.571
13	66	906.1	1.081	0.540
14	13	153.8	0.989	0.494
15	73	1017.9	0.916	0.458
16	24	261.2	0.796	0.398
17	54	675.7	0.712	0.356
Total effective mass percentage				90.182

Table 4.8 Sorted natural frequencies in the *Z* direction

No.	Mode	Frequency (Hz)	Effective mass	Effective mass percentage (%)
1	6	96	82.055	41.028
2	8	104.7	34.756	17.378
3	3	62.6	19.819	9.909
4	10	128.3	9.452	4.726
5	17	187.9	5.027	2.513
6	11	137	4.889	2.444
7	55	696.8	4.118	2.059
8	34	378.1	3.601	1.801
9	69	950.4	3.491	1.745
10	19	200.3	2.989	1.495
11	106	1672	2.397	1.198
12	9	119.8	2.258	1.129
13	13	153.8	2.066	1.033
14	12	152.5	1.843	0.921
15	108	1703	1.575	0.788
Total effective mass percentage				90.168

Table 4.9 The first 15 sorted natural frequencies for the whole satellite structure and their damping ratios

No.	Mode	Frequency (Hz)	Damping ratio	Effective mass percentage (%)			Percentage root square (%)
				X direction	Y direction	Z direction	
1	2	35	0.085	0.0003	68.507	0.010	68.507
2	1	33	0.086	63.711	0.001	0.048	63.711
3	6	96	0.068	2.156	0.018	41.028	41.084
4	8	104.7	0.066	4.807	5.71E − 06	17.378	18.031
5	3	62.6	0.076	3.429	0.00048	9.909	10.486
6	5	73.6	0.073	1.246	7.088	0.154	7.198
7	9	119.8	0.063	4.089	3.846	1.129	5.726
8	10	128.3	0.061	0.000195	0.002	4.726	4.726
9	17	187.9	0.052	0.632	0.846	2.513	2.726
10	11	137	0.059	0.026	0.007	2.444	2.444
11	40	471.7	0.030	2.394	0.030	0.224	2.404
12	55	696.8	0.022	0.007	0.031	2.059	2.059
13	34	378.1	0.035	0.335	0.002	1.801	1.832
14	69	950.4	0.017	0.090	0.048	1.745	1.748
15	28	311.3	0.039	1.510	0.033	0.048	1.511

4.11.1 Definition of Harmonic Response Analysis

Any sustained cyclic load will produce a sustained cyclic response (a harmonic response) in a structural system. Harmonic response analysis gives the ability to predict the sustained dynamic behavior of any structure, thus enabling to verify whether or not structural designs will successfully overcome resonance, fatigue, and other harmful effects of forced vibrations. Harmonic response analysis is a technique used to determine the steady-state response of a linear structure to loads that vary sinusoidally (harmonically) with time. This analysis technique calculates only steady-state forced vibrations of a structure. Transient vibrations, which occur at the beginning of the excitation, are not accounted for in a harmonic response analysis.

The general idea of harmonic response analysis is to calculate the structure's response at several frequencies, and obtain a graph of some response quantity (usually displacements) versus frequency. "Peak" responses are then identified on the graph and stresses are reviewed at those peak frequencies. Peak harmonic response occurs at forcing frequencies that match the natural frequencies of the structure (resonance frequencies). Therefore, before performing harmonic analysis, the natural frequencies of the structure should be first determined through a modal solution. Amplitudes of cyclic load stresses at each resonance frequency are calculated for the structure. These peak stresses are needed to calculate fatigue damage due to mechanical vibrations during transportation and launch vehicle flight.

4.11.2 Small Sat Dynamic Vibration Loads

By reviewing the satellite mechanical loads represented in Appendix A, there are two types of dynamic vibrations acting on the satellite structure during transportation and LV flight. The first type is harmonic oscillations which are characterized by the amplitude of vibroaccelerations and frequency. These sinusoidal vibrations influence the satellite interface in three mutually perpendicular directions during air transportation. They have a frequency band from 5 to 310 Hz along a duration of 750 min. During launch vehicle flight, the satellite/LV interface is affected by sinusoidal vibrations in the lateral and axial directions. The frequency band in the lateral direction ranges from 2 to 15 Hz over 300 s, while in the axial direction it ranges from 5 to 20 Hz over 100 s.

Random vibrations are the second type of mechanical dynamic vibrations acting on the satellite interface. They are characterized by acceleration power spectral density and the duration of influence. The values of amplitude and spectral densities are given in the extreme octave points. The random vibrations' parameters are defined for rail and road transportation in three perpendicular directions. In rail transportation, the frequency band is from 0 to 100 Hz. The duration is specified corresponding to rail transportation distance, which is approximately

10,000 km for Small Sat. Hence, the total duration of random vibrations during rail transportation is 200 h. In road transportation, the frequency band is from 4 to 43 Hz. The duration is also specified corresponding to road transportation distance, which is approximately 600 km for Small Sat. Hence, the total duration of random vibrations during road transportation is 18 h.

During launch, random vibrations act on the satellite/LV interface with approximately equal intensity of vibroaccelerations in each of the three selected mutually perpendicular directions. There are two cases during LV flight which have the same frequency band from 20 to 2,000 Hz, but influence different acceleration power spectral densities and have unequal durations. The first case is during Liftoff; the LV flight segment where $M = 1$, and maximum dynamic pressure. Its random vibrations act only for 35 s with acceleration root-mean-square (rms) equal to 6.5 g. The second case is during 1st, 2nd, and 3rd stage burn. The total duration in this case is 831 s and the acceleration rms is 3.6 g.

To predict the satellite's response to random vibrations of its mounting interface, a mathematical concept called Fourier transform is used. The Fourier transform can be considered a Fourier series for a function having an infinite period. This concept is discussed in Appendix B. It is used to calculate the vibration-acceleration amplitude of the equivalent sinusoidal harmonic vibration for random vibrations of road and rail transportation and launch vehicle flight. Converting random vibrations to equivalent sinusoidal harmonic vibrations eases numerical calculation of fatigue damage due to dynamic vibration. Figures 4.39, 4.40, 4.41, 4.42, 4.43, 4.44, 4.45 and 4.46 show the dynamic vibration loads acting on the satellite interface during transportation and LV flight.

4.11.3 Performing Harmonic Analysis

The procedure for a harmonic analysis consists of the following steps:

1. Build the model: The entire finite-element model of the satellite beside the substructural superelement modules is used during harmonic analysis. These models are built in Sect. 4.1.
2. Apply the loads and define boundary conditions: A harmonic analysis assumes that any applied load varies harmonically (sinusoidally) with time. To completely specify harmonic loads for the entire finite-element model of the satellite, three pieces of information are required: the amplitude, the resonance frequency, and the damping ratio. The amplitude is the maximum value (peak) of the harmonic load which is the maximum acceleration value specified for each cycle. The resonance frequency and its damping ratio are calculated from modal analysis of the entire satellite structure model. By reviewing the ANSYS package for harmonic analysis, it is found that the harmonic load is fully reversed sinusoidal cycling with zero mean value, without taking into account the gravity effect in any direction.

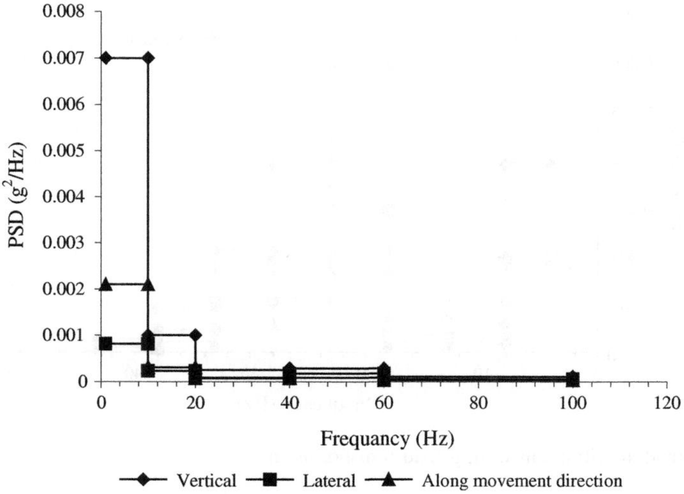

Fig. 4.39 Random vibrations during rail transportation

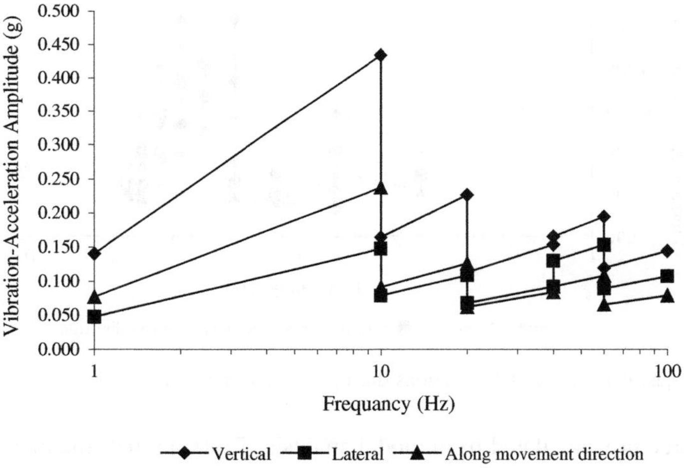

Fig. 4.40 Equivalent sinusoidal vibrations during rail transportation

3. As mentioned before, peak harmonic response occurs at forcing frequencies that match the natural frequencies of the structure. Therefore, amplitudes of the cyclic loading (acceleration) are calculated at the satellite's natural frequencies for each loading phase. These calculated values are used to perform harmonic response analysis for Small Sat structure individually in each loading phase for different forcing frequencies matched with the natural frequencies of the satellite. Section 4.10 gives the most important natural frequencies of Small

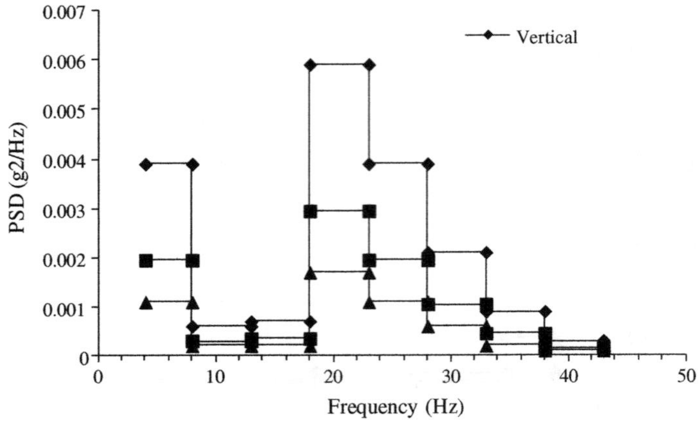

Fig. 4.41 Random vibrations during road transportation

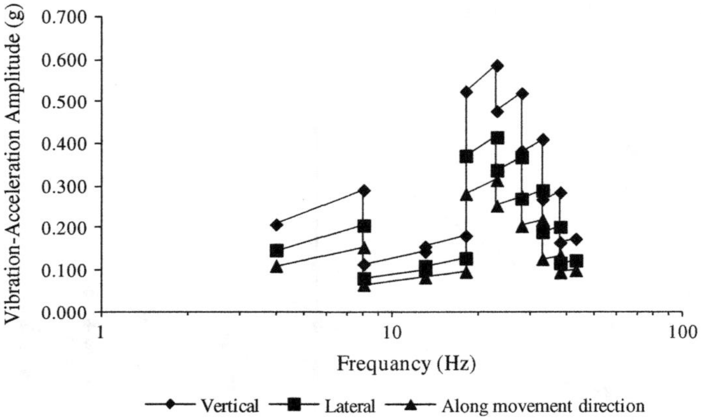

Fig. 4.42 Equivalent sinusoidal vibrations during road transportation

Sat's structure calculated from modal analysis. Table 4.9 lists the first 15 sorted natural frequencies for the whole satellite structure and their damping ratios, which are to be used in harmonic response analysis.

4. Vibration acceleration amplitudes are specified for all loading phases in three perpendicular axes as presented in Appendix A. Tables A.3, A.5, A.7, and A.12 give the equivalent sinusoidal acceleration amplitudes for rail, road, air transportation, and two launch phases, respectively. Variation of the vibration-acceleration amplitude within every frequency band is linear on logarithmic scale of frequency. So a linear interpolation is performed to calculate the acceleration value at each resonance frequency. The following equation is used to calculate the vibration-acceleration amplitude (A) in (g) at a resonance frequency (f) in (Hz):

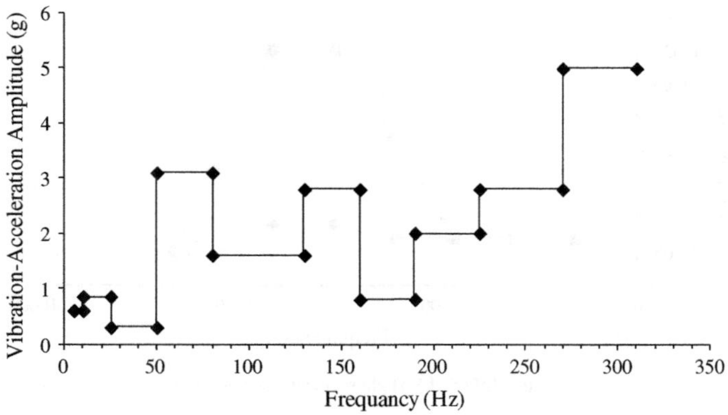

Fig. 4.43 Sinusoidal vibrations for the three mutually perpendicular directions during air transportation

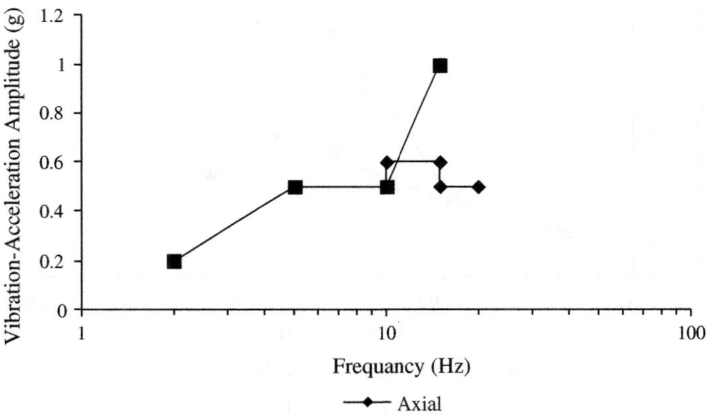

Fig. 4.44 Axial and lateral sinusoidal vibrations during flight of the launch vehicle

$$A(f) = \frac{A_2 - A_1}{\log(f_2) - \log(f_1)} (\log(f) - \log(f1)) + A_1$$

5. Where: f_1 & f_2 are the lower and upper values of the frequency band, respectively in (Hz), and A_1 & A_2 are the corresponding accelerations in (g).
6. Table 4.10 shows the vibration-acceleration amplitude at the first 15 sorted natural frequencies of the satellite structure for all dynamic loading phases. Vibrations in each phase act simultaneously, and are generated from the same source, that is why they affect the satellite structure in all directions in the same time. Displacement boundary conditions are the DOF constraints usually specified at model boundaries to define rigid support points. The same boundary conditions applied for the satellite during static and modal analyses are reapplied for harmonic response analysis.

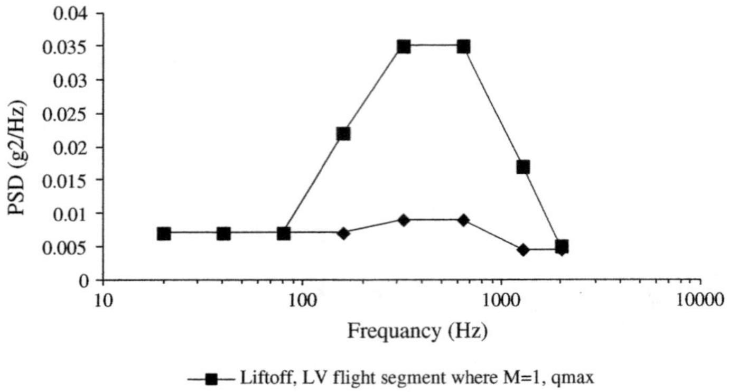

Fig. 4.45 Random vibrations in the three mutually perpendicular directions during launch

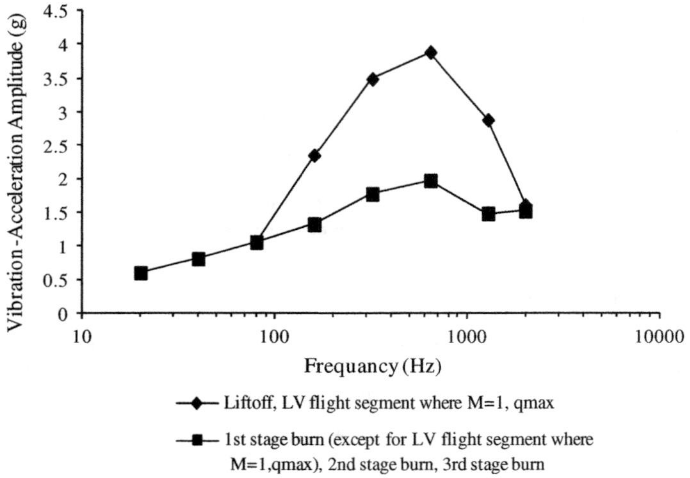

Fig. 4.46 Equivalent sinusoidal vibrations in the three mutually perpendicular directions during launch

7. Set solution controls and solve the analysis: a harmonic analysis type is defined during performing the analysis on ANSYS package. The default settings that will work well for many structural harmonic analyses are selected. An individual fully reversed harmonic analysis is performed for the entire satellite model in each harmonic loading phase, and at every selected structural resonance frequency. Then each superelement module is resolved to calculate the fully reversed stress amplitudes in all DOF in the superelement for all analysis sets. The harmonic analysis is performed by the full method which is the default and the easiest one on ANSYS. It uses the full system matrices to calculate the harmonic response (no matrix reduction). Solution listing format is chosen to be real and imaginary parts. This option determines how the harmonic solution is

Table 4.10 Vibration-acceleration amplitude at the first 15 sorted natural frequencies of the satellite structure for all dynamic loading phases

Loading phase	Frequency (Hz)	Vibration-acceleration Amplitude (g)		
		Vertical "z"	Lateral "y"	Mov. Dir. "x"
Rail transportation	35	0.145	0.087	0.08
	33	0.142	0.085	0.078
	96	0.14	0.108	0.077
	62.6	0.121	0.092	0.066
	73.6	0.128	0.098	0.07
	35	0.286	0.195	0.13
	33	0.346	0.242	0.173
Road transportation	35	0.3	0.3	0.3
	33	0.3	0.3	0.3
	96	1.6	1.6	1.6
	104.7	1.6	1.6	1.6
Air transportation	62.6	3.1	3.1	3.1
	73.6	3.1	3.1	3.1
	119.8	1.6	1.6	1.6
	128.3	1.6	1.6	1.6
	187.9	0.8	0.8	0.8
	137	2.8	2.8	2.8
	Vertical "z"	Lateral "y"	Lateral "x"	
	35	0.771	0.771	0.771
	33	0.753	0.753	0.753
	96	1.406	1.406	1.406
	104.7	3.759	3.759	3.759
	62.6	0.974	0.974	0.974
	73.6	1.033	1.033	1.033
Launch (phase I)	119.8	1.813	1.813	1.813
	128.3	1.941	1.941	1.941
	187.9	2.615	2.615	2.615
	137	2.063	2.063	2.063
	471.7	3.709	3.709	3.709
	696.8	3.759	3.759	3.759
	378.1	3.584	3.584	3.584
	950.4	3.312	3.312	3.312
	311.3	3.444	3.444	3.444
Launch (phase II)	35	0.771	0.771	0.771
	33	0.753	0.753	0.753
	96	1.133	1.133	1.133
	104.7	1.909	1.909	1.909
	62.6	0.974	0.974	0.974
	73.6	1.033	1.033	1.033
	119.8	1.216	1.216	1.216
	128.3	1.242	1.242	1.242
	187.9	1.429	1.429	1.429
	137	1.267	1.267	1.267
	471.7	1.458	1.458	1.458
	696.8	1.909	1.909	1.909
	378.1	1.233	1.233	1.233
	950.4	1.692	1.692	1.692
	311.3	1.751	1.751	1.751

The three perpendicular axes (x, y, and z) for each phase align with the design axes of the entire satellite model
g = gravity acceleration = 9.81 m/s^2

listed in the printed output. Therefore, for each analysis set, the expansion pass of every superelement structural module is done twice to calculate the real and imaginary parts of the solution.

8. Review the results: The results data for fully reversed harmonic analysis of the entire model and the expansion pass for each superelement module are the same as the data for a basic structural analysis. All results are complex in nature, and are stored in terms of real and imaginary parts because damping is defined.

4.11.4 Harmonic Response Analysis Results

As mentioned before, harmonic response analysis [2] is performed to calculate peak stresses in the structural modules due to harmonic inertia loads. These maximum stresses are used in fatigue damage calculations. Therefore, the output results from the harmonic analysis must serve the fatigue analysis. The harmonic tension stresses are responsible for dynamic fatigue failure. So the output results from the harmonic analysis are the equivalent amplitude stresses (Von Mises) in each structural module. For Small Sat harmonic analysis, the equivalent amplitude stress at each point is calculated from the formula:

$$\sigma_e = \sqrt{(\sigma_R)^2 + (\sigma_{\text{Imag}})^2}$$

where: σ_R and σ_{Imag} are the real and imaginary equivalent amplitude stress at the point.

Harmonic loads do not act individually on structural elements, but they operate simultaneously with quasi-static loads. Quasi-static loads comprise both static and dynamic loads, and are applied at a frequency sufficiently below the first natural frequency of the structure. Therefore, the quasi-static loads are independent of time or very slowly, so that the dynamic response of the structure due to the dynamic component is not significant.

The maximum stresses affecting the structure are a combination of the equivalent amplitude stresses due to fully reversed harmonic loads and the equivalent stresses from the static components of the quasi-static loads. Since the finite-element model of Small Sat is linear, then superposition is applicable to calculate the maximum stresses in each structural module. The equivalent stresses from the static components of the quasi-static loads can be considered as the mean stresses for the fatigue analysis due to dynamic loads. The satellite mechanical loads given in Appendix A, Tables A.1, A.6, and A.8 show the quasi-static loads affecting the satellite during road, rail, and air transportation and launch, respectively. The mean stresses in each structural module can be calculated by performing static analysis for the entire satellite model. The load factors used for the static analysis are specified as follows:

Table 4.11 Maximum Von Mises equivalent stresses and yield margins of safety for structural modules during all dynamic phases

Structural module	Maximum equivalent stress (MPa)	Yield margin of safety "MS_y"
Base plate	122.816	0.22
Mounting plate	68.883	1.18
Basis plate	147.05	0.02
Basis walls	146.082	0.03
Upper frame	129.159	0.16
Lower frame	118.846	0.26

- For road, rail, and air transportation: $n_z = 1$ (vertically down)
- For launch: $n_z = 7.8$ (axially down); $n_l = 0.2$ (laterally)

Only the maximum regular static components of the quasi-static loads are taken into account. For air transportation, the mean static stresses are calculated for only flight condition.

The nodal stresses for both harmonic analysis and mean loads static analysis calculated and are listed for all DOF of each structural module. Tables 4.11, 4.12, 4.13, 4.14, 4.15 and 4.16 list the harmonic response results due to various dynamic loading phases for the satellite structural modules. They represent the total equivalent maximum stresses and their locations for each dynamic loading phase. Figures 4.47, 4.48, 4.49, 4.50, 4.51 and 4.52 show the locations of these critical points in each structural module. Table 4.11 lists the maximum Von Mises equivalent stresses and their yield margins of safety MS_y, in each structural module of Small Sat during all dynamic phases (Table 4.17).

4.12 Fatigue Damage Calculations Due to Dynamic Vibrations

In this section, fatigue damage is calculated for Small Sat structure due to mechanical dynamic vibrations [3]. The results of both modal and harmonic analyses are used to estimate fatigue damage of each satellite structure module. The main purpose of this analysis is to investigate whether the satellite structure will withstand all mechanical dynamic loads encountered during road, rail, and air transportation and launch vehicle flight without significant accumulated damage.

4.12.1 Definition of Fatigue Analysis

Fatigue is the phenomenon in which damage accumulates due to repetitive application of loads that may be well below the yield point. The process is dangerous because a single application of the load will not produce static failure, which deceivingly shows a safe structure. However, as loading cycles repeat, structural damage accumulates leading to catastrophic failure. In fatigue analysis,

Table 4.12 Harmonic response results due to various dynamic loading phases for base plate module

Loading phase	Frequency (Hz)	Critical point	Equivalent mean static stress (MPa)	Equivalent real amplitude stress (MPa)	Equivalent imaginary amplitude stress (MPa)	Equivalent amplitude stress (MPa)	Total equivalent maximum stress (MPa)
	35	3	4.772	3.243	8.427	9.030	13.802
	33	2	4.780	3.079	8.464	9.007	13.787
Rail transportation	96	2	4.780	5.956	2.101	6.316	11.096
	62.6	3	4.772	1.657	0.207	1.669	6.442
	73.6	3	4.772	1.588	0.389	1.635	6.407
Road transportation	35	4	4.611	4.552	17.282	17.871	22.483
	33	2	4.780	8.389	20.851	22.475	27.255
	35	3	4.772	11.300	30.760	32.770	37.542
	33	2	4.780	9.759	31.743	33.209	37.989
	96	1	4.784	2.362	49.567	49.623	54.407
	104.7	1	4.784	21.681	21.888	30.808	35.592
	62.6	3	4.772	54.648	5.804	54.955	59.728
Air transportation	73.6	3	4.772	46.457	9.999	47.521	52.293
	119.8	4	4.611	4.787	14.469	15.240	19.852
	128.3	4	4.611	8.289	7.046	10.879	15.491
	187.9	4	4.611	5.153	4.380	6.763	11.374
	137	4	4.611	12.742	10.832	16.724	21.335
	35	2	39.439	19.062	79.178	81.440	120.879
	33	2	39.439	24.502	79.696	83.377	122.816
	96	1	35.185	2.075	43.545	43.594	78.779
	104.7	1	35.185	21.180	21.382	30.096	65.281
	62.6	3	34.961	17.174	1.824	17.271	52.232
	73.6	3	34.961	15.486	3.333	15.841	50.802
	119.8	2	39.439	12.383	2.466	12.626	52.065

(continued)

Table 4.12 (continued)

Loading phase	Frequency (Hz)	Critical point	Equivalent mean static stress (MPa)	Equivalent real amplitude stress (MPa)	Equivalent imaginary amplitude stress (MPa)	Equivalent amplitude stress (MPa)	Total equivalent maximum stress (MPa)
Launch (phase I)	128.3	5	36.938	7.885	7.901	11.163	48.101
	187.9	4	34.073	16.692	14.189	21.908	55.981
	137	5	36.938	7.404	7.419	10.482	47.420
	471.7	2	39.439	0.807	3.986	4.067	43.506
	696.8	2	39.439	0.333	0.976	1.032	40.471
	378.1	2	39.439	2.388	2.679	3.588	43.027
	950.4	2	39.439	0.368	2.212	2.242	41.681
	311.3	2	39.439	3.634	1.754	4.035	43.474
Launch (phase II)	35	2	39.439	19.062	79.178	81.440	120.879
	33	2	39.439	24.502	79.696	83.377	122.816
	96	1	35.185	1.673	35.097	35.137	70.322
	104.7	2	39.439	14.335	11.308	18.258	57.697
	62.6	3	34.961	17.174	1.824	17.271	52.232
	73.6	3	34.961	15.486	3.333	15.841	50.802
	119.8	2	39.439	8.305	1.654	8.468	47.907
	128.3	2	39.439	4.904	1.570	5.149	44.588
	187.9	5	36.938	7.193	7.208	10.183	47.121
	137	2	39.439	4.413	1.413	4.634	44.073
	471.7	2	39.439	0.317	1.567	1.599	41.038
	696.8	2	39.439	0.169	0.496	0.524	39.963
	378.1	2	39.439	0.822	0.922	1.235	40.674
	950.4	2	39.439	0.188	1.130	1.145	40.584
	311.3	2	39.439	1.848	0.892	2.052	41.491

Table 4.13 Harmonic response results due to various dynamic loading phases for mounting plate module

Loading phase	Frequency (Hz)	Critical point	Equivalent mean static stress (MPa)	Equivalent real amplitude stress (MPa)	Equivalent imaginary amplitude stress (MPa)	Equivalent amplitude stress (MPa)	Total equivalent maximum stress (MPa)
	35	7	2.634	1.503	1.496	2.121	4.755
	33	7	2.634	0.395	1.950	1.989	4.624
Rail transportation	96	10	0.337	3.961	6.947	7.997	8.334
	62.6	7	2.634	1.128	2.048	2.338	4.973
	73.6	6	2.594	2.009	1.324	2.406	5.000
Road transportation	35	7	2.634	2.615	2.624	3.705	6.339
	33	7	2.634	0.910	4.444	4.536	7.170
	35	7	2.634	4.839	5.847	7.590	10.224
	33	7	2.634	0.764	7.809	7.846	10.480
	96	10	0.337	7.684	35.121	35.952	36.289
	104.7	9	0.557	16.246	10.543	19.367	19.924
Air transportation	62.6	7	2.634	38.342	51.425	64.145	66.780
	73.6	10	0.337	58.712	35.376	68.546	68.883
	119.8	10	0.337	10.253	26.641	28.546	28.883
	128.3	10	0.337	22.405	6.477	23.322	23.660
	187.9	10	0.337	13.927	4.026	14.497	14.835
	137	10	0.337	34.440	9.956	35.850	36.188
	35	7	19.848	12.439	15.030	19.510	39.358
	33	7	19.848	1.918	19.605	19.699	39.547
	96	8	20.769	7.269	12.223	14.221	34.990
	104.7	7	19.848	1.549	8.210	8.355	28.203
	62.6	7	19.848	12.049	16.161	20.158	40.006

(continued)

Table 4.13 (continued)

Loading phase	Frequency (Hz)	Critical point	Equivalent mean static stress (MPa)	Equivalent real amplitude stress (MPa)	Equivalent imaginary amplitude stress (MPa)	Equivalent amplitude stress (MPa)	Total equivalent maximum stress (MPa)
Launch (phase I)	73.6	6	19.520	17.104	12.513	21.192	40.712
	119.8	10	2.984	11.620	30.194	32.353	35.337
	128.3	10	2.984	27.175	7.856	28.288	31.272
	187.9	10	2.984	45.118	13.043	46.965	49.949
	137	10	2.984	25.517	7.377	26.562	29.546
	471.7	6	19.520	2.883	1.979	3.497	23.017
	696.8	8	20.769	0.525	2.769	2.818	23.587
	378.1	6	19.520	1.970	6.706	6.989	26.509
	950.4	8	20.769	0.937	2.236	2.424	23.193
	311.3	8	20.769	0.864	3.687	3.787	24.556
Launch (phase II)	35	7	19.848	12.439	15.030	19.510	39.358
	33	7	19.848	1.918	19.605	19.699	39.547
	96	8	20.769	5.859	9.852	11.463	32.232
	104.7	7	19.848	1.155	6.120	6.228	26.076
	62.6	7	19.848	12.049	16.161	20.158	40.006
	73.6	6	19.520	17.104	12.513	21.192	40.712
	119.8	10	2.984	7.793	20.251	21.699	24.683
	128.3	8	20.769	1.773	1.532	2.343	23.112
	187.9	10	2.984	24.790	7.166	25.805	28.789
	137	8	20.769	1.596	1.378	2.109	22.878
	471.7	8	20.769	0.568	0.091	0.575	21.344
	696.8	8	20.769	0.267	1.406	1.431	22.200
	378.1	8	20.769	1.458	0.682	1.610	22.379
	950.4	8	20.769	0.478	1.142	1.238	22.007
	311.3	8	20.769	0.440	1.875	1.926	22.695

Table 4.14 Harmonic response results due to various dynamic loading phases for basis plate module

Loading phase	Frequency (Hz)	Critical point	Equivalent mean static stress (MPa)	Equivalent real amplitude stress (MPa)	Equivalent imaginary amplitude stress (MPa)	Equivalent amplitude stress (MPa)	Total equivalent maximum stress (MPa)
	35	11	7.960	3.153	6.808	7.503	15.462
	33	11	7.960	1.009	6.390	6.469	14.429
Rail transportation	96	13	6.790	2.294	11.921	12.140	18.930
	62.6	11	7.960	2.719	1.340	3.031	10.991
	73.6	11	7.960	1.581	2.512	2.968	10.928
Road transportation	35	11	7.960	5.524	14.044	15.091	23.051
	33	11	7.960	3.283	15.867	16.203	24.162
	35	11	7.960	10.347	24.771	26.845	34.805
	33	11	7.960	5.280	24.512	25.074	33.034
	96	13	6.790	19.599	69.044	71.772	78.562
	104.7	13	6.790	50.753	19.585	54.401	61.191
Air transportation	62.6	13	6.790	92.773	55.055	107.879	114.669
	73.6	12	2.265	123.090	76.232	144.784	147.050
	119.8	13	6.790	25.088	21.651	33.139	39.929
	128.3	11	7.960	10.722	19.506	22.259	30.218
	187.9	11	7.960	6.665	12.125	13.836	21.795
	137	11	7.960	16.482	29.985	34.216	42.176
	35	11	60.907	26.597	63.674	69.006	129.913
	33	11	60.907	13.256	61.542	62.953	123.860
	96	11	60.907	7.215	55.007	55.478	116.385
	104.7	13	52.727	49.579	19.132	53.142	105.869
	62.6	11	60.907	26.063	10.804	28.214	89.121
	73.6	11	60.907	14.302	24.022	27.957	88.864

(continued)

Table 4.14 (continued)

Loading phase	Frequency (Hz)	Critical point	Equivalent mean static stress (MPa)	Equivalent real amplitude stress (MPa)	Equivalent imaginary amplitude stress (MPa)	Equivalent amplitude stress (MPa)	Total equivalent maximum stress (MPa)
Launch (phase I)	119.8	13	52.727	28.434	24.538	37.558	90.285
	128.3	11	60.907	13.004	23.659	26.997	87.904
	187.9	11	60.907	21.591	39.280	44.823	105.730
	137	11	60.907	12.211	22.216	25.351	86.258
	471.7	11	60.907	6.271	17.568	18.654	79.561
	696.8	11	60.907	8.758	5.835	10.523	71.430
	378.1	11	60.907	3.411	8.993	9.618	70.525
	950.4	11	60.907	7.225	25.927	26.915	87.822
	311.3	11	60.907	5.421	3.630	6.524	67.431
Launch (phase II)	35	11	60.907	26.597	63.674	69.006	129.913
	33	11	60.907	13.256	61.542	62.953	123.860
	96	11	60.907	5.816	44.336	44.716	105.623
	104.7	13	52.727	36.959	14.262	39.615	92.342
	62.6	11	60.907	26.063	10.804	28.214	89.121
	73.6	11	60.907	14.302	24.022	27.957	88.864
	119.8	11	60.907	9.522	16.415	18.977	79.884
	128.3	11	60.907	8.324	15.144	17.281	78.188
	187.9	11	60.907	11.863	21.583	24.628	85.535
	137	11	60.907	7.492	13.630	15.553	76.460
	471.7	11	60.907	2.465	6.907	7.334	68.241
	696.8	11	60.907	4.448	2.963	5.345	66.252
	378.1	11	60.907	1.174	3.094	3.309	64.216
	950.4	11	60.907	3.690	13.243	13.748	74.655
	311.3	11	60.907	2.757	1.846	3.318	64.225

Table 4.15 Harmonic response results due to various dynamic loading phases for basis walls module

Loading phase	Frequency (Hz)	Critical point	Equivalent mean static stress (MPa)	Equivalent real amplitude stress (MPa)	Equivalent imaginary amplitude stress (MPa)	Equivalent amplitude stress (MPa)	Total equivalent maximum stress (MPa)
	35	16	3.686	2.244	3.391	4.067	7.753
	33	16	3.686	0.120	3.551	3.553	7.239
Rail transportation	96	17	4.094	4.751	0.879	4.831	8.925
	62.6	17	4.094	3.665	1.978	4.165	8.258
	73.6	17	4.094	3.182	3.540	4.760	8.854
Road transportation	35	16	3.686	3.846	6.622	7.657	11.344
	33	16	3.686	0.189	8.624	8.626	12.312
	35	16	3.686	7.470	12.578	14.629	18.315
	33	16	3.686	0.415	13.743	13.749	17.435
	96	14	2.900	4.299	38.699	38.937	41.837
	104.7	14	2.900	25.304	16.547	30.234	33.134
Air transportation	62.6	17	4.094	116.060	75.623	138.524	142.617
	73.6	17	4.094	95.394	105.170	141.989	146.082
	119.8	15	3.253	7.427	18.036	19.505	22.759
	128.3	15	3.253	16.330	7.561	17.996	21.249
	187.9	15	3.253	10.151	4.700	11.186	14.439
	137	15	3.253	25.102	11.623	27.663	30.916
	35	16	28.338	19.201	32.333	37.605	65.943
	33	16	28.338	1.042	34.504	34.520	62.858
	96	14	22.355	3.777	33.998	34.207	56.562
	104.7	18	24.238	24.238	16.021	29.054	53.292
	62.6	17	31.656	36.473	23.765	43.532	75.188
	73.6	17	31.656	31.798	35.057	47.330	78.986

(continued)

Table 4.15 (continued)

Loading phase	Frequency (Hz)	Critical point	Equivalent mean static stress (MPa)	Equivalent real amplitude stress (MPa)	Equivalent imaginary amplitude stress (MPa)	Equivalent amplitude stress (MPa)	Total equivalent maximum stress (MPa)
Launch (phase I)	119.8	15	25.297	8.418	20.442	22.107	47.404
	128.3	15	25.297	19.807	9.171	21.827	47.124
	187.9	15	25.297	32.885	15.226	36.239	61.536
	137	15	25.297	18.599	8.612	20.496	45.793
	471.7	17	31.656	7.350	4.934	8.853	40.509
	696.8	17	31.656	2.783	3.765	4.682	36.338
	378.1	17	31.656	9.240	7.594	11.960	43.616
	950.4	17	31.656	0.124	1.481	1.486	33.142
	311.3	17	31.656	2.364	3.031	3.843	35.499
Launch (phase II)	35	16	28.338	19.201	32.333	37.605	65.943
	33	16	28.338	1.042	34.504	34.520	62.858
	96	14	22.355	3.044	27.402	27.571	49.926
	104.7	18	24.238	18.068	11.943	21.658	45.896
	62.6	17	31.656	36.473	23.765	43.532	75.188
	73.6	17	31.656	31.798	35.057	47.330	78.986
	119.8	15	25.297	5.646	13.710	14.827	40.124
	128.3	17	31.656	7.808	4.210	8.871	40.527
	187.9	15	25.297	18.069	8.366	19.911	45.208
	137	17	31.656	11.410	5.283	12.574	37.871
	471.7	17	31.656	2.890	1.940	3.481	35.137
	696.8	17	31.656	1.413	1.912	2.378	34.034
	378.1	17	31.656	3.179	2.613	4.115	35.771
	950.4	17	31.656	0.063	0.756	0.759	32.415
	311.3	17	31.656	1.202	1.541	1.955	33.611

Table 4.16 Harmonic response results due to various dynamic loading phases for upper frame module

Loading phase	Frequency (Hz)	Critical point	Equivalent mean static stress (MPa)	Equivalent real amplitude stress (MPa)	Equivalent imaginary amplitude stress (MPa)	Equivalent amplitude stress (MPa)	Total equivalent maximum stress (MPa)
	35	22	3.450	0.739	10.974	10.999	14.449
	33	23	3.506	5.089	7.510	9.072	12.578
Rail transportation	96	21	5.447	0.735	7.913	7.947	13.394
	62.6	20	5.596	3.156	1.392	3.449	9.045
	73.6	20	5.596	0.903	0.259	0.939	6.535
Road transportation	35	22	3.450	1.347	25.148	25.184	28.634
	33	23	3.506	14.138	20.637	25.015	28.521
	35	22	3.450	3.156	37.759	37.891	41.341
	33	23	3.506	17.244	26.856	31.916	35.422
	96	21	5.447	1.752	54.606	54.634	60.081
	104.7	21	5.447	25.878	22.112	34.038	39.485
Air transportation	62.6	19	5.144	108.990	59.167	124.014	129.159
	73.6	19	5.144	38.849	83.548	92.139	97.283
	119.8	23	3.506	7.220	47.178	47.727	51.233
	128.3	23	3.506	28.475	25.893	38.487	41.993
	187.9	23	3.506	17.700	16.095	23.924	27.430
	137	23	3.506	43.772	39.802	59.162	62.668
	35	22	27.517	8.113	97.060	97.398	124.915
	33	23	27.976	43.294	67.427	80.130	108.106
	96	21	40.770	1.539	47.972	47.997	88.767
	104.7	21	40.770	25.280	21.600	33.251	74.021
	62.6	19	40.288	34.252	18.594	38.974	79.262
	73.6	19	40.288	12.950	27.849	30.713	71.001

(continued)

Table 4.16 (continued)

Loading phase	Frequency (Hz)	Critical point	Equivalent mean static stress (MPa)	Equivalent real amplitude stress (MPa)	Equivalent imaginary amplitude stress (MPa)	Equivalent amplitude stress (MPa)	Total equivalent maximum stress (MPa)
Launch (phase I)	119.8	23	27.976	8.183	53.469	54.092	82.068
	128.3	23	27.976	34.538	31.406	46.682	74.658
	187.9	23	27.976	57.342	52.142	77.504	105.480
	137	23	27.976	32.431	29.490	43.834	71.810
	471.7	20	43.409	7.511	3.752	8.396	51.805
	696.8	20	43.409	6.996	3.129	7.663	51.072
	378.1	20	43.409	4.235	12.570	13.264	56.673
	950.4	20	43.409	0.636	3.947	3.998	47.407
	311.3	21	40.770	6.998	2.005	7.280	48.050
Launch (phase II)	35	22	27.517	8.113	97.060	97.398	124.915
	33	23	27.976	43.294	67.427	80.130	108.106
	96	21	40.770	1.241	38.666	38.686	79.456
	104.7	21	40.770	18.845	16.102	24.787	65.557
	62.6	19	40.288	34.252	18.594	38.974	79.262
	73.6	19	40.288	12.950	27.849	30.713	71.001
	119.8	23	27.976	5.488	35.862	36.280	64.256
	128.3	19	40.288	16.998	10.151	19.798	60.086
	187.9	23	27.976	31.507	28.650	42.585	70.561
	137	19	40.288	15.298	9.136	17.819	58.107
	471.7	20	43.409	2.953	1.475	3.301	46.710
	696.8	20	43.409	3.553	1.589	3.892	47.301
	378.1	20	43.409	1.457	4.325	4.563	47.972
	950.4	20	43.409	0.325	2.016	2.042	45.451
	311.3	20	43.409	1.401	1.000	1.721	45.130

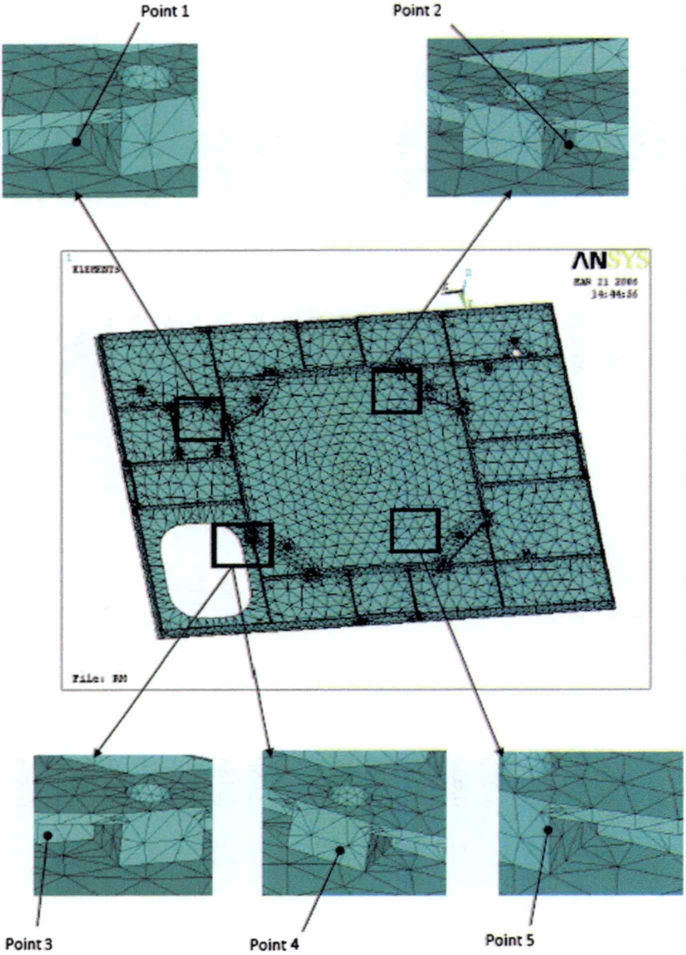

Fig. 4.47 Critical points in base plate module [6]

an empirical approach is used to estimate the incremental fatigue damage for each cycle of load. Fatigue analysis does not predict when a crack will start, or how fast it will grow. It simply predicts how many cycles at a particular stress cycle are typically needed to induce failure, assuming the material has no initial cracks. Fatigue damage occurs more rapidly with higher stresses, so fatigue cracks almost always begin at discontinuities, where there is stress concentration.

Fatigue is an important issue when a part undergoes high stress concentration or a large number of loading cycles. Without stress concentration, most metals can withstand about a thousand fully reversing cycles at stress levels of ±80 % of their ultimate tensile strength. Stress concentration may produce localized stresses

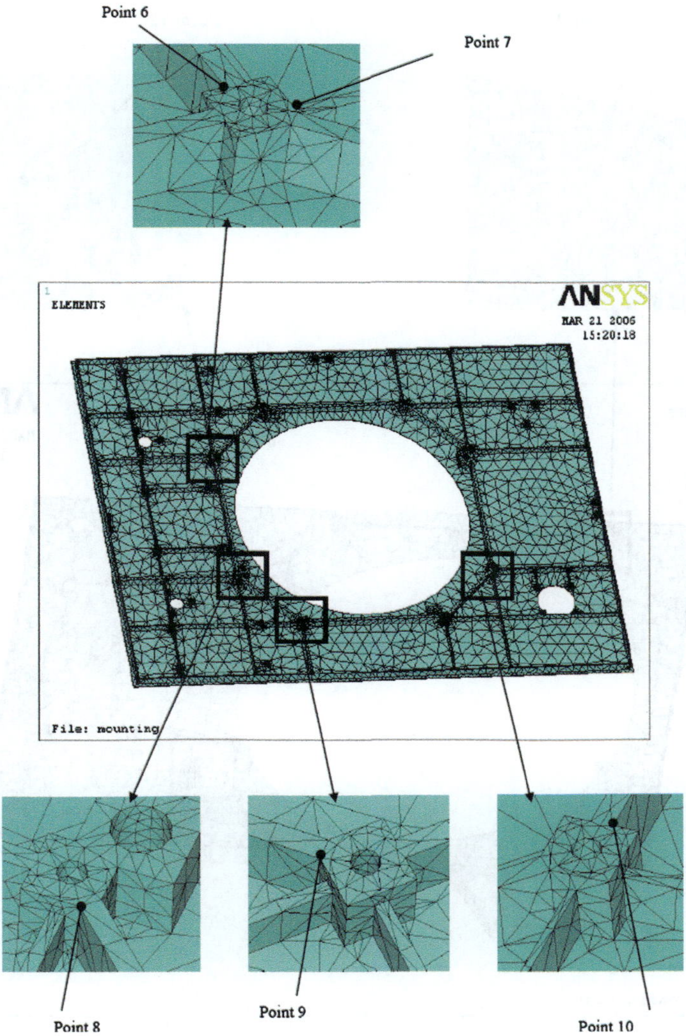

Fig. 4.48 Critical points in mounting plate module

above the material's elastic limit, which may lead to fatigue failure only after a few loading cycles. The main factors that contribute to fatigue failure include:

- Number of load cycles experienced
- Range of stress experienced in each load cycle
- Mean stress experienced in each load cycle

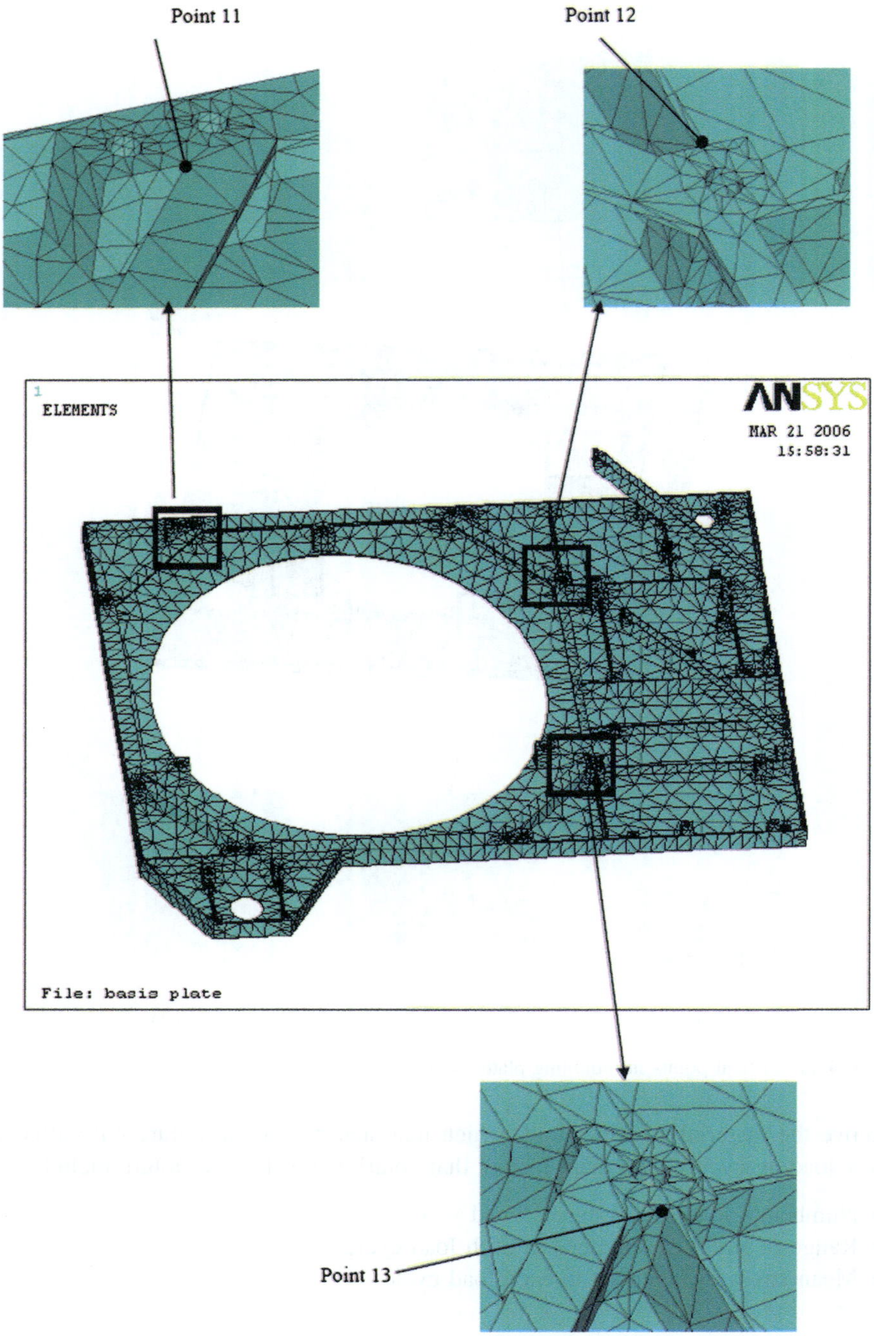

Fig. 4.49 Critical points in basis plate module

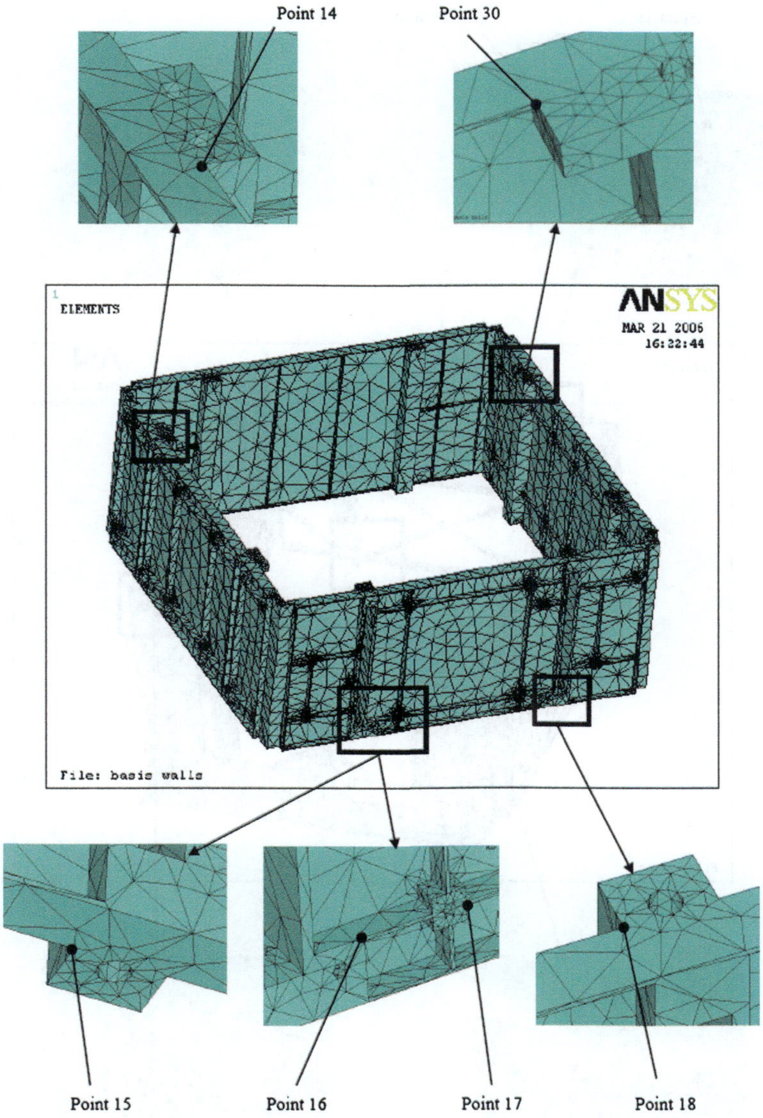

Fig. 4.50 Critical points in basis walls module

4.12.1.1 S–N Curves

Well before a microstructural understanding of fatigue processes was developed, engineers had developed empirical means of quantifying the fatigue process and designing against it. Perhaps the most important concept is the S–N diagram, in which a constant cyclic stress amplitude S is applied to a specimen for a number of loading cycles *N* until the specimen fails. Millions of cycles may be required to

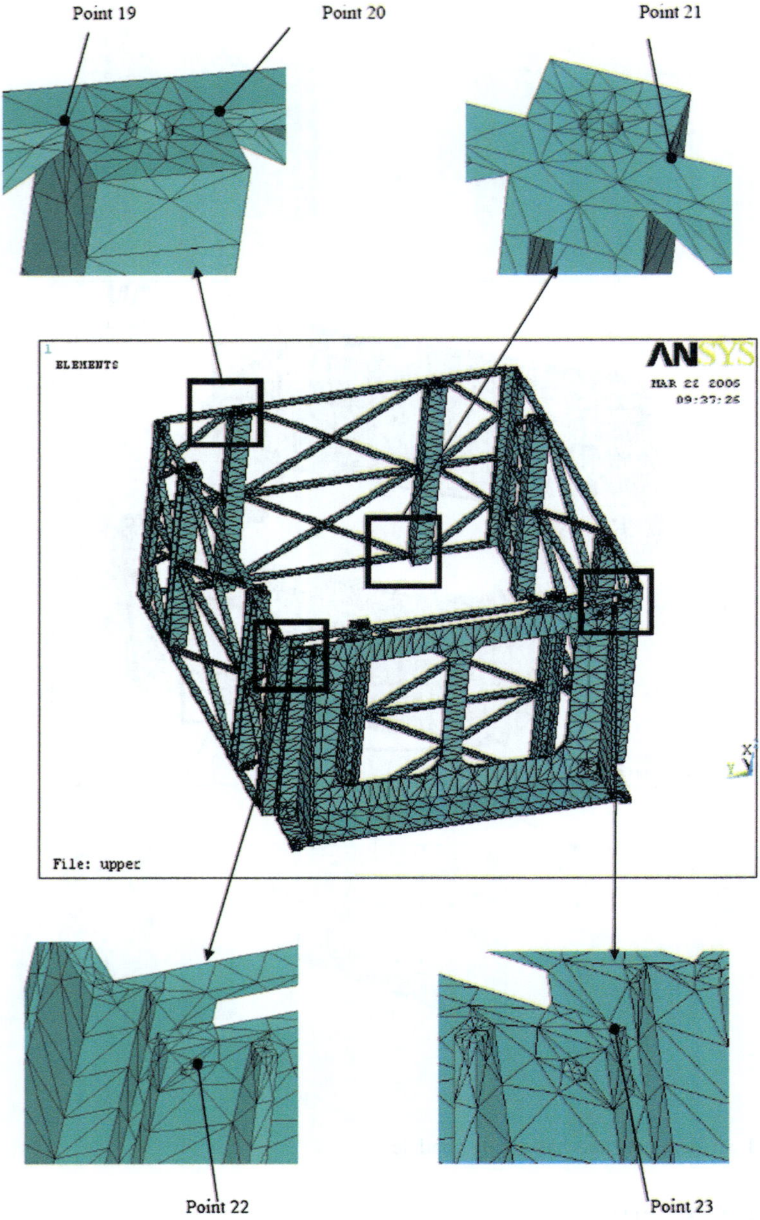

Fig. 4.51 Critical points in upper frame module

cause failure at lower loading levels, so the abscissa in usually plotted logarithmically.

In some materials, particularly ferrous alloys, the S–N curve flattens out so that below a certain endurance limit σ_{end} failure does not occur no matter how long the

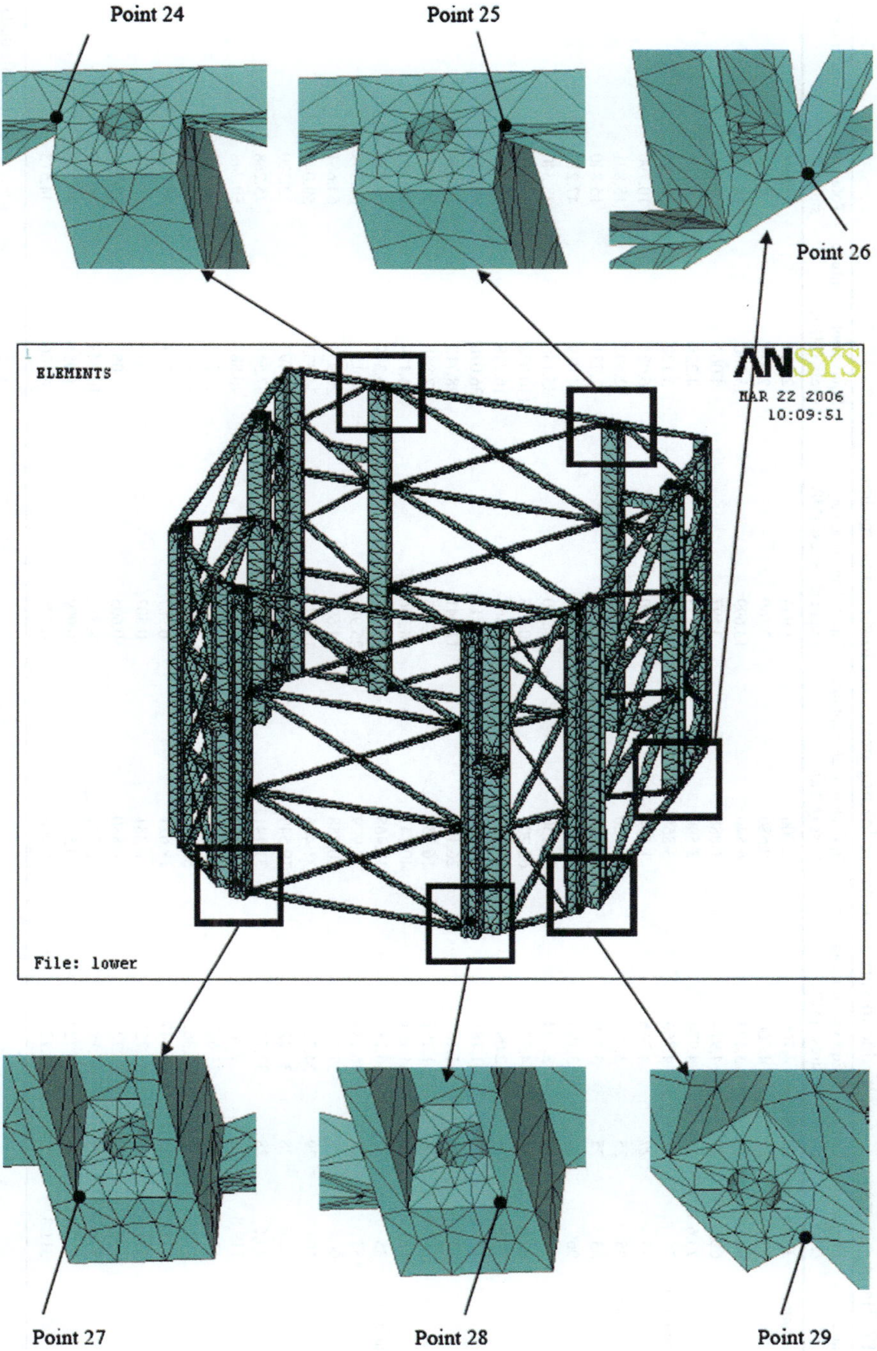

Fig. 4.52 Critical points in lower frame module

Table 4.17 Harmonic response results due to various dynamic loading phases for lower frame module

Loading phase	Frequency (Hz)	Critical point	Equivalent mean static stress (MPa)	Equivalent real amplitude stress (MPa)	Equivalent imaginary amplitude stress (MPa)	Equivalent amplitude stress (MPa)	Total equivalent maximum stress (MPa)
Rail transportation	35	27	4.766	2.200	1.934	2.929	7.694
	33	27	4.766	0.680	2.560	2.649	7.414
	96	29	3.194	6.540	13.699	15.180	18.374
	62.6	28	4.974	1.888	3.597	4.062	9.036
	73.6	28	4.974	3.590	2.233	4.228	9.201
Road transportation	35	27	4.766	3.868	3.354	5.120	9.886
	33	27	4.766	1.554	5.820	6.024	10.790
Air transportation	35	25	3.109	8.420	9.497	12.692	15.801
	33	25	3.109	1.544	12.637	12.731	15.840
	96	29	3.194	9.781	42.918	44.019	47.212
	104.7	29	3.194	32.150	14.200	35.146	38.340
	62.6	28	4.974	62.207	90.816	110.078	115.052
	73.6	24	2.098	87.431	77.368	116.748	**118.846**
	119.8	26	1.181	9.826	34.712	36.076	37.257
	128.3	26	1.181	26.112	11.941	28.713	29.894
	187.9	26	1.181	16.231	7.423	17.848	19.029
	137	26	1.181	40.141	18.356	44.139	45.320
Launch (phase I)	35	27	36.189	17.465	19.727	26.347	62.536
	33	27	36.189	2.929	25.985	26.150	62.339
	96	28	37.727	10.758	24.692	26.934	64.661
	104.7	29	24.714	31.406	13.872	34.333	59.047
	62.6	28	37.727	19.549	28.540	34.593	72.320
	73.6	28	37.727	31.141	20.965	37.541	75.268
	119.8	26	9.362	11.136	39.341	40.887	50.248
	128.3	25	23.647	19.830	7.420	21.173	44.820
	187.9	26	9.362	52.583	24.047	57.821	67.183
	137	25	23.647	18.621	6.967	19.881	43.528
	471.7	28	37.727	0.381	0.507	0.634	38.361
	696.8	28	37.727	0.880	0.605	1.067	38.794
	378.1	27	36.189	5.221	8.812	10.243	46.432
	950.4	28	37.727	0.527	0.909	1.051	38.778
	311.3	28	37.727	0.400	6.775	6.786	44.513

(continued)

Table 4.17 (continued)

Loading phase	Frequency (Hz)	Critical point	Equivalent mean static stress (MPa)	Equivalent real amplitude stress (MPa)	Equivalent imaginary amplitude stress (MPa)	Equivalent amplitude stress (MPa)	Total equivalent maximum stress (MPa)
Launch (phase II)	35	27	36.189	17.465	19.727	26.347	62.536
	33	27	36.189	2.929	25.985	26.150	62.339
	96	28	37.727	8.671	19.902	21.709	59.436
	104.7	29	24.714	23.412	10.341	25.594	50.308
	62.6	28	37.727	19.549	28.540	34.593	72.320
	73.6	28	37.727	31.141	20.965	37.541	75.268
	119.8	29	24.714	4.539	15.683	16.327	41.041
	128.3	28	37.727	1.460	0.644	1.596	39.323
	187.9	25	23.647	18.090	6.768	19.314	42.961
	137	28	37.727	1.314	0.579	1.436	39.163
	471.7	28	37.727	0.150	0.199	0.249	37.976
	696.8	28	37.727	0.447	0.307	0.542	38.269
	378.1	28	37.727	1.532	1.915	2.452	40.179
	950.4	28	37.727	0.269	0.464	0.537	38.264
	311.3	28	37.727	0.204	3.445	3.451	41.178

load is cycled. Obviously, the structure should be sized to keep the stresses below σ_{end} by a suitable safety factor if cyclic loads are to be withstood. For some other materials such as aluminum, no endurance limit exists, and the planned lifetime of the structure must be less than the failure point on the S–N curve.

4.12.1.2 Effect of Mean Load

The fully reversed fatigue loading varies sinusoidally from tension to compression with zero mean stress. Of course, not all actual loading applications involve fully reversed stress cycling. A more general sort of fatigue testing adds a mean stress σ_m on which a sinusoidal cycle is superimposed, as shown in Fig. 4.51. Such a cycle can be described in several ways, a common one is to state the alternating stress σ_{alt} and the stress ratio $R = \sigma_{\mathrm{Min}}/\sigma_{\mathrm{Max}}$. Accordingly, for loading from zero to maximum stress, the stress ratio R is $0/\sigma_{\mathrm{Max}} = 0$. For fully reversed loading, σ_{Min} is equal and opposite in sign to σ_{Max}, in which case $R = -1$.

A very substantial amount of testing is required to obtain an S–N curve for the simple case of fully reversed loading, and it will usually be impractical to determine whole families of curves for every combination of mean and alternating stress. There are a number of solutions for this difficulty. For convenience, the fatigue data are presented in constant life or modified Goodman diagrams as shown in Fig. 4.52 [8]. Here, life for any given stress ratio can be found by the intercept of appropriate life with the diagonal lines. It is usually necessary to interpolate between life lines for actual values.

4.12.2 Cumulative Fatigue Damage

When the cyclic load level varies during the fatigue process, a cumulative damage model is often hypothesized. To illustrate, take the lifetime to be N_1 cycles at a stress level σ_1 and N_2 at σ_2. If damage is assumed to accumulate at a constant rate during fatigue and a number of cycles n_1 are applied at stress σ_1, where $n_1 < N_1$, then the fraction of lifetime consumed will be n_1/N_1. To determine how many additional cycles the specimen will survive at stress σ_2, an additional fraction of life will be available such that the sum of the two fractions equals one:

$$\frac{n_1}{N_1} + \frac{n_2}{N_2} = 1$$

Solving for the remaining permissible cycles at σ_2:

$$n_2 = N_2\left(1 - \frac{n_1}{N_1}\right)$$

The generalization of this approach is called Miner's Law, and can be written

$$\sum \frac{n_j}{N_j} = 1$$

where n_j is number of cycles applied at a load corresponding to a lifetime of N_j.

4.12.3 Performing a Fatigue Damage Calculation

The procedure of a fatigue damage calculation of Small Sat structure due to mechanical dynamic vibrations consists of the following steps:

1. Run a modal analysis to determine the resonance frequencies of the satellite structure. This step is done in Sect. 4.10.
2. Run a harmonic analysis to calculate the fully reserved stress distribution amplitudes at each selected resonance frequency (the first 15 sorted modes) for every loading phase during transportation and launch. This step is done in Sect. 4.11.
3. Run a static analysis of the satellite model to determine the mean stress distribution of the cyclic stress during transportation and launch. This step is done in Sect. 4.9.4.
4. Define the most critical points at each structural module where there are maximum resultant stresses during any loading phase. These points are represented in Figs. 4.47, 4.48, 4.49 and 4.50. Be sure that the maximum resultant stress at any critical point does not exceed the yield stress of AMg6 aluminum alloy (150 MPa).
5. Calculate the number of cycles n_j corresponding to each loading phase and selected resonance frequency. It is calculated from:

$$n_j = \sum f_j t_j$$

6. Where f_j is the resonance frequency and t_j is the corresponding time duration of applied load in each phase.
7. The satellite mechanical loads in Appendix A, Tables A.3, A.5, A.7, and A.12 represent the durations of dynamic vibrations affecting the satellite during rail, road, and air transportation and launch vehicle flight, respectively. The duration of each loading phase has been mentioned before in Sect. 4.11.2. If there is more than one selected resonance frequency located in the same frequency band in any loading phase, the corresponding time duration should be equally divided among the common frequencies. Table 4.18 lists the number of cycles n_j corresponding to each loading phase and selected resonance frequency.
8. Calculate the time life cycles N_j corresponding to each loading phase and selected resonance frequency. This is calculated directly for any given stress ratio by reading from the typical constant life diagram corresponding to the material of the primary structural modules, which is AMg6 aluminum alloy. Unfortunately, this alloy has not available fatigue behavior data, but it is equivalent to 6061-T6

Table 4.18 Number of cycles n_j corresponding to each loading phase and selected resonance frequency

Loading phase	Frequency band (Hz)	Total duration (sec)	Frequency, f_j (Hz)	Duration, t_j (sec)	Number of cycles, n_j
Rail transportation	0–100	720,000	35	144000	5040000
			33	144000	4752000
			96	144000	13852800
			62.6	144000	9014400
			73.6	144000	10612800
Road Transportation	4–43	64,800	35	32400	1134000
			33	32400	1069200
Air transportation	25–50	5,400	35	2700	94500
			33	2700	89100
	50–80	4,320	62.6	2160	135216
			73.6	2160	159192
	80–130	6,120	96	1530	147186
			104.7	1530	160191
			119.8	1530	183294
			128.3	1530	196299
	130–160	3,420	137	3420	468540
	160–190	3,240	187.9	3240	608796
Launch (phase I)	20–2,000	35	35	2.333	81.67
			33	2.333	77
			96	2.333	224.47
			104.7	2.333	244.3
			62.6	2.333	146.07
			73.6	2.333	171.97
			119.8	2.333	279.53
			128.3	2.333	299.37
			187.9	2.333	438.43
			137	2.333	319.67
			471.7	2.333	1100.63
			696.8	2.333	1625.87
			378.1	2.333	882.23
			950.4	2.333	2217.6
			311.3	2.333	726.37
Launch (phase II)	20–2,000	831	35	55.4	1939
			33	55.4	1828.2
			96	55.4	5329.48
			104.7	55.4	5800.38
			62.6	55.4	3468.04
			73.6	55.4	4082.98
			119.8	55.4	6636.92
			128.3	55.4	7107.82
			187.9	55.4	10409.66
			137	55.4	7589.8
			471.7	55.4	26132.18
			696.8	55.4	38602.72
			378.1	55.4	20946.74
			950.4	55.4	52652.16
			311.3	55.4	17246.02

aluminum alloy plates with 0.25 – 2 inch thickness. Figure 4.55 shows the typical constant life diagram for fatigue behavior of 6061-T6 aluminum alloy. The fatigue data of this alloy are presented for unnotched various wrought products under room temperature without specification for surface finish. The stresses in the fatigue diagram are presented in (Ksi), so all stress values in (MPa) from harmonic and mean static analyses must be converted into (Ksi) values. Using the formula:

$$\sigma(\text{ksi}) = \frac{\sigma(\text{MPa})}{6.89}$$

The following empirical stress equation provides an appropriate approximation for the fatigue behavior data of wrought products made of 6061-T6 aluminum alloy at room temperature [9]:

$$\log N_j = a_1 - a_2 \log[\sigma_{\text{Max}}(1 - R)^n - a_3]$$

where:

- N_j: life time cycles
- σ_{Max}: maximum value of stress in the stress cycle (ksi)
- R: stress ratio $R = \sigma_{\text{Min}}/\sigma_{\text{Max}}$
- a_1, a_2, a_3, and n are empirical constants having the following values:
- $a_1 = 20.68$, $a_2 = 9.84$, $a_3 = 0$, $n = 0.63$

For accurate results, the previous equation is used to calculate the life time cycles N_j corresponding to each dynamic loading phase and resonance frequency at the most critical points of Small Sat structure.

Calculate cumulative fatigue damage at the critical points by the relation:

$$\sum D_j = \sum \frac{n_j}{N_j}$$

If fatigue damage accumulation is more than one $\left(\sum D_j > 1\right)$, proper changes have to be made either for the section dimensions or the type of material used in design.

4.12.4 Fatigue Damage Calculation Results

The cumulative fatigue damage is calculated for the most critical points located in the satellite structure. These points are specified during performing the harmonic response analysis of Small Sat structure. By reviewing Sect. 4.9.4, there are 29 critical points distributed on the whole satellite structure. The cyclic stress loading acting on each critical point is specified from the harmonic response analysis for all dynamic loading phases. Table 4.19 lists the cumulative fatigue damage values for each critical point calculated according to the criterion mentioned before.

Table 4.19 Cumulative fatigue damage values for each critical point due to mechanical dynamic vibrations

Structural module	Critical points	Cumulative fatigue damage
Base plate	1	9.25E − 06
	2	1.01E − 04
	3	**1.27E − 04**
	4	9.17E − 05
	5	7.38E − 07
Mounting plate	6	1.46E − 04
	7	1.49E − 04
	8	5.94E − 05
	9	1.64E − 09
	10	**1.65E − 04**
Basis plate	11	4.41E − 03
	12	**2.68E − 01**
	13	1.68E − 02
Basis walls	14	8.24E − 07
	15	1.09E − 03
	16	9.74E − 02
	17	**3.85E − 01**
	18	4.01E − 06
Upper frame	**19**	**5.77E − 02**
	20	2.16E − 02
	21	2.55E − 05
	22	2.34E − 04
	23	2.80E − 04
Lower frame	24	3.45E − 02
	25	3.55E − 02
	26	2.08E − 05
	27	2.31E − 02
	28	**4.19E − 02**
	29	2.46E − 04

It clarifies that point 17 located at the basis walls module is the most affected point in the whole satellite structure by mechanical dynamic vibrations as it has the maximum value of cumulative fatigue damage index (0.385). Tables 4.20, 4.21, 4.22, 4.23, 4.24 and 4.25 explain the dynamic fatigue damage calculation for only the most critical point located in each structural module. They list the fatigue calculations criterion and the cumulative damage index for points 3, 10, 12, 17, 19, and 28 located on the base, mounting, basis plate, basis walls, upper, and lower frame module, respectively.

After performing the fatigue damage analysis for the entire satellite structure and determining the most critical points affected by dynamic vibrations in several loading phases, the main question is how are the first 15 sorted natural frequencies of the satellite structure sufficient to give accurate fatigue calculations or not? To answer this question, the fatigue damage is summed up for each natural frequency in all dynamic loading phases. Figures 4.53, 4.54, 4.55, 4.56, 4.57, 4.58, 4.59, 4.60 and 4.61 show the total accumulated fatigue damage for the most critical points

Table 4.20 Harmonic and dynamic fatigue damage analysis results for critical point No. 3 located on the base plate module

Loading phase	Frequency (Hz)	Static analysis		Harmonic analysis				Fatigue analysis							
		Mean stress		Real	Imag	Amp. stress		Max stress		Min stress		R	N (cycles)	n (cycles)	Damage
		MPa	Ksi	Mpa	MPa	MPa	Ksi	MPa	Ksi	MPa	Ksi				
Rail transportation	35	4.772	0.693	3.243	8.427	9.030	1.311	13.802	2.003	−4.258	−0.618	−0.308	9.708E + 16	5.040E + 06	5.191E − 11
	33	4.772	0.693	1.554	8.430	8.572	1.244	13.344	1.937	−3.800	−0.551	−0.285	1.516E + 17	4.752E + 06	3.135E − 11
	96	4.772	0.693	1.506	5.593	5.792	0.841	10.565	1.533	−1.020	−0.148	−0.097	4.028E + 18	1.385E + 07	3.439E − 12
	62.6	4.772	0.693	1.657	0.207	1.669	0.242	6.442	0.935	3.103	0.450	0.482	5.457E + 22	9.014E + 06	1.652E − 16
	73.6	4.772	0.693	1.588	0.389	1.635	0.237	6.407	0.930	3.137	0.455	0.490	6.333E + 22	1.061E + 07	1.676E − 16
Road transportation	35	4.772	0.693	5.417	16.658	17.517	2.542	22.289	3.235	−12.744	−1.850	−0.572	2.789E + 14	1.134E + 06	4.067E − 09
	33	4.772	0.693	4.669	20.690	21.210	3.078	25.982	3.771	−16.438	−2.386	−0.633	4.874E + 13	1.069E + 06	2.194E − 08
Air transportation	35	4.772	0.693	11.300	30.760	32.770	4.756	37.542	5.449	−27.998	−4.064	−0.746	8.604E + 11	9.450E + 04	1.098E − 07
	33	4.772	0.693	6.456	32.151	32.793	4.759	37.565	5.452	−28.020	−4.067	−0.746	8.548E + 11	8.910E + 04	1.042E − 07
	96	4.772	0.693	10.336	42.252	43.498	6.313	48.270	7.006	−38.726	−5.621	−0.802	5.954E + 10	1.472E + 05	2.472E − 06
	104.7	4.772	0.693	8.834	19.846	21.723	3.153	26.496	3.846	−16.951	−2.460	−0.640	3.913E + 13	1.602E + 05	4.093E − 09
	62.6	4.772	0.693	54.648	5.804	54.955	7.976	59.728	8.669	−50.183	−7.283	−0.840	6.435E + 09	1.352E + 05	2.101E − 05
	73.6	4.772	0.693	46.457	9.999	47.521	6.897	52.293	7.590	−42.748	−6.204	−0.817	2.571E + 10	1.592E + 05	6.192E − 06
	119.8	4.772	0.693	5.734	13.758	14.905	2.163	19.677	2.856	−10.133	−1.471	−0.515	1.194E + 15	1.833E + 05	1.535E − 10
	128.3	4.772	0.693	8.328	5.794	10.145	1.472	14.918	2.165	−5.373	−0.780	−0.360	3.554E + 16	1.963E + 05	5.523E − 12
	187.9	4.772	0.693	5.177	3.602	6.306	0.915	11.079	1.608	−1.534	−0.223	−0.138	2.001E + 18	6.088E + 05	3.043E − 13
	137	4.772	0.693	12.802	8.907	15.596	2.263	20.368	2.956	−10.823	−1.571	−0.531	7.956E + 14	4.685E + 05	5.889E − 10
	35	34.961	5.074	29.047	79.069	84.236	12.226	**119.197**	17.300	−49.275	−7.152	−0.413	3.682E + 07	8.167E + 01	2.218E − 06
Launch (phase I)	33	34.961	5.074	16.209	80.722	82.333	11.950	117.294	17.024	−47.372	−6.876	−0.404	4.499E + 07	7.700E + 01	1.712E − 06
	96	34.961	5.074	9.080	37.119	38.214	5.546	73.175	10.620	−3.253	−0.472	−0.044	2.922E + 10	2.245E + 02	7.682E − 09
	104.7	34.961	5.074	8.630	19.387	21.221	3.080	56.182	8.154	13.740	1.994	0.245	2.932E + 12	2.443E + 02	8.333E − 11
	62.6	34.961	5.074	17.174	1.824	17.271	2.507	52.232	7.581	17.690	2.568	0.339	1.371E + 13	1.461E + 02	1.066E − 11
	73.6	34.961	5.074	15.486	3.333	15.841	2.299	50.802	7.373	19.120	2.775	0.376	2.592E + 13	1.720E + 02	6.636E − 12
	119.8	34.961	5.074	6.499	15.593	16.893	2.452	51.854	7.526	18.068	2.622	0.348	1.614E + 13	2.795E + 02	1.732E − 11
	128.3	34.961	5.074	10.101	7.028	12.305	1.786	47.266	6.860	22.656	3.288	0.479	1.613E + 14	2.994E + 02	1.856E − 12
	187.9	34.961	5.074	16.770	11.668	20.430	2.965	55.391	8.039	14.531	2.109	0.262	3.907E + 12	4.384E + 02	1.122E − 10
	137	34.961	5.074	9.485	6.599	11.555	1.677	46.516	6.751	23.406	3.397	0.503	2.525E + 14	3.197E + 02	1.266E − 12
	471.7	34.961	5.074	2.024	1.649	2.611	0.379	37.572	5.453	32.350	4.695	0.861	5.554E + 18	1.101E + 03	1.982E − 16
	696.8	34.961	5.074	1.218	0.302	1.255	0.182	36.216	5.256	33.706	4.892	0.931	5.953E + 20	1.626E + 03	2.731E − 18
	378.1	34.961	5.074	0.868	5.224	5.295	0.769	40.256	5.843	29.666	4.306	0.737	5.389E + 16	8.822E + 02	1.637E − 14
	950.4	34.961	5.074	0.426	1.161	1.237	0.180	36.198	5.254	33.724	4.895	0.932	6.529E + 20	2.218E + 03	3.396E − 18

(continued)

Table 4.20 (continued)

Loading phase	Frequency (Hz)	Static analysis		Harmonic analysis				Fatigue analysis							
		Mean stress		Real	Imag	Amp. stress		Max stress		Min stress		R	N (cycles)	n (cycles)	Damage
		MPa	Ksi	Mpa	MPa	MPa	Ksi	MPa	Ksi	MPa	Ksi				
Launch (phase II)	311.3	34.961	5.074	2.156	2.177	3.064	0.445	38.025	5.519	31.897	4.629	0.839	1.970E + 18	7.264E + 02	3.688E − 16
	35	34.961	5.074	29.047	79.069	84.236	12.226	**119.197**	17.300	−49.275	−7.152	−0.413	3.682E + 07	1.939E + 03	5.266E − 05
	33	34.961	5.074	16.209	80.722	82.333	11.950	117.294	17.024	−47.372	−6.876	−0.404	4.499E + 07	1.828E + 03	4.064E − 05
	96	34.961	5.074	7.319	29.918	30.800	4.470	65.761	9.544	4.161	0.604	0.063	1.641E + 11	5.329E + 03	3.247E − 08
	104.7	34.961	5.074	6.433	14.452	15.819	2.296	50.780	7.370	19.142	2.778	0.377	2.617E + 13	5.800E + 03	2.216E − 10
	62.6	34.961	5.074	17.174	1.824	17.271	2.507	52.232	7.581	17.690	2.568	0.339	1.371E + 13	3.468E + 03	2.530E − 10
	73.6	34.961	5.074	15.486	3.333	15.841	2.299	50.802	7.373	19.120	2.775	0.376	2.592E + 13	4.083E + 03	1.576E − 10
	119.8	34.961	5.074	4.359	10.458	11.330	1.644	46.291	6.719	23.631	3.430	0.510	2.903E + 14	6.637E + 03	2.286E − 11
	128.3	34.961	5.074	6.466	4.498	7.877	1.143	42.838	6.217	27.084	3.931	0.632	3.666E + 15	7.108E + 03	1.939E − 12
	187.9	34.961	5.074	9.214	6.411	11.225	1.629	46.186	6.703	23.736	3.445	0.514	3.100E + 14	1.041E + 04	3.358E − 11
	137	34.961	5.074	5.819	4.049	7.089	1.029	42.050	6.103	27.872	4.045	0.663	7.538E + 15	7.590E + 03	1.007E − 12
	471.7	34.961	5.074	0.796	0.648	1.026	0.149	35.987	5.223	33.935	4.925	0.943	2.119E + 21	2.613E + 04	1.233E − 17
	696.8	34.961	5.074	0.619	0.153	0.637	0.093	35.598	5.167	34.324	4.982	0.964	4.225E + 22	3.860E + 04	9.137E − 19
	378.1	34.961	5.074	0.299	1.797	1.822	0.264	36.783	5.339	33.139	4.810	0.901	5.582E + 19	2.095E + 04	3.753E − 16
	950.4	34.961	5.074	0.218	0.593	0.632	0.092	35.593	5.166	34.329	4.982	0.965	4.470E + 22	5.265E + 04	1.178E − 18
	311.3	34.961	5.074	1.097	1.107	1.558	0.226	36.519	5.300	33.403	4.848	0.915	1.509E + 20	1.725E + 04	1.143E − 16
						Maximum stress (MPa)		119.197					Cumulative fatigue damage		1.27E − 04

Table 4.21 Harmonic and dynamic fatigue damage analysis results for critical point No. 10 located on the mounting plate module

Loading phase	Frequency (Hz)	Static analysis		Harmonic analysis									Fatigue analysis		
		Mean stress		Real	Imag	Amp. stress		Max stress		Min stress		R	N (cycles)	n (cycles)	Damage
		MPa	Ksi	Mpa	MPa	MPa	Ksi	MPa	Ksi	MPa	Ksi				
Rail transportation	35	0.337	0.049	0.613	0.612	0.866	0.126	1.204	0.175	−0.529	−0.077	−0.439	1.432E + 27	5.040E + 06	3.519E − 21
	33	0.337	0.049	0.098	1.007	1.012	0.147	1.350	0.196	−0.675	−0.098	−0.500	3.592E + 26	4.752E + 06	1.323E − 20
	96	0.337	0.049	3.961	6.947	7.997	1.161	8.334	1.210	−7.659	−1.112	−0.919	1.294E + 18	1.385E + 07	1.071E − 11
	62.6	0.337	0.049	0.887	0.494	1.015	0.147	1.352	0.196	−0.678	−0.098	−0.501	3.502E + 26	9.014E + 06	2.574E − 20
	73.6	0.337	0.049	1.964	1.169	2.286	0.332	2.623	0.381	−1.948	−0.283	−0.743	2.047E + 23	1.061E + 07	5.184E − 17
Road transportation	35	0.337	0.049	0.979	0.912	1.339	0.194	1.676	0.243	−1.001	−0.145	−0.597	2.887E + 25	1.134E + 06	3.928E − 20
	33	0.337	0.049	0.251	2.171	2.186	0.317	2.523	0.366	−1.848	−0.268	−0.733	3.114E + 23	1.069E + 06	3.433E − 18
Air transportation	35	0.337	0.049	2.429	2.401	3.415	0.496	3.753	0.545	−3.078	−0.447	−0.820	4.614E + 21	9.450E + 04	2.048E − 17
	33	0.337	0.049	0.433	4.001	4.024	0.584	4.362	0.633	−3.687	−0.535	−0.845	9.648E + 20	8.910E + 04	9.235E − 17
	96	0.337	0.049	7.684	35.121	35.952	5.218	36.289	5.267	−35.614	−5.169	−0.981	5.481E + 11	1.472E + 05	2.685E − 07
	104.7	0.337	0.049	5.758	13.010	14.227	2.065	14.565	2.114	−13.890	−2.016	−0.954	4.766E + 15	1.602E + 05	3.361E − 11
	62.6	0.337	0.049	28.572	22.827	36.571	5.308	36.908	5.357	−36.233	−5.259	−0.982	4.636E + 11	1.352E + 05	2.917E − 07
	73.6	0.337	0.049	58.712	35.376	68.546	9.949	**68.883**	9.998	−68.209	−9.900	−0.990	9.730E + 08	1.592E + 05	1.636E − 04
	119.8	0.337	0.049	10.253	26.641	28.546	4.143	28.883	4.192	−28.208	−4.094	−0.977	5.258E + 12	1.833E + 05	3.486E − 08
	128.3	0.337	0.049	22.405	6.477	23.322	3.385	23.660	3.434	−22.985	−3.336	−0.971	3.805E + 13	1.963E + 05	5.159E − 09
	187.9	0.337	0.049	13.927	4.026	14.497	2.104	14.835	2.153	−14.160	−2.055	−0.955	3.967E + 15	6.088E + 05	1.534E − 10
	137	0.337	0.049	34.440	9.956	35.850	5.203	36.188	5.252	−35.513	−5.154	−0.981	5.635E + 11	4.685E + 05	8.315E − 07
Launch (phase I)	35	2.984	0.433	6.243	6.172	8.779	1.274	11.763	1.707	−5.795	−0.841	−0.493	2.069E + 17	8.167E + 01	3.947E − 16
	33	2.984	0.433	1.088	10.045	10.104	1.466	13.088	1.900	−7.120	−1.033	−0.544	5.870E + 16	7.700E + 01	1.312E − 15
	96	2.984	0.433	6.750	30.854	31.584	4.584	34.568	5.017	−28.600	−4.151	−0.827	1.460E + 12	2.245E + 02	1.537E − 10
	104.7	2.984	0.433	5.625	12.709	13.898	2.017	16.882	2.450	−10.914	−1.584	−0.646	3.219E + 15	2.443E + 02	7.590E − 14
	62.6	2.984	0.433	8.979	7.174	11.493	1.668	14.477	2.101	−8.509	−1.235	−0.588	1.830E + 16	1.461E + 02	7.983E − 15
	73.6	2.984	0.433	19.571	11.792	22.849	3.316	25.833	3.749	−19.865	−2.883	−0.769	3.138E + 13	1.720E + 02	5.481E − 12
	119.8	2.984	0.433	11.620	30.194	32.353	4.696	35.337	5.129	−29.369	−4.263	−0.831	1.161E + 12	2.795E + 02	2.407E − 10
	128.3	2.984	0.433	27.175	7.856	28.288	4.106	31.272	4.539	−25.304	−3.673	−0.809	4.165E + 12	2.994E + 02	7.187E − 11
	187.9	2.984	0.433	45.118	13.043	46.965	6.816	49.949	7.250	−43.981	−6.383	−0.881	3.268E + 10	4.384E + 02	1.342E − 08
	137	2.984	0.433	25.517	7.377	26.562	3.855	29.546	4.288	−23.578	−3.422	−0.798	7.566E + 12	3.197E + 02	4.225E − 11
	471.7	2.984	0.433	1.973	0.618	2.068	0.300	5.052	0.733	0.916	0.133	0.181	3.506E + 22	1.101E + 03	3.139E − 20
	696.8	2.984	0.433	7.060	2.846	7.612	1.105	10.597	1.538	−4.628	−0.672	−0.437	7.324E + 17	1.626E + 03	2.220E − 15
	378.1	2.984	0.433	8.175	11.654	14.235	2.066	17.219	2.499	−11.251	−1.633	−0.653	2.582E + 15	8.822E + 02	3.417E − 13
	950.4	2.984	0.433	1.452	1.627	2.180	0.316	5.164	0.750	0.804	0.117	0.156	2.331E + 22	2.218E + 03	9.515E − 20
	311.3	2.984	0.433	0.524	1.006	1.135	0.165	4.119	0.598	1.850	0.268	0.449	3.048E + 24	7.264E + 02	2.383E − 22

(continued)

Table 4.21 (continued)

Loading phase	Frequency (Hz)	Static analysis		Harmonic analysis									Fatigue analysis		
		Mean stress		Real	Imag	Amp. stress		Max stress		Min stress		R	N (cycles)	n (cycles)	Damage
		MPa	Ksi	Mpa	MPa	MPa	Ksi	MPa	Ksi	MPa	Ksi				
Launch (phase II)	35	2.984	0.433	6.243	6.172	8.779	1.274	11.763	1.707	−5.795	−0.841	−0.493	2.069E + 17	1.939E + 03	9.370E − 15
	33	2.984	0.433	1.088	10.045	10.104	1.466	13.088	1.900	−7.120	−1.033	−0.544	5.870E + 16	1.828E + 03	3.114E − 14
	96	2.984	0.433	5.441	24.868	25.456	3.695	28.440	4.128	−22.472	−3.262	−0.790	1.131E + 13	5.329E + 03	4.710E − 10
	104.7	2.984	0.433	4.193	9.474	10.360	1.504	13.344	1.937	−7.376	−1.071	−0.553	4.682E + 16	5.800E + 03	1.239E − 13
	62.6	2.984	0.433	8.979	7.174	11.493	1.668	14.477	2.101	−8.509	−1.235	−0.588	1.830E + 16	3.468E + 03	1.895E − 13
	73.6	2.984	0.433	19.571	11.792	22.849	3.316	25.833	3.749	−19.865	−2.883	−0.769	3.138E + 13	4.083E + 03	1.301E − 10
	119.8	2.984	0.433	7.793	20.251	21.699	3.149	24.683	3.582	−18.715	−2.716	−0.758	5.101E + 13	6.637E + 03	1.301E − 10
	128.3	2.984	0.433	17.394	5.028	18.106	2.628	21.090	3.061	−15.122	−2.195	−0.717	2.778E + 14	7.108E + 03	2.559E − 11
	187.9	2.984	0.433	24.790	7.166	25.805	3.745	28.789	4.178	−22.821	−3.312	−0.793	9.948E + 12	1.041E + 04	1.046E − 09
	137	2.984	0.433	15.655	4.526	16.296	2.365	19.280	2.798	−13.312	−1.932	−0.690	7.401E + 14	7.590E + 03	1.026E − 11
	471.7	2.984	0.433	0.776	0.243	0.813	0.118	3.797	0.551	2.171	0.315	0.572	3.234E + 25	2.613E + 04	8.080E − 22
	696.8	2.984	0.433	3.586	1.446	3.866	0.561	6.850	0.994	−0.882	−0.128	−0.129	2.390E + 20	3.860E + 04	1.615E − 16
	378.1	2.984	0.433	2.812	4.010	4.898	0.711	7.882	1.144	−1.913	−0.278	−0.243	3.313E + 19	2.095E + 04	6.322E − 16
	950.4	2.984	0.433	0.741	0.831	1.114	0.162	4.098	0.595	1.870	0.271	0.456	3.484E + 24	5.265E + 04	1.511E − 20
	311.3	2.984	0.433	0.266	0.512	0.577	0.084	3.561	0.517	2.407	0.349	0.676	3.422E + 26	1.725E + 04	5.040E − 23
					Maximum stress (MPa)			68.883					Cumulative fatigue damage		1.65E − 04

Table 4.22 Harmonic and dynamic fatigue damage analysis results for critical point No. 12 located on the basis plate module

Loading phase	Frequency (Hz)	Static analysis		Harmonic analysis									Fatigue analysis		
		Mean stress		Real	Imag	Amp. stress		Max stress		Min stress		R	N (cycles)	n (cycles)	Damage
		MPa	Ksi	Mpa	MPa	MPa	Ksi	MPa	Ksi	MPa	Ksi				
Rail transportation	35	2.265	0.329	0.882	2.631	2.774	0.403	5.040	0.731	−0.509	−0.074	−0.101	5.720E + 21	5.040E + 06	8.811E − 16
	33	2.265	0.329	1.540	0.943	1.806	0.262	4.071	0.591	0.459	0.067	0.113	1.782E + 23	4.752E + 06	2.667E − 17
	96	2.265	0.329	6.777	8.809	11.114	1.613	13.380	1.942	−8.849	−1.284	−0.661	3.000E + 16	1.385E + 07	4.617E − 10
	62.6	2.265	0.329	2.011	1.326	2.409	0.350	4.674	0.678	−0.143	−0.021	−0.031	1.807E + 22	9.014E + 06	4.989E − 16
	73.6	2.265	0.329	4.132	2.529	4.844	0.703	7.109	1.032	−2.579	−0.374	−0.363	5.163E + 19	1.061E + 07	2.056E − 13
Road transportation	35	2.265	0.329	1.499	6.533	6.703	0.973	8.968	1.302	−4.437	−0.644	−0.495	2.960E + 18	1.134E + 06	3.831E − 13
	33	2.265	0.329	4.368	3.623	5.675	0.824	7.940	1.152	−3.410	−0.495	−0.429	1.294E + 19	1.069E + 06	8.266E − 14
Air transportation	35	2.265	0.329	3.289	8.750	9.347	1.357	11.613	1.685	−7.082	−1.028	−0.610	1.470E + 17	9.450E + 04	6.430E − 13
	33	2.265	0.329	5.568	2.802	6.233	0.905	8.498	1.233	−3.968	−0.576	−0.467	5.649E + 18	8.910E + 04	1.577E − 14
	96	2.265	0.329	20.986	53.956	57.894	8.403	60.159	8.731	−55.628	−8.074	−0.925	4.539E + 09	1.472E + 05	3.243E − 05
	104.7	2.265	0.329	7.796	23.108	24.388	3.540	26.653	3.868	−22.122	−3.211	−0.830	1.869E + 13	1.602E + 05	8.569E − 09
	62.6	2.265	0.329	62.583	54.804	83.187	12.074	85.453	12.402	−80.922	−11.745	−0.947	1.337E + 08	1.352E + 05	1.011E − 03
	73.6	2.265	0.329	123.090	76.232	144.784	21.014	**147.050**	21.342	−142.519	−20.685	−0.969	5.969E + 05	1.592E + 05	2.667E − 01
	119.8	2.265	0.329	4.996	29.807	30.223	4.386	32.488	4.715	−27.958	−4.058	−0.861	2.405E + 12	1.833E + 05	7.621E − 08
	128.3	2.265	0.329	17.325	13.764	22.127	3.211	24.392	3.540	−19.862	−2.883	−0.814	4.718E + 13	1.963E + 05	4.160E − 09
	187.9	2.265	0.329	10.769	8.556	13.754	1.996	16.020	2.325	−11.489	−1.667	−0.717	4.156E + 15	6.088E + 05	1.465E − 10
	137	2.265	0.329	26.633	21.158	34.015	4.937	36.280	5.266	−31.749	−4.608	−0.875	7.734E + 11	4.685E + 05	6.058E − 07
Launch (phase I)	35	18.014	2.615	8.455	22.491	24.028	3.487	42.042	6.102	−6.014	−0.873	−0.143	3.901E + 12	8.167E + 01	2.094E − 11
	33	18.014	2.615	13.979	7.035	15.649	2.271	33.663	4.886	2.365	0.343	0.070	1.250E + 14	7.700E + 01	6.159E − 13
	96	18.014	2.615	18.437	47.400	50.859	7.382	68.873	9.996	−32.845	−4.767	−0.477	6.192E + 09	2.245E + 02	3.625E − 08
	104.7	18.014	2.615	7.616	22.574	23.824	3.458	41.838	6.072	−5.810	−0.843	−0.139	4.185E + 12	2.443E + 02	5.837E − 11
	62.6	18.014	2.615	19.667	17.223	26.142	3.794	44.156	6.409	−8.128	−1.180	−0.184	1.934E + 12	1.461E + 02	7.553E − 11
	73.6	18.014	2.615	41.029	25.411	48.261	7.004	66.275	9.619	−30.247	−4.390	−0.456	9.859E + 09	1.720E + 02	1.744E − 08
	119.8	18.014	2.615	5.662	33.782	34.253	4.971	52.267	7.586	−16.239	−2.357	−0.311	1.960E + 11	2.795E + 02	1.426E − 09
	128.3	18.014	2.615	21.014	16.694	26.838	3.895	44.852	6.510	−8.824	−1.281	−0.197	1.552E + 12	2.994E + 02	1.928E − 10
	187.9	18.014	2.615	34.888	27.717	44.558	6.467	62.572	9.082	−26.544	−3.853	−0.424	1.994E + 10	4.384E + 02	2.199E − 08
	137	18.014	2.615	19.733	15.676	25.202	3.658	43.216	6.272	−7.188	−1.043	−0.166	2.625E + 12	3.197E + 02	1.218E − 10
	471.7	18.014	2.615	9.245	13.515	16.375	2.377	34.389	4.991	1.639	0.238	0.048	8.735E + 13	1.101E + 03	1.260E − 11
	696.8	18.014	2.615	8.165	35.234	36.168	5.249	54.182	7.864	−18.154	−2.635	−0.335	1.227E + 11	1.626E + 03	1.325E − 08
	378.1	18.014	2.615	12.071	13.108	17.819	2.586	35.833	5.201	0.195	0.028	0.005	4.452E + 13	8.822E + 02	1.982E − 11
	950.4	18.014	2.615	6.515	13.300	14.810	2.150	32.824	4.764	3.204	0.465	0.098	1.929E + 14	2.218E + 03	1.150E − 11
	311.3	18.014	2.615	3.909	9.089	9.894	1.436	27.908	4.050	8.120	1.179	0.291	4.246E + 15	7.264E + 02	1.711E − 13

(continued)

Table 4.22 (continued)

Loading phase	Frequency (Hz)	Static analysis		Harmonic analysis									Fatigue analysis		
		Mean stress		Real	Imag	Amp. stress		Max stress		Min stress		R	N (cycles)	n (cycles)	Damage
		MPa	Ksi	Mpa	MPa	MPa	Ksi	MPa	Ksi	MPa	Ksi				
Launch (phase II)	35	18.014	2.615	8.455	22.491	24.028	3.487	42.042	6.102	−6.014	−0.873	−0.143	3.901E + 12	1.939E + 03	4.971E − 10
	33	18.014	2.615	13.979	7.035	15.649	2.271	33.663	4.886	2.365	0.343	0.070	1.250E + 14	1.828E + 03	1.462E − 11
	96	18.014	2.615	14.860	38.205	40.993	5.950	59.007	8.564	−22.979	−3.335	−0.389	4.139E + 10	5.329E + 03	1.288E − 07
	104.7	18.014	2.615	5.677	16.828	17.760	2.578	35.774	5.192	0.254	0.037	0.007	4.573E + 13	5.800E + 03	1.268E − 10
	62.6	18.014	2.615	19.667	17.223	26.142	3.794	44.156	6.409	−8.128	−1.180	−0.184	1.934E + 12	3.468E + 03	1.793E − 09
	73.6	18.014	2.615	41.029	25.411	48.261	7.004	66.275	9.619	−30.247	−4.390	−0.456	9.859E + 09	4.083E + 03	4.141E − 07
	119.8	18.014	2.615	3.798	22.657	22.973	3.334	40.987	5.949	−4.959	−0.720	−0.121	5.651E + 12	6.637E + 03	1.174E − 09
	128.3	18.014	2.615	13.451	10.686	17.179	2.493	35.193	5.108	0.835	0.121	0.024	5.965E + 13	7.108E + 03	1.192E − 10
	187.9	18.014	2.615	19.170	15.229	24.482	3.553	42.496	6.168	−6.468	−0.939	−0.152	3.339E + 12	1.041E + 04	3.117E − 09
	137	18.014	2.615	12.106	9.617	15.461	2.244	33.475	4.859	2.553	0.371	0.076	1.375E + 14	7.590E + 03	5.519E − 11
	471.7	18.014	2.615	3.635	5.314	6.438	0.934	24.452	3.549	11.576	1.680	0.473	9.858E + 16	2.613E + 04	2.651E − 13
	696.8	18.014	2.615	4.147	17.895	18.369	2.666	36.383	5.281	−0.355	−0.052	−0.010	3.489E + 13	3.860E + 04	1.107E − 09
	378.1	18.014	2.615	4.153	4.510	6.131	0.890	24.145	3.504	11.883	1.725	0.492	1.397E + 17	2.095E + 04	1.499E − 13
	950.4	18.014	2.615	3.328	6.793	7.564	1.098	25.578	3.712	10.450	1.517	0.409	3.079E + 16	5.265E + 04	1.710E − 12
	311.3	18.014	2.615	1.988	4.622	5.032	0.730	23.046	3.345	12.982	1.884	0.563	5.637E + 17	1.725E + 04	3.060E − 14
					Maximum stress (MPa)			147.050				Cumulative fatigue damage			2.68E − 01

Table 4.23 Harmonic and dynamic fatigue damage analysis results for critical point No. 17 located on the basis walls module

Loading phase	Frequency (Hz)	Static analysis		Harmonic analysis									Fatigue analysis		
		Mean stress		Real	Imag	Amp. stress		Max stress		Min stress		R	N (cycles)	n (cycles)	Damage
		MPa	Ksi	Mpa	MPa	MPa	Ksi	MPa	Ksi	MPa	Ksi				
Rail transportation	35	4.094	0.594	2.428	2.678	3.615	0.525	7.709	1.119	0.479	0.069	0.062	2.359E + 20	5.040E + 06	2.137E − 14
	33	4.094	0.594	0.585	3.034	3.090	0.448	7.184	1.043	1.004	0.146	0.140	8.070E + 20	4.752E + 06	5.889E − 15
	96	4.094	0.594	4.751	0.879	4.831	0.701	8.925	1.295	−0.738	−0.107	−0.083	2.292E + 19	1.385E + 07	6.044E − 13
	62.6	4.094	0.594	3.665	1.978	4.165	0.604	8.258	1.199	−0.071	−0.010	−0.009	7.635E + 19	9.014E + 06	1.181E − 13
	73.6	4.094	0.594	3.182	3.540	4.760	0.691	8.854	1.285	−0.666	−0.097	−0.075	2.589E + 19	1.061E + 07	4.100E − 13
Road transportation	35	4.094	0.594	4.214	5.063	6.588	0.956	10.681	1.550	−2.494	−0.362	−0.233	1.744E + 18	1.134E + 06	6.504E − 13
	33	4.094	0.594	1.151	7.204	7.295	1.059	11.389	1.653	−3.202	−0.465	−0.281	7.331E + 17	1.069E + 06	1.458E − 12
Air transportation	35	4.094	0.594	7.994	10.119	12.896	1.872	16.990	2.466	−8.802	−1.278	−0.518	5.003E + 15	9.450E + 04	1.889E − 11
	33	4.094	0.594	1.224	11.912	11.975	1.738	16.069	2.332	−7.881	−1.144	−0.490	9.701E + 15	8.910E + 04	9.185E − 12
	96	4.094	0.594	15.049	0.788	15.070	2.187	19.163	2.781	−10.976	−1.593	−0.573	1.229E + 15	1.472E + 05	1.198E − 10
	104.7	4.094	0.594	8.042	0.925	8.095	1.175	12.188	1.769	−4.001	−0.581	−0.328	3.007E + 17	1.602E + 05	5.328E − 13
	62.6	4.094	0.594	116.060	75.623	138.524	20.105	142.617	20.699	−134.430	−19.511	−0.943	8.776E + 05	1.352E + 05	1.541E − 01
	73.6	4.094	0.594	95.394	105.170	141.989	20.608	**146.082**	21.202	−137.895	−20.014	−0.944	6.899E + 05	1.592E + 05	2.307E − 01
	119.8	4.094	0.594	5.681	6.476	8.614	1.250	12.708	1.844	−4.520	−0.656	−0.356	1.756E + 17	1.833E + 05	1.044E − 12
	128.3	4.094	0.594	10.057	5.423	11.426	1.658	15.520	2.253	−7.332	−1.064	−0.472	1.473E + 16	1.963E + 05	1.333E − 11
	187.9	4.094	0.594	6.251	3.371	7.102	1.031	11.196	1.625	−3.009	−0.437	−0.269	9.214E + 17	6.088E + 05	6.607E − 13
	137	4.094	0.594	15.460	8.336	17.564	2.549	21.658	3.143	−13.470	−1.955	−0.622	3.044E + 14	4.685E + 05	1.539E − 09
Launch (phase I)	35	31.656	4.594	20.549	26.011	33.149	4.811	64.805	9.406	−1.493	−0.217	−0.023	1.098E + 11	8.167E + 01	7.438E − 10
	33	31.656	4.594	3.074	29.908	30.066	4.364	61.722	8.958	1.590	0.231	0.026	2.401E + 11	7.700E + 01	3.206E − 10
	96	31.656	4.594	13.220	0.692	13.238	1.921	44.894	6.516	18.418	2.673	0.410	1.237E + 14	2.245E + 02	1.815E − 12
	104.7	31.656	4.594	7.856	0.904	7.907	1.148	39.563	5.742	23.749	3.447	0.600	4.780E + 15	2.443E + 02	5.111E − 14
	62.6	31.656	4.594	36.473	23.765	43.532	6.318	75.188	10.913	−11.876	−1.724	−0.158	1.180E + 10	1.461E + 02	1.238E − 08
	73.6	31.656	4.594	31.798	35.057	47.330	6.869	78.986	11.464	−15.674	−2.275	−0.198	5.873E + 09	1.720E + 02	2.928E − 08
	119.8	31.656	4.594	6.438	7.339	9.763	1.417	41.419	6.011	21.893	3.178	0.529	1.095E + 15	2.795E + 02	2.553E − 13
	128.3	31.656	4.594	12.198	6.578	13.858	2.011	45.514	6.606	17.798	2.583	0.391	8.856E + 13	2.994E + 02	3.381E − 12
	187.9	31.656	4.594	20.252	10.921	23.009	3.339	54.665	7.934	8.647	1.255	0.158	1.962E + 12	4.384E + 02	2.235E − 10
	137	31.656	4.594	11.454	6.177	13.013	1.889	44.669	6.483	18.643	2.706	0.417	1.400E + 14	3.197E + 02	2.283E − 12
	471.7	31.656	4.594	7.350	4.934	8.853	1.285	40.509	5.879	22.803	3.310	0.563	2.177E + 15	1.101E + 03	5.055E − 13
	696.8	31.656	4.594	2.783	3.765	4.682	0.679	36.338	5.274	26.974	3.915	0.742	1.679E + 17	1.626E + 03	9.685E − 15
	378.1	31.656	4.594	9.240	7.594	11.960	1.736	43.616	6.330	19.696	2.859	0.452	2.577E + 14	8.822E + 02	3.423E − 12
	950.4	31.656	4.594	0.124	1.481	1.486	0.216	33.142	4.810	30.170	4.379	0.910	2.889E + 20	2.218E + 03	7.675E − 18
	311.3	31.656	4.594	2.364	3.031	3.843	0.558	35.499	5.152	27.813	4.037	0.783	6.212E + 17	7.264E + 02	1.169E − 15

(continued)

Table 4.23 (continued)

Loading phase	Frequency (Hz)	Static analysis		Harmonic analysis									Fatigue analysis		
		Mean stress		Real	Imag	Amp. stress		Max stress		Min stress		R	N (cycles)	n (cycles)	Damage
		MPa	Ksi	Mpa	MPa	MPa	Ksi	MPa	Ksi	MPa	Ksi				
Launch (phase II)	35	31.656	4.594	20.549	26.011	33.149	4.811	64.805	9.406	−1.493	−0.217	−0.023	1.098E + 11	1.939E + 03	1.766E − 08
	33	31.656	4.594	3.074	29.908	30.066	4.364	61.722	8.958	1.590	0.231	0.026	2.401E + 11	1.828E + 03	7.613E − 09
	96	31.656	4.594	10.656	0.558	10.671	1.549	42.327	6.143	20.985	3.046	0.496	5.832E + 14	5.329E + 03	9.138E − 12
	104.7	31.656	4.594	5.856	0.674	5.895	0.856	37.551	5.450	25.761	3.739	0.686	3.572E + 16	5.800E + 03	1.624E − 13
	62.6	31.656	4.594	36.473	23.765	43.532	6.318	75.188	10.913	−11.876	−1.724	−0.158	1.180E + 10	3.468E + 03	2.939E − 07
	73.6	31.656	4.594	31.798	35.057	47.330	6.869	78.986	11.464	−15.674	−2.275	−0.198	5.873E + 09	4.083E + 03	6.952E − 07
	119.8	31.656	4.594	4.318	4.922	6.548	0.950	38.204	5.545	25.108	3.644	0.657	1.748E + 16	6.637E + 03	3.796E − 13
	128.3	31.656	4.594	7.808	4.210	8.871	1.287	40.527	5.882	22.785	3.307	0.562	2.147E + 15	7.108E + 03	3.310E − 12
	187.9	31.656	4.594	11.128	6.000	12.642	1.835	44.298	6.429	19.014	2.760	0.429	1.727E + 14	1.041E + 04	6.027E − 11
	137	31.656	4.594	7.027	3.789	7.984	1.159	39.640	5.753	23.672	3.436	0.597	4.472E + 15	7.590E + 03	1.697E − 12
	471.7	31.656	4.594	2.890	1.940	3.481	0.505	35.137	5.100	28.175	4.089	0.802	1.192E + 18	2.613E + 04	2.192E − 14
	696.8	31.656	4.594	1.413	1.912	2.378	0.345	34.034	4.940	29.278	4.249	0.860	1.421E + 19	3.860E + 04	2.717E − 15
	378.1	31.656	4.594	3.179	2.613	4.115	0.597	35.771	5.192	27.541	3.997	0.770	3.956E + 17	2.095E + 04	5.295E − 14
	950.4	31.656	4.594	0.063	0.756	0.759	0.110	32.415	4.705	30.897	4.484	0.953	2.017E + 22	5.265E + 04	2.610E − 18
	311.3	31.656	4.594	1.202	1.541	1.955	0.284	33.611	4.878	29.701	4.311	0.884	5.012E + 19	1.725E + 04	3.441E − 16
					Maximum stress (MPa)			146.082					Cumulative fatigue damage		3.85E − 01

Table 4.24 Harmonic and dynamic fatigue damage analysis results for critical point No. 19 located on the upper frame module

Loading phase	Frequency (Hz)	Static analysis		Harmonic analysis									Fatigue analysis		
		Mean stress		Real	Imag	Amp. stress		Max stress		Min stress		R	N (cycles)	n (cycles)	Damage
		MPa	Ksi	Mpa	MPa	MPa	Ksi	MPa	Ksi	MPa	Ksi				
Rail transportation	35	5.144	0.747	2.380	4.470	5.064	0.735	10.208	1.482	0.081	0.012	0.008	1.051E + 19	5.040E + 06	4.797E − 13
	33	5.144	0.747	1.378	3.568	3.824	0.555	8.969	1.302	1.320	0.192	0.147	9.589E + 19	4.752E + 06	4.956E − 14
	96	5.144	0.747	0.810	7.736	7.778	1.129	12.923	1.876	−2.634	−0.382	−0.204	3.111E + 17	1.385E + 07	4.453E − 11
	62.6	5.144	0.747	3.425	1.595	3.779	0.548	8.923	1.295	1.366	0.198	0.153	1.053E + 20	9.014E + 06	8.563E − 14
	73.6	5.144	0.747	1.304	0.281	1.334	0.194	6.478	0.940	3.811	0.553	0.588	2.151E + 23	1.061E + 07	4.933E − 17
Road transportation	35	5.144	0.747	4.200	9.764	10.629	1.543	15.774	2.289	−5.485	−0.796	−0.348	2.173E + 16	1.134E + 06	5.220E − 11
	33	5.144	0.747	4.018	9.385	10.209	1.482	15.353	2.228	−5.064	−0.735	−0.330	3.079E + 16	1.069E + 06	3.472E − 11
Air transportation	35	5.144	0.747	7.669	15.934	17.684	2.567	22.828	3.313	−12.539	−1.820	−0.549	2.410E + 14	9.450E + 04	3.921E − 10
	33	5.144	0.747	5.033	13.474	14.383	2.088	19.528	2.834	−9.239	−1.341	−0.473	1.531E + 15	8.910E + 04	5.818E − 11
	96	5.144	0.747	5.397	35.986	36.389	5.281	41.533	6.028	−31.244	−4.535	−0.752	3.111E + 11	1.472E + 05	4.730E − 07
	104.7	5.144	0.747	14.261	12.808	19.168	2.782	24.313	3.529	−14.024	−2.035	−0.577	1.163E + 14	1.602E + 05	1.378E − 09
	62.6	5.144	0.747	108.990	59.167	124.014	17.999	**129.159**	18.746	−118.870	−17.253	−0.920	2.500E + 06	1.352E + 05	5.409E − 02
	73.6	5.144	0.747	38.849	83.548	92.139	13.373	97.283	14.119	−86.994	−12.626	−0.894	4.425E + 07	1.592E + 05	3.597E − 03
	119.8	5.144	0.747	10.568	27.857	29.794	4.324	34.939	5.071	−24.650	−3.578	−0.706	2.017E + 12	1.833E + 05	9.089E − 08
	128.3	5.144	0.747	21.894	13.075	25.501	3.701	30.645	4.448	−20.357	−2.955	−0.664	8.528E + 12	1.963E + 05	2.302E − 08
	187.9	5.144	0.747	13.610	8.128	15.852	2.301	20.996	3.047	−10.707	−1.554	−0.510	6.438E + 14	6.088E + 05	9.456E − 10
	137	5.144	0.747	33.656	20.099	39.201	5.690	44.345	6.436	−34.056	−4.943	−0.768	1.545E + 11	4.685E + 05	3.033E − 06
Launch (phase I)	35	40.288	5.847	19.714	40.957	45.455	6.597	85.743	12.444	−5.167	−0.750	−0.060	5.596E + 09	8.167E + 01	1.459E − 08
	33	40.288	5.847	12.637	33.829	36.112	5.241	76.400	11.089	4.176	0.606	0.055	3.546E + 10	7.700E + 01	2.171E − 09
	96	40.288	5.847	4.742	31.614	31.968	4.640	72.256	10.487	8.320	1.208	0.115	9.250E + 10	2.245E + 02	2.427E − 09
	104.7	40.288	5.847	13.931	12.512	18.725	2.718	59.013	8.565	21.563	3.130	0.365	5.325E + 12	2.443E + 02	4.588E − 11
	62.6	40.288	5.847	34.252	18.594	38.974	5.657	79.262	11.504	1.314	0.191	0.017	1.933E + 10	1.461E + 02	7.555E − 09
	73.6	40.288	5.847	12.950	27.849	30.713	4.458	71.001	10.305	9.575	1.390	0.135	1.264E + 11	1.720E + 02	1.361E − 09
	119.8	40.288	5.847	11.977	31.571	33.767	4.901	74.055	10.748	6.521	0.947	0.088	6.024E + 10	2.795E + 02	4.640E − 09
	128.3	40.288	5.847	26.556	15.859	30.931	4.489	71.219	10.337	9.357	1.358	0.131	1.196E + 11	2.994E + 02	2.503E − 09
	187.9	40.288	5.847	44.090	26.330	51.353	7.453	91.641	13.301	−11.065	−1.606	−0.121	2.062E + 09	4.384E + 02	2.127E − 07
	137	40.288	5.847	24.936	14.892	29.044	4.215	69.332	10.063	11.244	1.632	0.162	1.948E + 11	3.197E + 02	1.641E − 09
	471.7	40.288	5.847	6.017	5.337	8.043	1.167	48.331	7.015	32.245	4.680	0.667	2.076E + 15	1.101E + 03	5.301E − 13
	696.8	40.288	5.847	3.310	5.985	6.839	0.993	47.127	6.840	33.449	4.855	0.710	6.215E + 15	1.626E + 03	2.616E − 13
	378.1	40.288	5.847	1.182	6.592	6.697	0.972	46.985	6.819	33.591	4.875	0.715	7.159E + 15	8.822E + 02	1.232E − 13
	950.4	40.288	5.847	1.043	3.323	3.483	0.505	43.771	6.353	36.805	5.342	0.841	5.335E + 17	2.218E + 03	4.157E − 15
	311.3	40.288	5.847	3.027	1.363	3.320	0.482	43.608	6.329	36.968	5.365	0.848	7.280E + 17	7.264E + 02	9.978E − 16

(continued)

Table 4.24 (continued)

Loading phase	Frequency (Hz)	Static analysis		Harmonic analysis									Fatigue analysis		
		Mean stress		Real	Imag	Amp. stress		Max stress		Min stress		R	N (cycles)	n (cycles)	Damage
		MPa	Ksi	Mpa	MPa	MPa	Ksi	MPa	Ksi	MPa	Ksi				
Launch (phase II)	35	40.288	5.847	19.714	40.957	45.455	6.597	85.743	12.444	−5.167	−0.750	−0.060	5.596E + 09	1.939E + 03	3.465E − 07
	33	40.288	5.847	12.637	33.829	36.112	5.241	76.400	11.089	4.176	0.606	0.055	3.546E + 10	1.828E + 03	5.156E − 08
	96	40.288	5.847	3.822	25.481	25.766	3.740	66.054	9.587	14.522	2.108	0.220	4.883E + 11	5.329E + 03	1.091E − 08
	104.7	40.288	5.847	10.385	9.327	13.959	2.026	54.247	7.873	26.329	3.821	0.485	4.470E + 13	5.800E + 03	1.298E − 10
	62.6	40.288	5.847	34.252	18.594	38.974	5.657	79.262	11.504	1.314	0.191	0.017	1.933E + 10	3.468E + 03	1.794E − 07
	73.6	40.288	5.847	12.950	27.849	30.713	4.458	71.001	10.305	9.575	1.390	0.135	1.264E + 11	4.083E + 03	3.231E − 08
	119.8	40.288	5.847	8.033	21.175	22.647	3.287	62.935	9.134	17.641	2.560	0.280	1.296E + 12	6.637E + 03	5.123E − 09
	128.3	40.288	5.847	16.998	10.151	19.798	2.873	60.086	8.721	20.490	2.974	0.341	3.529E + 12	7.108E + 03	2.014E − 09
	187.9	40.288	5.847	24.225	14.467	28.216	4.095	68.504	9.943	12.072	1.752	0.176	2.435E + 11	1.041E + 04	4.275E − 08
	137	40.288	5.847	15.298	9.136	17.819	2.586	58.107	8.433	22.469	3.261	0.387	7.662E + 12	7.590E + 03	9.906E − 10
	471.7	40.288	5.847	2.365	2.098	3.162	0.459	43.450	6.306	37.126	5.388	0.854	9.977E + 17	2.613E + 04	2.619E − 14
	696.8	40.288	5.847	1.681	3.040	3.474	0.504	43.762	6.351	36.814	5.343	0.841	5.427E + 17	3.860E + 04	7.114E − 14
	378.1	40.288	5.847	0.407	2.268	2.304	0.334	42.592	6.182	37.984	5.513	0.892	7.631E + 18	2.095E + 04	2.745E − 15
	950.4	40.288	5.847	0.533	1.697	1.779	0.258	42.067	6.106	38.509	5.589	0.915	3.969E + 19	5.265E + 04	1.326E − 15
	311.3	40.288	5.847	1.540	0.693	1.688	0.245	41.976	6.092	38.600	5.602	0.920	5.531E + 19	1.725E + 04	3.118E − 16
						Maximum stress (MPa)		129.159					Cumulative fatigue damage		5.77E − 02

Table 4.25 Harmonic and dynamic fatigue damage analysis results for critical point No. 28 located on the lower frame module

Loading phase	Frequency (Hz)	Static analysis		Harmonic analysis								Fatigue analysis			
		Mean stress		Real	Imag	Amp. stress		Max stress		Min stress		R	N (cycles)	n (cycles)	Damage
		MPa	Ksi	Mpa	MPa	MPa	Ksi	MPa	Ksi	MPa	Ksi				
Rail transportation	35	4.974	0.722	2.409	1.085	2.642	0.383	7.616	1.105	2.332	0.338	0.306	1.722E + 21	5.040E + 06	2.927E − 15
	33	4.974	0.722	1.187	2.108	2.419	0.351	7.393	1.073	2.555	0.371	0.346	3.314E + 21	4.752E + 06	1.434E − 15
	96	4.974	0.722	1.782	3.589	4.007	0.582	8.981	1.303	0.966	0.140	0.108	7.144E + 19	1.385E + 07	1.939E − 13
	62.6	4.974	0.722	1.888	3.597	4.062	0.590	9.036	1.311	0.911	0.132	0.101	6.419E + 19	9.014E + 06	1.404E − 13
	73.6	4.974	0.722	3.590	2.233	4.228	0.614	9.201	1.335	0.746	0.108	0.081	4.692E + 19	1.061E + 07	2.262E − 13
Road transportation	35	4.974	0.722	4.247	1.370	4.463	0.648	9.436	1.370	0.511	0.074	0.054	3.061E + 19	1.134E + 06	3.705E − 14
	33	4.974	0.722	2.971	4.403	5.312	0.771	10.285	1.493	−0.338	−0.049	−0.033	7.601E + 18	1.069E + 06	1.407E − 13
Air transportation	35	4.974	0.722	7.493	4.814	8.906	1.293	13.880	2.014	−3.932	−0.571	−0.283	1.036E + 17	9.450E + 04	9.119E − 13
	33	4.974	0.722	2.897	8.854	9.316	1.352	14.290	2.074	−4.342	−0.630	−0.304	7.052E + 16	8.910E + 04	1.263E − 12
	96	4.974	0.722	12.246	28.107	30.659	4.450	35.633	5.172	−25.685	−3.728	−0.721	1.572E + 12	1.472E + 05	9.362E − 08
	104.7	4.974	0.722	8.959	8.786	12.548	1.821	17.522	2.543	−7.575	−1.099	−0.432	5.296E + 15	1.602E + 05	3.025E − 11
	62.6	4.974	0.722	62.207	90.816	110.078	15.977	115.052	16.698	−105.105	−15.255	−0.914	7.975E + 06	1.352E + 05	1.696E − 02
	73.6	4.974	0.722	93.423	62.894	112.621	16.346	**117.595**	17.067	−107.647	−15.624	−0.915	6.392E + 06	1.592E + 05	2.490E − 02
	119.8	4.974	0.722	1.082	2.613	2.828	0.411	7.802	1.132	2.145	0.311	0.275	1.034E + 21	1.833E + 05	1.773E − 16
	128.3	4.974	0.722	1.881	0.829	2.055	0.298	7.029	1.020	2.918	0.424	0.415	1.094E + 22	1.963E + 05	1.795E − 17
	187.9	4.974	0.722	1.169	0.515	1.278	0.185	6.251	0.907	3.696	0.536	0.591	3.194E + 23	6.088E + 05	1.906E − 18
	137	4.974	0.722	2.891	1.274	3.160	0.459	8.133	1.180	1.814	0.263	0.223	4.472E + 20	4.685E + 05	1.048E − 15
Launch (phase I)	35	37.727	5.476	19.261	12.373	22.893	3.323	60.620	8.798	14.834	2.153	0.245	1.389E + 12	8.167E + 01	5.879E − 11
	33	37.727	5.476	7.274	22.230	23.390	3.395	61.117	8.870	14.337	2.081	0.235	1.180E + 12	7.700E + 01	6.524E − 11
	96	37.727	5.476	10.758	24.692	26.934	3.909	64.661	9.385	10.793	1.567	0.167	4.009E + 11	2.245E + 02	5.599E − 10
	104.7	37.727	5.476	8.752	8.583	12.258	1.779	49.985	7.255	25.469	3.697	0.510	1.347E + 14	2.443E + 02	1.813E − 12
	62.6	37.727	5.476	19.549	28.540	34.593	5.021	72.320	10.496	3.134	0.455	0.043	5.652E + 10	1.461E + 02	2.584E − 09
	73.6	37.727	5.476	31.141	20.965	37.541	5.449	75.268	10.924	0.186	0.027	0.002	2.944E + 10	1.720E + 02	5.841E − 09
	119.8	37.727	5.476	1.226	2.962	3.206	0.465	40.933	5.941	34.521	5.010	0.843	1.139E + 18	2.795E + 02	2.454E − 16
	128.3	37.727	5.476	2.281	1.006	2.493	0.362	40.220	5.837	35.234	5.114	0.876	5.769E + 18	2.994E + 02	5.189E − 17
	187.9	37.727	5.476	3.787	1.670	4.139	0.601	41.866	6.076	33.588	4.875	0.802	2.152E + 17	4.384E + 02	2.037E − 15
	137	37.727	5.476	2.142	0.944	2.341	0.340	40.068	5.815	35.386	5.136	0.883	8.640E + 18	3.197E + 02	3.700E − 17
	471.7	37.727	5.476	0.381	0.507	0.634	0.092	38.361	5.568	37.093	5.384	0.967	3.337E + 22	1.101E + 03	3.298E − 20
	696.8	37.727	5.476	0.880	0.605	1.067	0.155	38.794	5.631	36.660	5.321	0.945	1.264E + 21	1.626E + 03	1.286E − 18
	378.1	37.727	5.476	4.451	5.566	7.127	1.034	44.854	6.510	30.600	4.441	0.682	5.763E + 15	8.822E + 02	1.531E − 13
	950.4	37.727	5.476	0.527	0.909	1.051	0.152	38.778	5.628	36.676	5.323	0.946	1.397E + 21	2.218E + 03	1.587E − 18

(continued)

Table 4.25 (continued)

Loading phase	Frequency (Hz)	Static analysis		Harmonic analysis									Fatigue analysis		
		Mean stress		Real	Imag	Amp. stress		Max stress		Min stress		R	N (cycles)	n (cycles)	Damage
		MPa	Ksi	Mpa	MPa	MPa	Ksi	MPa	Ksi	MPa	Ksi				
311.3	37.727	5.476	0.400	6.775	6.786	0.985	44.513	6.461	30.941	4.491	0.695	8.029E + 15	7.264E + 02	9.047E − 14	
Launch (phase II)	35	37.727	5.476	19.261	12.373	22.893	3.323	60.620	8.798	14.834	2.153	0.245	1.389E + 12	1.939E + 03	1.396E − 09
	33	37.727	5.476	7.274	22.230	23.390	3.395	61.117	8.870	14.337	2.081	0.235	1.180E + 12	1.828E + 03	1.549E − 09
	96	37.727	5.476	8.671	19.902	21.709	3.151	59.436	8.626	16.018	2.325	0.270	2.074E + 12	5.329E + 03	2.569E − 09
	104.7	37.727	5.476	6.524	6.398	9.138	1.326	46.865	6.802	28.589	4.149	0.610	1.053E + 15	5.800E + 03	5.510E − 12
	62.6	37.727	5.476	19.549	28.540	34.593	5.021	72.320	10.496	3.134	0.455	0.043	5.652E + 10	3.468E + 03	6.136E − 08
	73.6	37.727	5.476	31.141	20.965	37.541	5.449	75.268	10.924	0.186	0.027	0.002	2.944E + 10	4.083E + 03	1.387E − 07
	119.8	37.727	5.476	0.822	1.987	2.150	0.312	39.877	5.788	35.577	5.164	0.892	1.490E + 19	6.637E + 03	4.454E − 16
	128.3	37.727	5.476	1.460	0.644	1.596	0.232	39.323	5.707	36.131	5.244	0.919	9.952E + 19	7.108E + 03	7.142E − 17
	187.9	37.727	5.476	2.081	0.917	2.274	0.330	40.001	5.806	35.453	5.146	0.886	1.040E + 19	1.041E + 04	1.001E − 15
	137	37.727	5.476	1.314	0.579	1.436	0.208	39.163	5.684	36.291	5.267	0.927	1.941E + 20	7.590E + 03	3.910E − 17
	471.7	37.727	5.476	0.150	0.199	0.249	0.036	37.976	5.512	37.478	5.439	0.987	1.129E + 25	2.613E + 04	2.315E − 21
	696.8	37.727	5.476	0.447	0.307	0.542	0.079	38.269	5.554	37.185	5.397	0.972	8.859E + 22	3.860E + 04	4.358E − 19
	378.1	37.727	5.476	1.532	1.915	2.452	0.356	40.179	5.832	35.275	5.120	0.878	6.416E + 18	2.095E + 04	3.265E − 15
	950.4	37.727	5.476	0.269	0.464	0.537	0.078	38.264	5.554	37.190	5.398	0.972	9.444E + 22	5.265E + 04	5.575E − 19
	311.3	37.727	5.476	0.204	3.445	3.451	0.501	41.178	5.977	34.276	4.975	0.832	7.050E + 17	1.725E + 04	2.446E − 14
						Maximum stress (MPa)		117.595					Cumulative fatigue damage		4.19E − 02

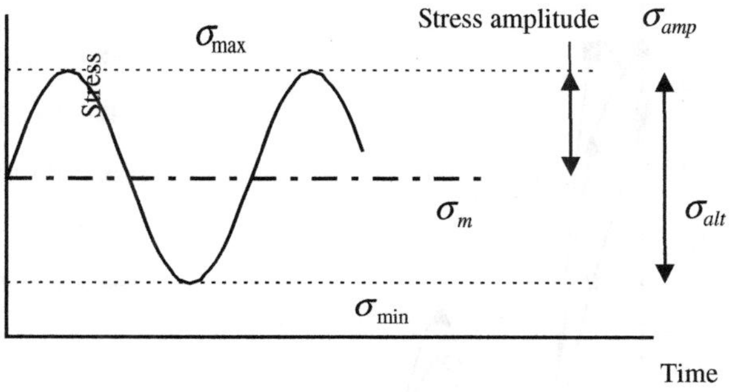

Fig. 4.53 Simultaneous mean and cyclic loading

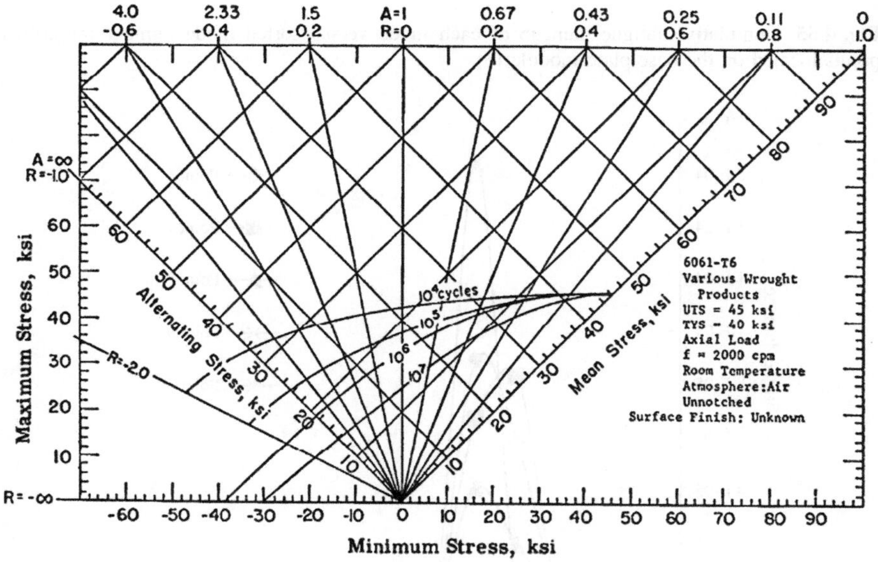

Fig. 4.54 Typical constant life diagram for fatigue behavior of various wrought products of 6061-T6 aluminum alloy [3]

located in each structural module. These results explain that the higher ranking frequency modes do not have a significant influence on the fatigue damage due to dynamic vibrations. The lower ranking modes have more than 99 % of the fatigue damage value at any critical point. Therefore, the fatigue calculation accuracy is acceptable for Small Sat structure.

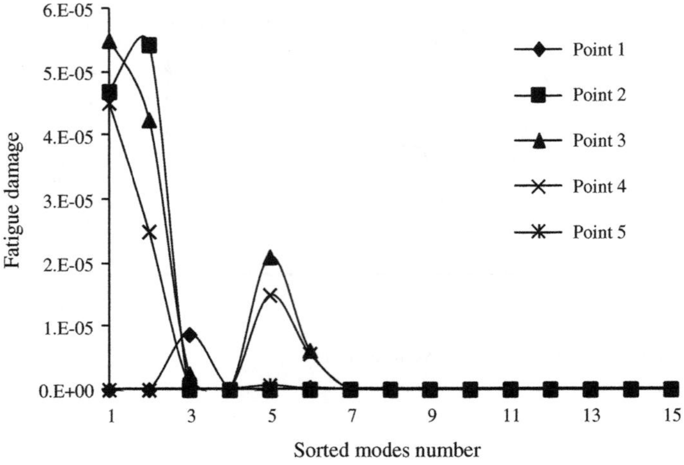

Fig. 4.55 Cumulative fatigue damage of each mode versus sorted mode number for critical points located on the base plate module [6]

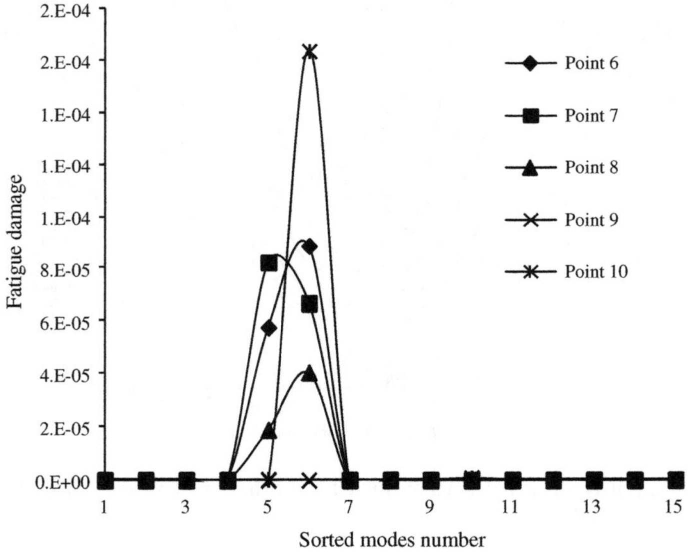

Fig. 4.56 Cumulative fatigue damage of each mode versus sorted mode number for critical points located on the mounting plate module

4.13 Spectrum Analysis

In this section, a spectrum analysis is performed for Small Sat structure [4]. The main purpose of this analysis is to provide investigate whether the satellite structure can withstand all dynamic random vibration loads encountered during

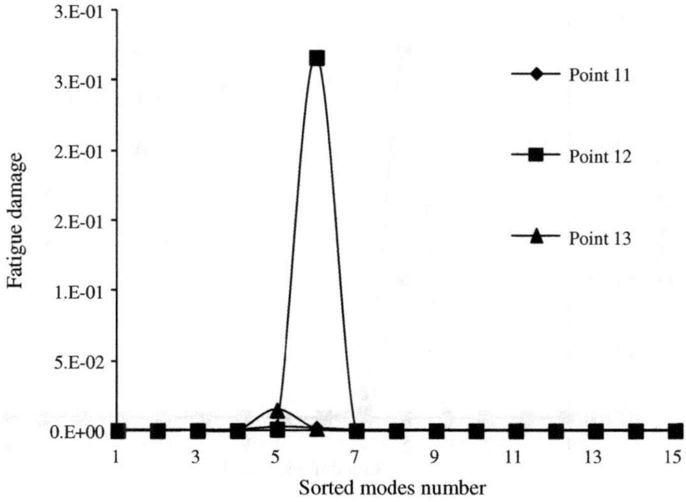

Fig. 4.57 Cumulative fatigue damage of each mode versus sorted mode number for critical points located on the basis plate module [6]

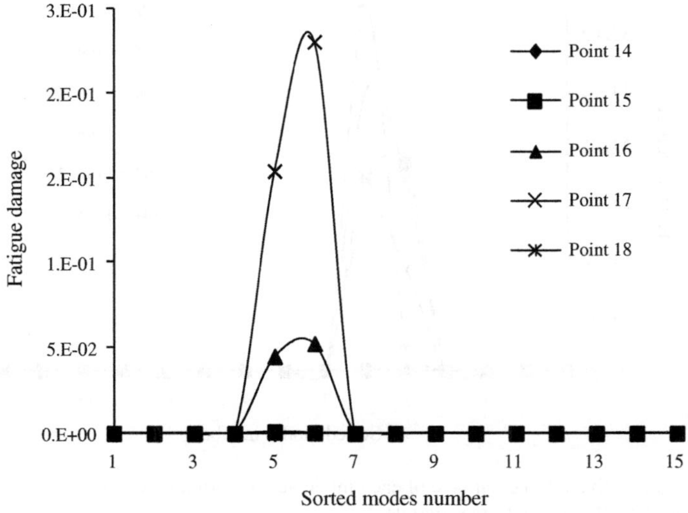

Fig. 4.58 Cumulative fatigue damage of each mode versus sorted mode number for critical points located on the basis walls module

transportation and launch. The entire finite-element model of the satellite is used during the spectrum analysis process to calculate displacements and stresses in the satellite structure modules.

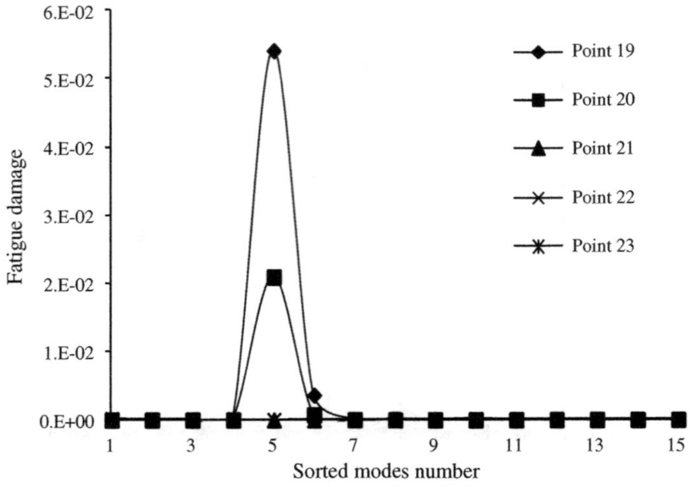

Fig. 4.59 Cumulative fatigue damage of each mode versus sorted mode number for critical points located on the upper frame module

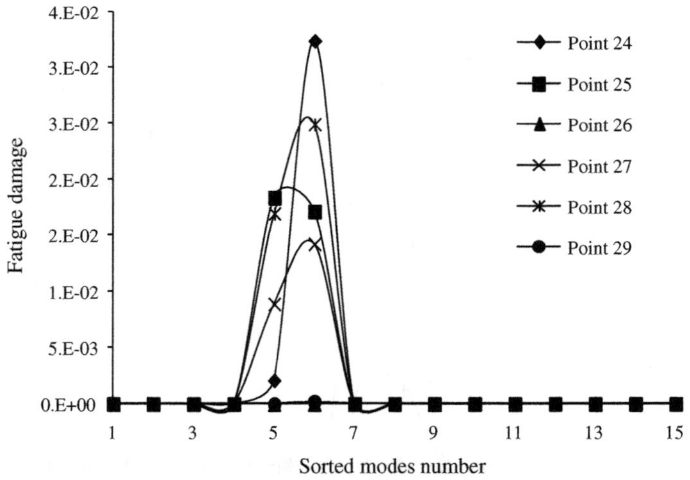

Fig. 4.60 Cumulative fatigue damage of each mode versus sorted mode number for critical point No. 28 located on the lower frame module

4.13.1 Definition of Spectrum Analysis

A spectrum analysis is one in which the results of a modal analysis with a known spectrum are used to calculate displacements and stresses in the model. It is mainly used in place of a time-history analysis to determine the response of structures to

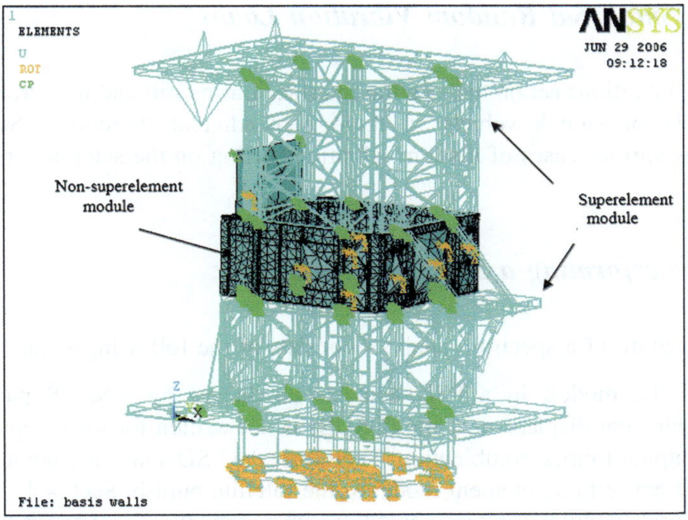

Fig. 4.61 The entire finite-element model of Small Sat used to carry out a PSD analysis for the basis walls module

random or time-dependent loading conditions such as earthquakes, wind loads, jet engine thrust, rocket motor vibrations, and so on. The spectrum is a graph of spectral values versus frequency, which captures the intensity and frequency content of time-history loads. Three types of spectra are available for a spectrum analysis:

- Response Spectrum
- Dynamic Design Analysis Method (DDAM)
- Power Spectral Density (PSD)

This section intends to study random vibrations acting on the satellite structure during transportation and launch. Therefore, the only spectrum type available in this study is the power spectral density (PSD), which is a statistical measure defined as the limiting mean-square value of a random variable. It is used in random vibration analyses in which the instantaneous magnitudes of the response can be specified only by probability distribution functions that show the probability of the magnitude taking a particular value. A PSD is a statistical measure of the response of a structure to random dynamic loading conditions. It is a graph of the PSD value versus frequency, where the PSD may be a displacement PSD, velocity PSD, acceleration PSD, or force PSD. Mathematically, the area under a PSD versus frequency curve is equal to the variance (square of the standard deviation of the response) of the variable.

4.13.2 Small Sat Random Vibration Loads

Random vibrations act on the satellite interface during rail and road transportation and flight of launch vehicle in three perpendicular directions. Section 4.11 describes various cases of random vibrations acting on the satellite structure.

4.13.3 Performing a Spectrum Analysis

The procedure of a spectrum analysis consists of the following steps:

1. Build the model: In a PSD analysis conducted by ANSYS package, the superelement displacement file (DSUB) is not written for load steps related to the output of mode combination. Therefore, the PSD analysis cannot conducted on the entire finite-element model of the satellite built in Sect. 4.1, because all structural modules are represented as superelements. To overcome this problem, six new finite-element models of the entire satellite are built to carry out the PSD analysis. Each one is specified to calculate stresses and deformations in a particular structural module represented as a nonsuperelement, while the rest are modeled using the superelement modules built before. The same procedure used to build the entire finite-element model in Sect. 4.1 is applied to build the new six models. As an example, Fig. 4.61 shows the entire finite-element model of Small Sat used to carry out a PSD analysis for the basis walls module.
2. Obtain the modal solution and expand the modes: The modal solution—natural frequencies and expanded mode shapes—is needed to calculate the spectrum solution. The procedure to obtain the modal solution is described in Sect. 4.10. However, the following additional points should be taken into account:
   - The Block Lanczos (default) is used to extract the modes.
   - The number of modes extracted should be sufficient to characterize the structure's response in the frequency range of interest.
   - The DOF where a base excitation spectrum is applied must be constrained.
   - Only expanded modes are used for the mode combination step.
   - Stress calculation should be requested during modal solution to calculate stresses caused by the spectrum.
3. *Obtain the spectrum solution*: To obtain the PSD spectrum solution, the database must contain the model data as well as the modal solution data. The spectrum type is defined as a PSD and stress calculation is specified ON. The PSD is defined as acceleration (g^2/Hz), and listed in a PSD versus frequency table. Then the PSD base excitation is applied only at nodes that were constrained in the modal analysis. It is followed by calculating the participation factor for the above PSD excitation. Then the output controls are defined which specifies the amount and form of output written to the results file. Up to three sets of solution quantities can be calculated: displacement solution, velocity

solution, and acceleration solution. Each one of these can be calculated relative to the base or in absolute value. Finally, the model is solved to calculate the spectrum results.

4. *Combine the Modes*: The modes can be combined in a separate solution phase. Only the PSD mode combination method is valid in a random vibration analysis. This method triggers calculation of the one-sigma displacements, stresses, etc., in the structure.
5. *Review the Results*: Results from a random vibration analysis are written to the structural results file. They consist of the following quantities:

- Expanded mode shapes from the modal analysis
- Static solution for base excitation

The following output, if mode combinations are requested:

- 1 σ displacement solution (displacements, stresses, strains, and forces)
- 1 σ velocity solution (velocities, stress velocities, strain velocities, and force velocities)
- 1 σ acceleration solution (accelerations, stress accelerations, strain accelerations, and force accelerations)

Nodal stress averaging may not be appropriate in a random vibration analysis because "stresses" here are not actual stresses, but stress statistics.

4.13.4 PSD Spectrum Analysis Results

As mentioned before, the PSD spectrum analysis is performed to calculate stresses and deformations in structural modules due to random vibration loading acting on the satellite during rail and road transportation and flight of launch vehicle. By reviewing the random vibration loads listed in Appendix A, the most severe power spectral density affecting the satellite structure is the base excitation during the first loading phase of launch. Figures 4.62, 4.63, 4.64, 4.65, 4.66, 4.67, 4.68, 4.69, 4.70, 4.71, 4.72 and 4.73 depict Von Mises stress and displacement distributions in the structural modules of Small Sat due to random vibrations in the first phase of launch. The obtained values of the maximum stresses and displacements are rms values representing the first standard deviation (1 σ) of the instantaneous response described by a normal distribution with zero mean. For random vibrations, the load at any given time has a 99.87 % probability of being less than this value (1 σ), but the structure must withstand the maximum load experienced during the total time of exposure to the random environment.

The maximum expected load during a random environment is a function of the number of positive relative maxima (number of cycles applied of a load corresponding to the random vibrations). With a single response frequency, this number equals the frequency multiplied by the exposure time. For wide band vibration, which includes a spectrum of response frequencies, the number of loading cycles

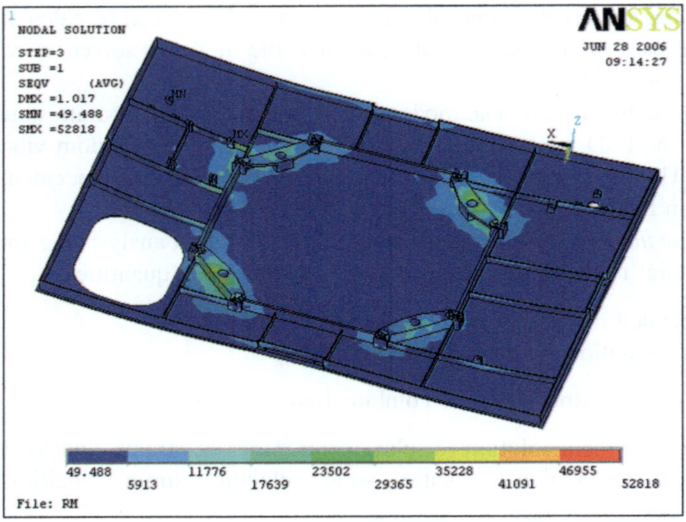

Fig. 4.62 Von Mises stress distribution for base plate module from launch random vibrations (phase I) [6]

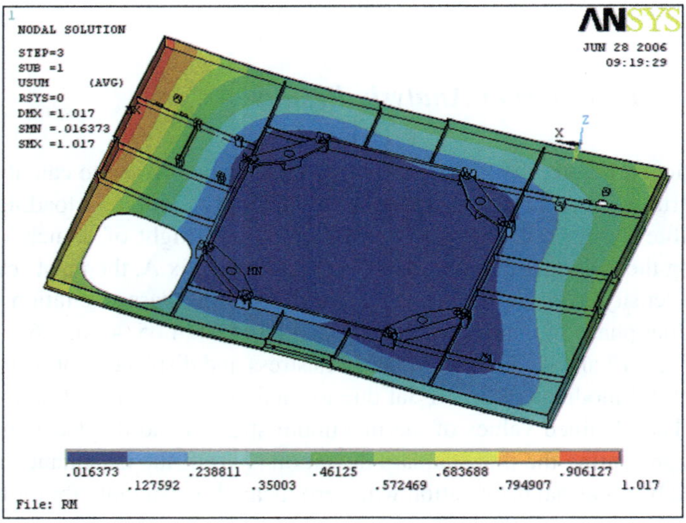

Fig. 4.63 Displacement distribution for base plate module from launch random vibrations (phase I)

can be approximated by multiplying an assumed primary frequency by the time span. The cumulative distribution function for the extreme value of random response is presented in the following formula [5]:

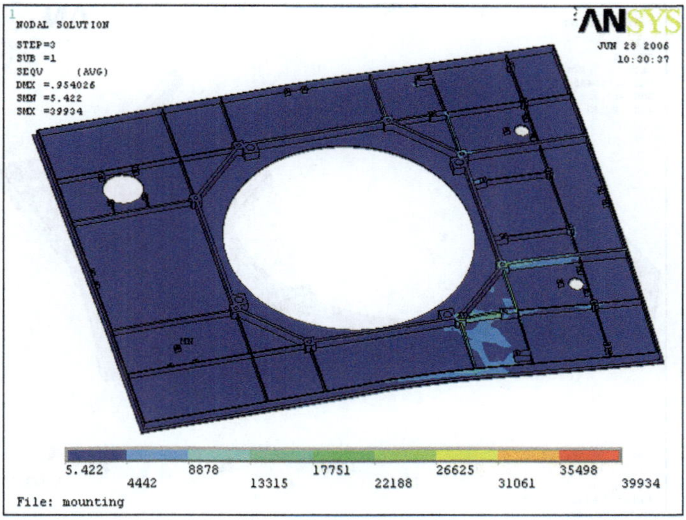

Fig. 4.64 Von Mises stress distribution for mounting plate module from launch random vibrations (phase I)

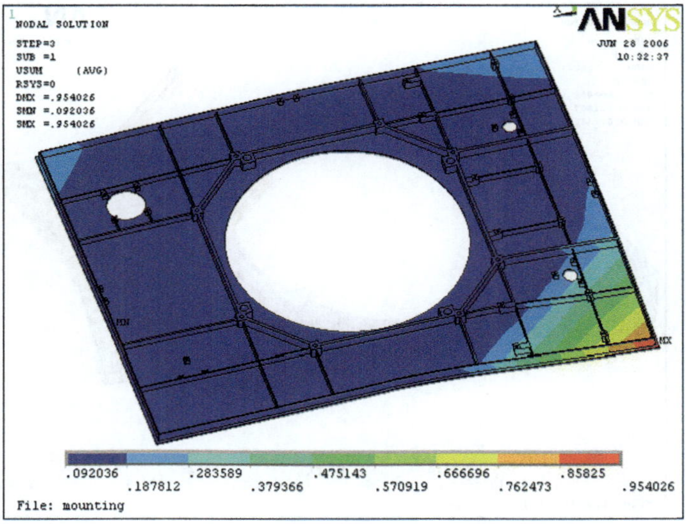

Fig. 4.65 Displacement distribution for mounting plate module from launch random vibrations (phase I)

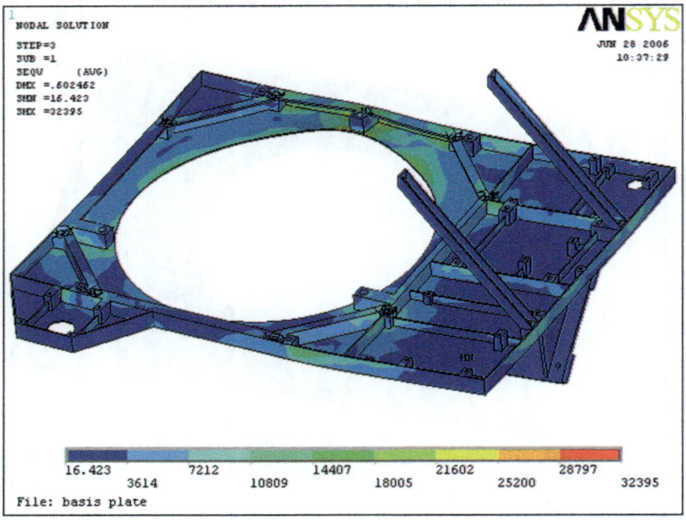

Fig. 4.66 Von Mises stress distribution for basis plate module from launch random vibrations (phase I)

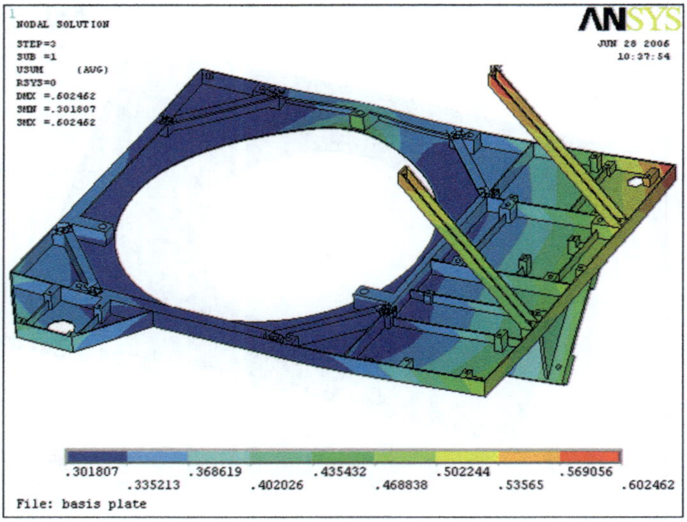

Fig. 4.67 Displacement distribution for basis plate module from launch random vibrations (phase I)

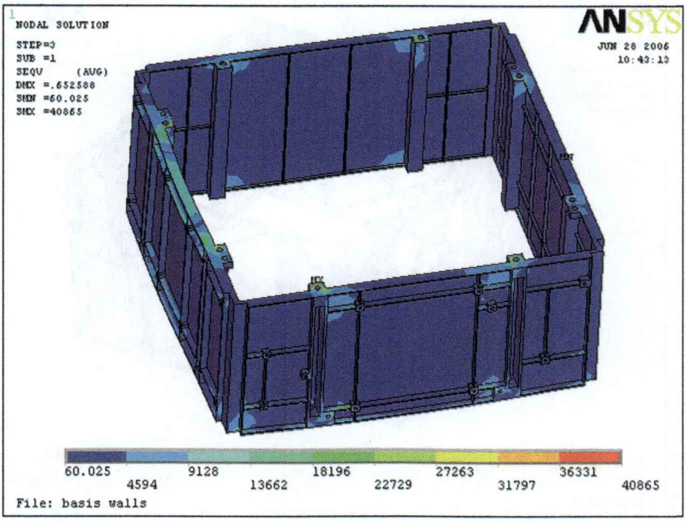

Fig. 4.68 Von Mises stress distribution for basis walls module from launch random vibrations (phase I)

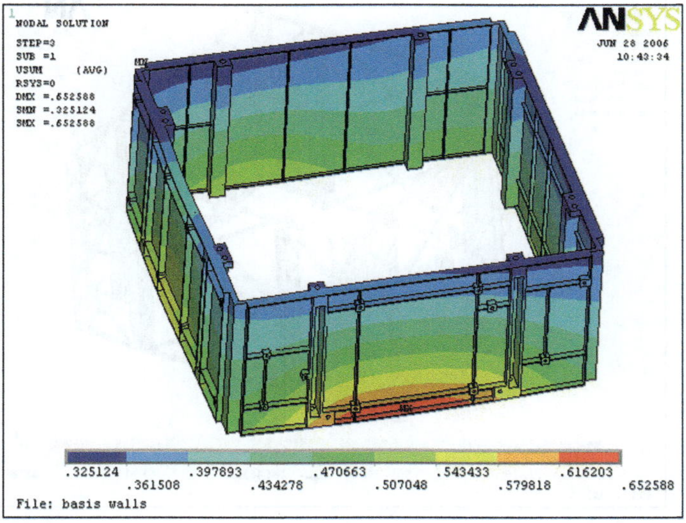

Fig. 4.69 Displacement distribution for basis walls module from launch random vibrations (phase I)

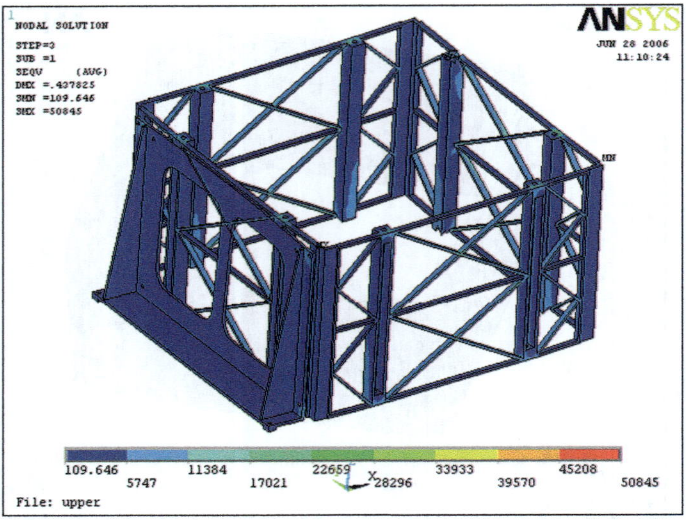

Fig. 4.70 Von Mises stress distribution for upper frame module from launch random vibrations (phase I)

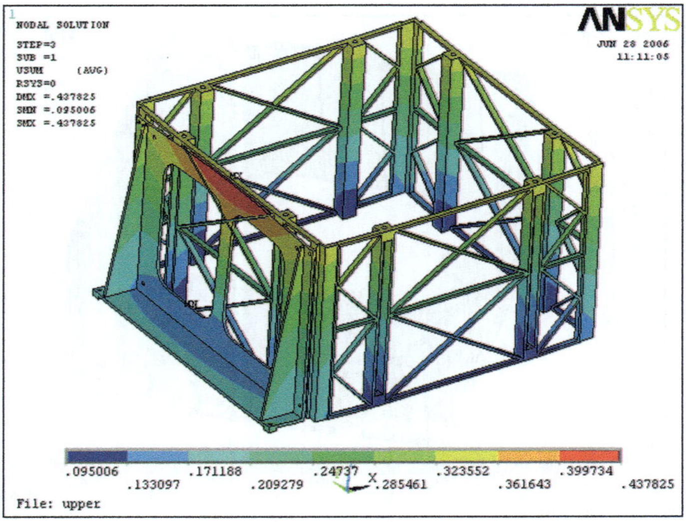

Fig. 4.71 Displacement distribution for upper frame module from launch random vibrations (phase I)

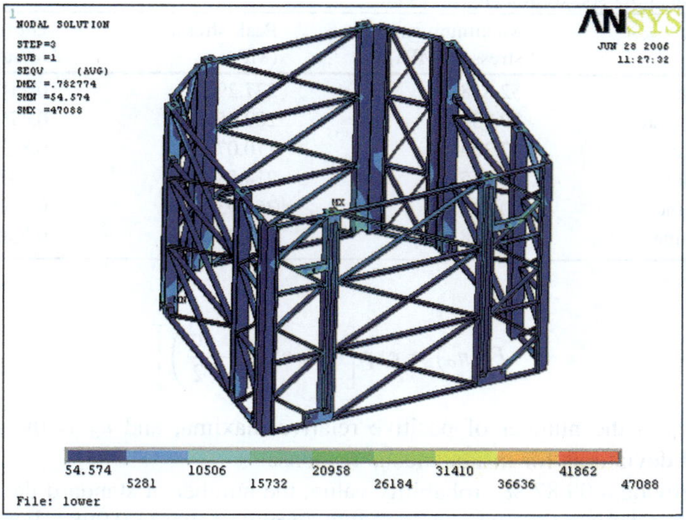

Fig. 4.72 Von Mises stress distribution for lower frame module from launch random vibrations (phase I)

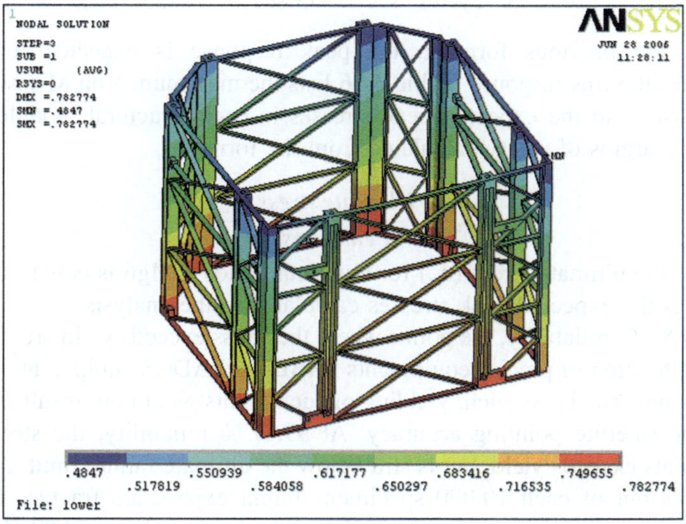

Fig. 4.73 Displacement distribution for lower frame module from launch random vibrations (phase I)

Table 4.26 PSD spectrum analysis results

Structural modules	Maximum rms stresses (MPa)	Peak stresses (MPa)	Ultimate margins of safety
Base plate	52.818	277.295	0.118
Mounting plate	39.934	209.654	0.479
Basis plate	32.395	170.074	0.823
Basis walls	40.865	214.541	0.445
Upper frame	50.845	266.936	0.161
Lower frame	47.088	247.212	0.254

$$F_e(n_\sigma) = \exp\left[-N_c \exp\left(-\frac{n_\sigma^2}{2}\right)\right] \tag{4.25}$$

where N_c is the number of positive relative maxima, and n_σ is the number of standard deviations for instantaneous response.

To provide a 99.87 % probability value, the number of standard deviation (n_σ) should be calculated to make the previous formula equal to 0.9987. For Small Sat, the primary natural frequency is 35 Hz, and the duration span of the first phase of launch is 35 s. Therefore, the number of loading cycles (N_c) can be estimated as follow:

$$N_c = 35(cycles/\sec) \times 35(\sec) = 1,225 cycles$$

From the previous formula, the peak response is expected to be about 5.25 times the rms response. Table 4.26 lists the maximum Von Mises equivalent rms stresses and the expected peak stresses in each structural module and their ultimate margins of safety calculated from the formula:

$$MS_u = \frac{Ultimate\ stress\ (\sigma_u)}{Peak\ stress\ (\sigma)} - 1$$

where the ultimate stress (σ_u) for aluminum alloy AMg6 is equal to 310 MPa, and (σ) is the expected peak stresses calculated in the analysis.

At 99.87 % reliability, the points where the stress exceeds yield are far from the points of fixation of precise equipments (MBEI and ADCS equipments) located in the basis unit block. As such, yielding of such points would not result in failure or affect the satellite pointing accuracy. At 99.87 % reliability, the stress level at some points exceeds yield, but is still below the ultimate failure limit. As such, 13 specimens out of each 10,000 specimens might experience fracture failure. To ensure that none of the satellite structure modules will fail at a 99.87 % reliability level, severe inspection rules must be applied during manufacturing, assembly, and testing. Such inspection would provide that the satellite structure would be among the 9,987 ones out of 10,000 that would not fail.

4.14 On-Orbit Thermal Deformation Analysis

In this section, a thermal deformation (thermoelastic) analysis for Small Sat structure due to on-orbit cyclic temperature is considered. It results in structural stresses and distortions due to thermal cyclic loading. The results of thermoelastic analysis are used to calculate fatigue damage due to on-orbit cyclic stresses. Moreover, it is used to check mounting accuracy of the precise equipments (MBIE and ADCS equipments) after on-orbit thermal deformation [6].

4.14.1 Definition of On-Orbit Thermoelastic Analysis

Temperature changes can affect structures in many ways. Most materials expand when heated and contract when cooled. In space, the orbiting satellite's temperature is neither uniform nor constant. As a result, satellite structures distort. The various materials that make up a satellite expand or contract by different amounts as temperatures change. Thus, they push and pull each other, resulting in stresses that can cause them to yield, rupture, or fatigue. High temperatures usually degrade material strength and stiffness, and if temperatures are high enough, materials can creep (deform permanently under sustained stress). Low temperatures often reduce ductility and fracture toughness. Thermal stresses develop when thermal expansion is restricted, which can happen in different ways,

- External constraints prevent free thermal expansion and develop boundary loads.
- Nonuniform temperatures generate thermal stresses when the resulting thermal strains are incompatible. Without external constraints, thermally caused internal tensile and compressive forces balance.
- Structures made of different materials can experience thermal stresses without external constraints even under uniform temperature. Differences in the materials' coefficients of thermal expansion produce incompatible strains, and result in thermal stresses. These stresses balance when no external constraints are present.

Thermally induced loads and stresses are limited by deformation. Once a material reaches its proportional limit or the structure begins to buckle, thermal stresses no longer increase in proportion to the change in temperature. Ductile materials seldom rupture or buckle from a single application of thermal stress, but they can fail in fatigue from the many cycles of thermal loading common to orbiting satellites.

Heat can transfer in three ways: conduction, convection, and radiation. Conduction is the transfer of heat within a material or between contacting materials, and results from the interaction of molecules and electrons. If one end of a body is hotter than the other, energy transfers from the hotter region to the cooler region. Convection is the transfer of heat through a fluid. Heat transfers from a solid through the fluidic boundary layer by conduction, and then fluid action distributes the heat. Thermal radiation is the transfer of energy by electromagnetic waves through a gas or the vacuum of space. Radiation is the driving mode of heat rejection in space; heat generated internally or absorbed from the environment (also by radiation) must eventually radiate to deep space.

Actual thermal-design problems for a satellite are complicated. The design problem typically must combine multiple modes of heat transfer with time-varying boundary conditions that require transient instead of steady-state solutions. To predict satellite's temperatures, the thermal analysis problem combines two types of heat transfer models. The first one is a thermal-radiation model to calculate external heating rates by simulating the external geometry of the satellite including surface properties. By subjecting this model to a simulated orbit, the output from this model consisting of; the environmental heating rates due to direct solar, albedo, and planetary emissions, and external radiation between satellite surfaces; becomes input for the second model.

The second thermal model uses a thermal analyzer. The satellite is modeled much the same as the structural analysis model with internal details, but the analysis is based on the finite difference method. Then the heating sources are defined. Finally, the model is solved to simulate the heat transfer paths of conduction, convection, and radiation within the satellite body. The thermal analyzer calculates temperatures at all nodes for steady state or transient conditions by solving energy equations. The thermal analysis process for on-orbit satellite structure is performed by the thermal analysis engineer.

4.14.2 Performing an On-Orbit Thermoelastic Analysis

The procedure of a thermal deformation (thermoelastic) analysis for Small Sat structure due to on-orbit cyclic temperature changes consists of the following steps:

1. Build the model: The most critical structural module in Small Sat regarding on-orbit thermal deformation is the basis unit block module. This module is very sensitive to thermal deformation, because it carries the precise equipments of the satellite. The rest of structural modules do not have severe restrictions on their equipment mounting accuracy. Therefore, only the basis unit block is studied. To speed up calculations, the substructure technique is not applied, and new simplified finite-element model of the entire satellite is built where fine structural details are ignored for all structural modules except the basis unit block module. The following points are taken into consideration during building this model:

   - The basis unit block module consists of the basis plate, four basis walls, two diagonal struts, and star sensor bracket.
   - The material is Aluminum alloy AMg6. Its properties are mentioned in Sect. 4.4. For thermal analysis, it has the following properties:
   - Thermal conductivity (K) = 117 W/m °C

- Coefficient of thermal expansion (α) = 24.7×10^{-6} m/m/ °C
- Assign the locations of mounting the high precise equipment to calculate thermal deformation, and hence mounting accuracy.
- Equipment support is designed to eliminate thermal loading as they allow sliding of the support under thermal expansion. Therefore, in the simplified thermal model, all satellite equipment are removed to avoid the appearance of any artificial thermal stresses that may result from improper representation of the sliding support.
- The basis unit block module is meshed according to the same criteria implemented in Sect. 4.3 for fine meshing, while the rest of structural modules are meshed with coarse meshing. ANSYS and SOLID92 elements are used in the meshing process.
- The nodes located at connection areas between different structural modules are coupled at all DOF.

2. Apply the loads and define boundary conditions: Thermal deformations are evaluated for the worst cases, which are obtained from the on-orbit thermal analysis for the satellite structure. Thermal satellite engineers should perform a complete on-orbit thermal analysis of the satellite, and supply input data for surface temperature gradients of different satellite structure modules. This data are the input data needed to perform an on-orbit thermal deformation analysis. For Small Sat, the satellite structure is solved thermally under on-orbit cyclic thermal loads. The results of this data are obtained from the research project titled "Prediction of satellite structure life on-orbit under thermal fatigue effect", National Authority of Remote Sensing and Space Science, NARSS, Egypt, 2006 [6]. Input data for surface temperature gradients of different satellite structure modules of Small Sat are listed in Table 4.26. It contains both maximum and minimum average temperatures for each structural module (Table 4.27).
   Displacement boundary conditions are the DOF constraints usually specified at model boundaries to define the structural support points. During on-orbit operation, the satellite is totally free without any fixation points. To conduct a thermoelastic analysis, displacement boundary conditions must be defined for the related model. Therefore, to simulate satellite thermal deformation due to on-orbit cycling, the satellite model must be constrained by rigid support points. Selection of these points' locations and their fixation manner should provide minimum effect on satellite deformation. For the Small Sat model represented at Fig. 4.61, the support points are selected at the plane of the star sensor mounting with its bracket. This plane is selected, because the mounting accuracy of all precise equipment installed at the basis unit block is measured relative to the star sensor bracket. Displacement boundary conditions are applied by fixing one of the four star sensor points of fixation (on bracket) in the "*X*" direction, the next point in the "*Y*" and one of the other in the "*Z*".
3. Set solution controls and solve the analysis: By reviewing the input data listed in Table 4.26, some of the satellite structural modules are divided into more than one division according to their position. Each division has its maximum

Table 4.27 Input data for surface temperatures of different satellite structure modules

Satellite structure module	Corner or face	Maximum temperature (°C)	Minimum temperature (°C)
Base plate	I	−5.3	−10.1
	II	−1.8	−6.2
	III	7.2	4.9
	IV	5.4	−1.6
Mounting plate	I	22	11.9
	II	4.5	−7.3
	III	21.2	9.4
	IV	18.6	7.5
Basis plate	I	20.3	16
	II	5.5	1.4
	III	16.3	10.6
	IV	21.7	15.9
Basis walls	I-II	26.3	25
	II-III	25.7	24.4
	III-IV	31.7	30.5
	IV-I	25.4	24.5
Upper frame	I-II	36.1	35
	II-III	28.5	29.6
	III-IV	32.5	31.7
	IV-I	36.3	35.2
Lower frame	I-II	18.5	16.3
	II-III	12	5.9
	III-IV	17.6	12.8
	IV-I	13.5	7.2
	II	14.7	10.3
	III	15.2	8
Diagonal strut	I	14.3	13.2
	IV	16.6	15.5
Star sensor bracket		18.2	16.9

and minimum average surface temperature. Before conducting a thermoelastic analysis, a thermal analysis process is performed to calculate the temperature distribution through all solid elements and nodes. The following points are taken into consideration during thermal analysis:

- The thermal analysis carried out by ANSYS package is based on conduction only, as the effect of radiation should be taken care of by the thermal control engineer.
- This process is performed twice for the worst cases; maximum and minimum average surface temperatures.
- The entire F.E. model shown in Fig. 4.74 is used to carry out the thermal analysis. The solid element used to mesh the model must be converted from structural elements into thermal elements. Displacement boundary conditions are not needed in this case.

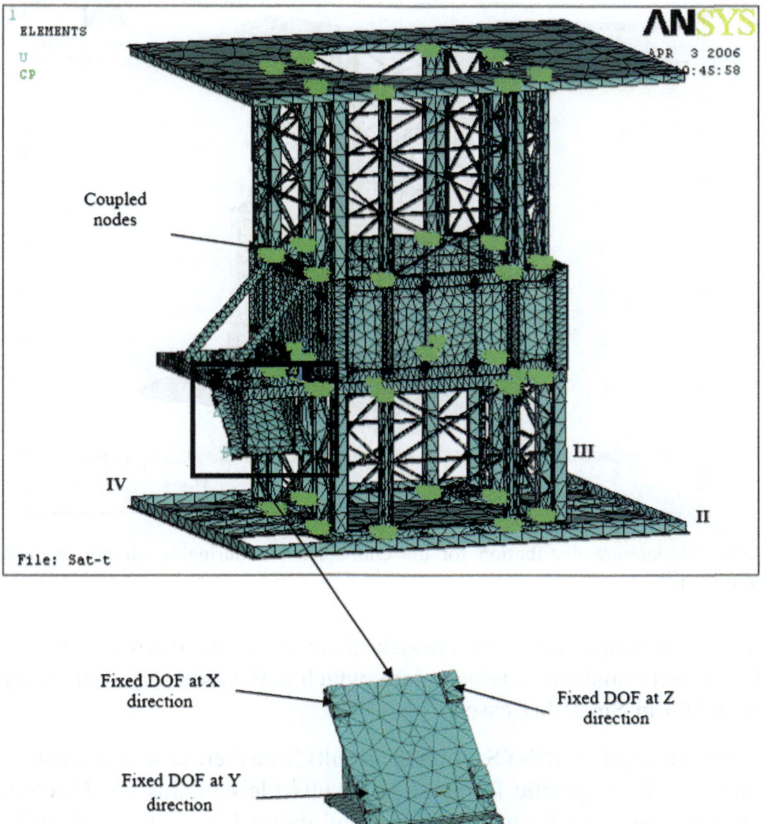

Fig. 4.74 The simplified entire finite-element model used in the on-orbit thermal deformation analysis [7]

- The output data from the thermal analysis are listed into a specific format file for the entire F.E. model of the satellite. It is used in the structure deformation "thermoelastic" analysis as applied loads.
  A thermoelastic analysis is conducted to the entire satellite model based on the output results from the thermal analysis. The following points are taken into consideration during thermoelastic analysis:
- The analysis is performed twice for the worst cases; maximum and minimum average surface temperatures.
- The entire F.E. model shown in Fig. 4.74 is used to carry out the thermoelastic analysis. The solid element used to mesh the model must be converted from thermal elements into structural elements. Displacement boundary conditions are applied as mentioned before, see Fig. 4.74.
- Thermal strains are given by "$\alpha.\ (T - T_{ref})$", where α is the coefficient of thermal expansion, T is the local element temperature, and T_{ref} is the reference

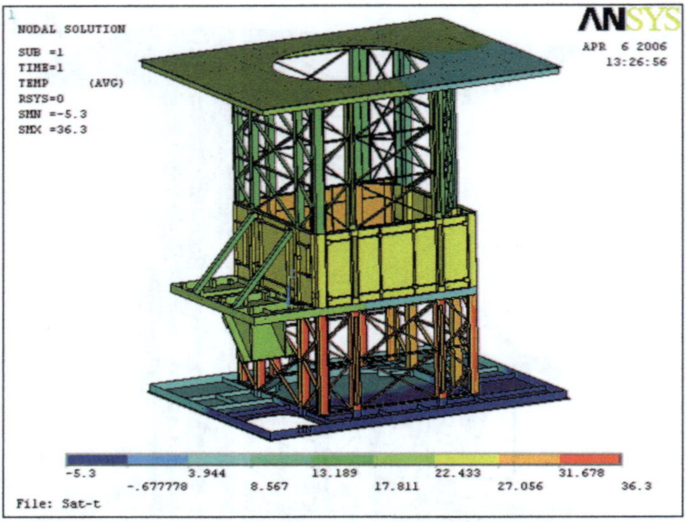

Fig. 4.75 Temperature distribution for the entire satellite during maximum average surface temperatures [7]

(ambient) temperature. In thermoelastic analysis, the reference temperature is the zero thermal stress temperature, which is the assembly room temperature (=20 °C) in Small Sat case.

4. Review the results: ANSYS writes the results from thermal analysis to the thermal results file in a specific format. The results file contains the following data: primary dataset (nodal temperatures) and derived dataset (nodal and element thermal fluxes, nodal and element thermal gradients, element heat flow rates, nodal reaction heat flow rates, … etc.). The data format for thermoelastic analysis is identical to the data for a basic structural analysis. They consist of the following data: primary dataset (nodal displacements), derived dataset (nodal and element stresses, nodal and element strains, element forces, nodal reaction forces, … etc.).

4.14.3 Thermal Analysis Results

Temperature distribution for the entire satellite is determined as a result of the thermal analysis. Figures 4.75 and 4.76 show the temperature distributions (°C) for the entire satellite during maximum and minimum average surface temperatures.

4.14.4 On-Orbit Thermal Deformation Analysis Results

Stress–strain state of the basis unit block structure module is determined as a result of the on-orbit thermoelastic analysis for both worst cases, where maximum and

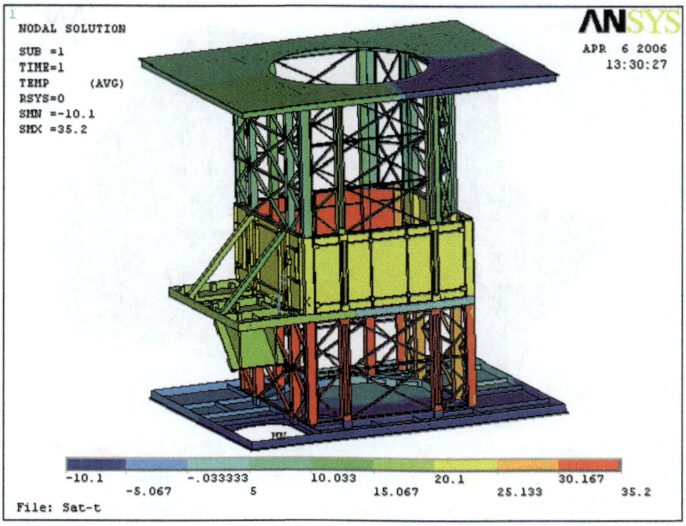

Fig. 4.76 Temperature distribution for the entire satellite during minimum average surface temperatures [7]

minimum average on-orbit surface temperatures are considered. The diagrams of equivalent stress (10^3 MPa) and displacement (mm) distributions in the basis unit block module are shown in Figs. 4.77, 4.78, 4.79 and 4.80. The stress values are determined according to Von Mises criterion. Displacements are relative to the points of the star sensor attachment to its bracket, which are the fixation points for the entire model of Small Sat structure during thermoelastic analysis. The maximum stress values (σ_e) in the basis unit block module in each on-orbit thermal case and their equivalent yield margins of safety (MS_y) are given in Table 4.28.

4.15 Mounting Accuracy Due to On-Orbit Thermal Deformation

From the static point of view, the basis unit block module is safe under thermal loading, because the yield margin of safety has a positive value in both design cases. But this is not enough to decide that the design is satisfactory. Satisfactory performance of the satellite requires accurate prediction of thermal deformations to verify pointing and alignment accuracy requirements for sensors (Mounting Accuracy). Therefore, it is important to calculate the angular positioning deviations for all high precise equipment (MBEI and ADCS devices) relative to the star sensor due to on-orbit thermal deformation. These values must not exceed the limiting deviations specified for mounting precise equipment. Table 4.29 lists the limiting angular positioning deviations for the most precise equipment relative to the star sensor, which are derived from the structure requirements [7].

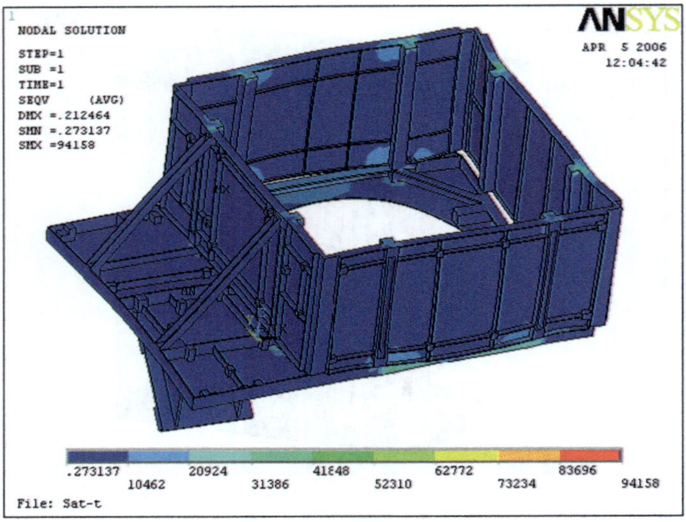

Fig. 4.77 Basis unit block stress distribution due to on-orbit thermal deformation under maximum average surface temperatures [7]

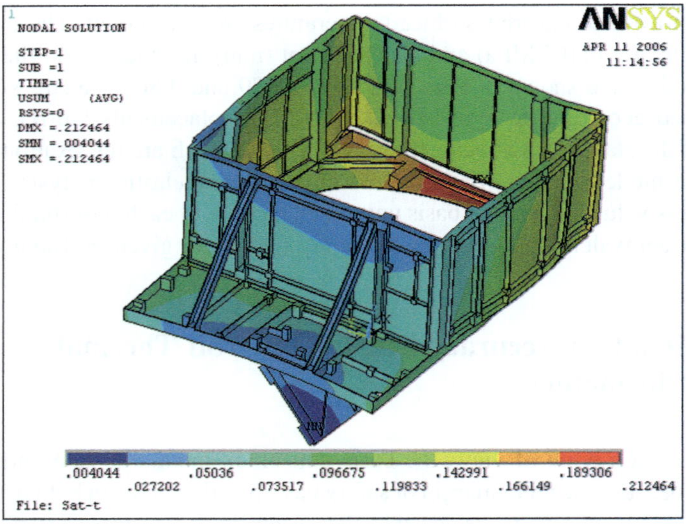

Fig. 4.78 Basis unit block displacement due to on-orbit thermal deformation under maximum average surface temperatures [7]

The angular positioning deviation between any two equipments is calculated by measuring the deviation angle between normal vectors for their mounting planes. The deviation angle is determined by subtracting the measured angle after on-orbit thermal deformation from the initial angle before deformation. The criterion used to calculate the angular positioning deviation between precise equipment and the

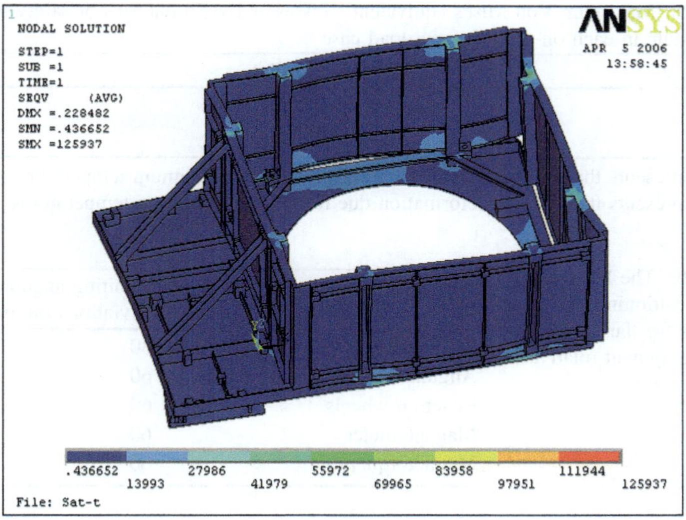

Fig. 4.79 Basis unit block stress distribution due to on-orbit thermal deformation under minimum average surface temperatures [7]

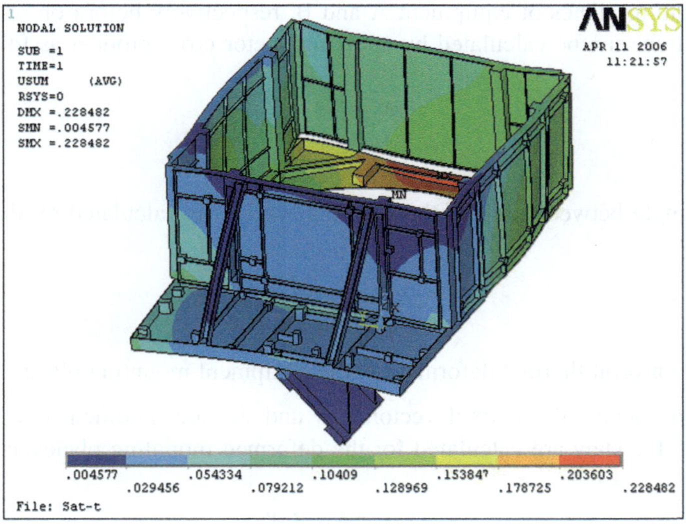

Fig. 4.80 Basis unit block displacement due to on-orbit thermal deformation under minimum average surface temperatures [7]

star sensor is explained below. Figure 4.81 explains the criterion used to calculate the angular positioning deviation between any two equipments.

All precise equipment mounted on the basis unit block are fixed through three or four fixation points. The mounting plane can be defined by only two vectors

Table 4.28 Maximum Von Mises equivalent stresses and yield margins of safety for basis unit block module in each on-orbit thermal load case

Design case	σ_e (MPa)	MS_y
Case 1	94.158	0.59
Case 2	125.937	0.19

Case 1 represents the thermal deformation due to on-orbit maximum temperatures
Case 2 represents the thermal deformation due to on-orbit minimum temperatures

Table 4.29 The limiting angular positioning deviations for the most precise equipment relative to the star sensor

Equipment	Limiting angular positioning deviations (arcmin)
MBEI (optical-mechanical unit)	30
Angular velocity meters (gyro)	60
Reaction wheels	60
Magnetometer	60
Magnetorquers	60

connecting at least three fixation points. In case of equipment A, $\vec{v}_{1A}$ and $\vec{v}_{2A}$ identify its mounting plane before on-orbit thermal deformation. $\vec{v}_{1B}$ and $\vec{v}_{2B}$ identify the mounting plane of equipment B. The normal vectors, $\vec{v}_{nA}$ and $\vec{v}_{nB}$ to the mounting planes of equipment A and B, respectively before on-orbit thermal deformation, can be calculated by applying vector cross product as follows:

$$\vec{v}_{nA} = \vec{v}_{1A} \times \vec{v}_{2A}$$

$$\vec{v}_{nB} = \vec{v}_{1B} \times \vec{v}_{2B}$$

The angle between both of these normal vectors is calculated by the formula:

$$\theta = \cos^{-1}\left(\frac{\vec{v}_{nA} \cdot \vec{v}_{nB}}{\left|\vec{v}_{nA}\right| \cdot \left|\vec{v}_{nB}\right|}\right)$$

After on-orbit thermal deformation, the equipment mounting planes are usually deformed. Hence, the normal vectors $\vec{v}_{nA}$ and $\vec{v}_{nB}$ are modified to $\vec{v}^{\,\prime}_{nA}$ and $\vec{v}^{\,\prime}_{nB}$, respectively. They are calculated for the deformed mounting planes as follows:

$$\vec{v}^{\,\prime}_{nA} = \vec{v}^{\,\prime}_{1A} \times \vec{v}^{\,\prime}_{2A}$$

$$\vec{v}^{\,\prime}_{nB} = \vec{v}^{\,\prime}_{1B} \times \vec{v}^{\,\prime}_{2B}$$

where: $\vec{v}^{\,\prime}_{1A}$ and $\vec{v}^{\,\prime}_{2A}$ and $\vec{v}^{\,\prime}_{1B}$ and $\vec{v}^{\,\prime}_{2B}$ identify the mounting planes of equipment *A* and *B*, respectively, after on-orbit thermal deformation. These vectors are calculated from the displacement deformation results of on-orbit thermal deformation analysis in the *X*, *Y*, and *Z* directions.

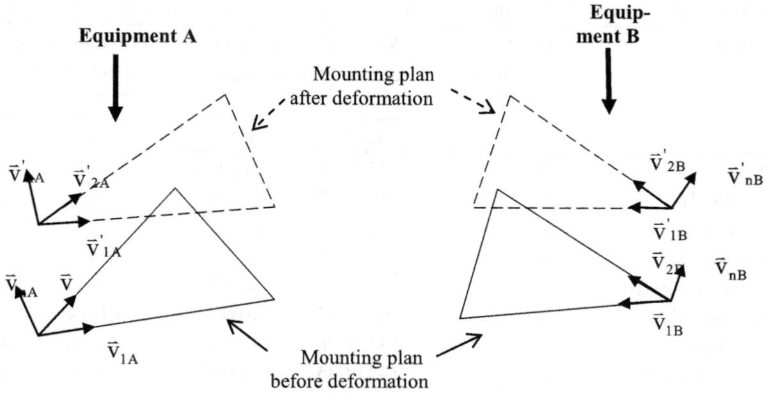

Fig. 4.81 Angular positioning deviation between two equipments [7]

The modified angle between normal vectors to the deformed mounting plane is calculated as:

$$\theta' = \cos^{-1}\left(\frac{\vec{v}'_{nA}\cdot\vec{v}'_{nB}}{\left|\vec{v}'_{nA}\right|\cdot\left|\vec{v}'_{nB}\right|}\right)$$

The angular positioning deviation angle between equipment *A* and *B* is calculated by:

$$\theta_{\text{dev}} = \theta' - \theta$$

For Small Sat, the angular positioning deviation angle θ_{dev} (arcmin) is calculated for the precise equipment mounted on the basis unit block relative to the star sensor. Table 4.30 lists the values of the angular positioning deviation angles due to on-orbit thermal deformation relative to the star sensor. It shows the results for both maximum and minimum on-orbit temperatures. By comparing these results with the limiting values listed in Table 4.29, it is found that the performance of the satellite is not significantly affected by on-orbit thermal deformation under maximum or minimum temperatures.

4.16 Fatigue Damage Calculation Due to On-Orbit Thermal Cyclic Loading

One of the most important results of on-orbit thermal deformation analysis is to evaluate the fatigue damage due to on-orbit cyclic thermal stresses. The ductile material (AMg6 aluminum alloy) used to manufacture the satellite structure modules does not rupture or buckle from a single application of thermal stress,

Table 4.30 Angular positioning deviation angles for precise equipment relative to the star sensor

Equipment	Angle before deformation θ (deg)	Modified angle after deformation θ' (deg)		Angular positioning deviations angle θ_{dev} (arcmin)	
		Case1	Case 2	Case1	Case 2
MBEI	136	135.9871	135.9955	−0.777	−0.269
AVM, gyro M_x	90	89.9982	90.0063	−0.105	0.379
AVM, gyro M_y	46	46.0042	46.0117	0.252	0.7002
AVM, gyro M_z	44	43.9885	43.9908	−0.691	−0.5512
AVM, skewed gyro	44	44.0233	44.0238	1.3996	1.428
Reaction wheels M_x	90	90.0127	90.0068	0.7614	0.40734
Reaction wheels M_y -1	134	134.0076	133.9993	0.4549	−0.0395
Reaction wheels M_y -2	134	134.0035	133.9959	0.2078	−0.2446
Reaction wheels M_z	44	44.0221	44.0202	1.3242	1.2117
Magnetometer	136	135.9947	135.9979	−0.3168	−0.1239
Magnetorquers	44	43.9828	43.9892	−1.0341	−0.6458

Note
Case 1 represents thermal deformation due to on-orbit maximum temperatures
Case 2 represents thermal deformation due to on-orbit minimum temperatures

which is clear from the previous analysis. Therefore, the material can fail in fatigue from the many cycles of on-orbit thermal deformation loading. The results data of thermoelastic analysis are used to evaluate the thermal fatigue damage for the basis unit block.

By reviewing the results data of thermal deformation analysis and the thermal stress distributions in the basis unit block module under maximum and minimum temperatures (Figs. 4.75 and 4.78), it is found that the maximum thermal stress occurs at the same location under both cases. This location is specified as "Point 30" in Fig. 4.50. The entire satellite structure is affected by cyclic thermal stresses along the operation life time of the satellite. However, Point 30 is the location most severely affected by fatigue damage due to on-orbit thermal cyclic loading.

The total fatigue damage at any point located in the basis unit block module, including Point 30, is the dynamic fatigue damage due to mechanical vibrations during transportation and launch plus the thermal fatigue damage due to on-orbit cyclic thermal stresses. Therefore, to calculate the overall fatigue damage for the basis unit block, dynamic fatigue damage must be known for Point 30 based on the fatigue analysis of dynamic vibrations discussed at Sect. 4.12. Moreover, thermal fatigue damage must be calculated for the critical points located in the basis plate module and basis walls module. Points 11 through 18 represent the locations of the maximum resultant stresses during mechanical loading phases.

By applying the same method used to calculate dynamic fatigue damage in Sect. 4.12, the dynamic fatigue damage results for Point 30 are given in Table 4.31. For thermal fatigue damage calculation, the maximum and minimum cyclic thermal stresses for the critical points are taken from the thermoelastic analysis results. The time life cycles (N_{th}) corresponding to each given stress ratio is calculated directly by reading from the typical constant life diagram related to the equivalent material of AMg6 aluminum alloy shown in Fig. 4.55, and based on the same criterion described in Sect. 4.12. The number of cycles (n_{th}) corresponding to the operational life time of the satellite is calculated by first calculating the satellite draconian period (T_{dr}), which is the time interval for completing one orbital cycle (revolution), calculated from the following relationship:

$$\tau_{dr} = \frac{2\pi}{\sqrt{\mu_e}} a^{\frac{3}{2}} \left\{ 1 - \frac{\varepsilon_e}{\mu_e a^2 (1-e^2)^2} \left[\begin{array}{c} \left(2 - \frac{5}{2}\sin^2(i)\right) \frac{(1-e^2)^{\frac{3}{2}}}{(1+e\cdot\cos(u-\omega))^2} + \\ \frac{(1+e\cdot\cos(u-\omega))^3}{(1-e^2)} (1 - 3\sin^2(i)\sin^2(u)) \end{array} \right] \right\}$$

where:

- a: the major semiaxis ($= R_e + h$)
- R_e : the mean earth radius, ($= 6378.14$ km)
- h: the satellite altitude
- μ_e : the gravitational constant of the earth, ($= 3.986005 \times 10^5$ km^3/s^2)
- ε_e: the earth oblateness parameter, ($= 2.6333 \times 10^{10}$ km^5/s^2)
- i: inclination
- e: eccentricity
- u: the argument of latitude
- ω: the argument of perigee

For Small Sat, the following parameter values are taken in the calculations, $h = 668$ km; $i = 98.085°$; $e = 0.001$; $u = 0$; $\omega = 0$. By substituting these values into the previous equation, the draconian period for Small Sat is found to be:

$T_{\text{dr}} = 5881.9$ s

The number of cycles (n) corresponding to the operational life time is then:

$n_{\text{th}} = \frac{T_{\text{opr}}}{T_{\text{dr}}}$

where: T_{opr} is the operational life time of the satellite, which equals to 5 years.

$T_{\text{opr}} = 5 \times 365 \times 24 \times 60 \times 60 = 157.86 \times 10^6$ s

Hence, $n_{\text{th}} = 26808$ cycles

The thermal damage at the critical points is calculated by the relation:

$$D_{\text{thermal}} = \frac{n_{\text{th}}}{N_{\text{th}}}$$

The overall fatigue damage at the critical points is calculated by the following relation:

$$D_{\text{overall}} = D_{\text{dynamic}} + D_{\text{thermal}}$$

Table 4.32 lists thermal and overall fatigue damage calculations for the critical points located in the basis unit block module.

Table 4.31 Harmonic and dynamic fatigue damage analysis results for critical point No. 30 located in the basis unit block module

Loading Phase	Frequency (Hz)	Static analysis		Harmonic analysis									Fatigue analysis		
		Mean Stress		Real	Imag	Amp. Stress		Max Stress		Min Stress		R	N (cycles)	n (cycles)	Damage
		MPa	Ksi	MPa	MPa	MPa	Ksi	MPa	Ksi	MPa	Ksi				
Rail transportation	35	0.443	0.064	0.211	0.527	0.568	0.082	1.012	0.147	−0.125	−0.018	−0.124	3.68E + 28	5.04E + 06	1.37E − 22
	33	0.443	0.064	0.164	0.651	0.671	0.097	1.114	0.162	−0.228	−0.033	−0.204	9.22E + 27	4.75E + 06	5.15E − 22
	96	0.443	0.064	1.042	0.466	1.141	0.166	1.584	0.23	−0.698	−0.101	−0.441	9.52E + 25	1.39E + 07	1.46E − 19
	62.6	0.443	0.064	0.075	0.024	0.079	0.011	0.522	0.076	0.365	0.053	0.699	8.66E + 34	9.01E + 06	1.04E − 28
	73.6	0.443	0.064	0.145	0.131	0.196	0.028	0.639	0.093	0.247	0.036	0.387	1.44E + 32	1.06E + 07	7.37E − 26
Road transportation	35	0.443	0.064	0.308	0.972	1.019	0.148	1.463	0.212	−0.576	−0.084	−0.394	2.57E + 26	1.13E + 06	4.42E − 21
	33	0.443	0.064	0.443	1.49	1.555	0.226	1.998	0.29	−1.111	−0.161	−0.556	6.02E + 24	1.07E + 06	1.78E − 19
Air transportation	35	0.443	0.064	0.955	1.94	2.163	0.314	2.606	0.378	−1.719	−0.25	−0.66	2.96E + 23	9.45E + 04	3.20E − 19
	33	0.443	0.064	0.504	2.51	2.56	0.372	3.004	0.436	−2.117	−0.307	−0.705	6.19E + 22	8.91E + 04	1.44E − 18
	96	0.443	0.064	0.995	6.567	6.642	0.964	7.085	1.028	−6.199	−0.9	−0.875	7.38E + 18	1.47E + 05	1.99E − 14
	104.7	0.443	0.064	4.212	3.551	5.509	0.8	5.952	0.864	−5.066	−0.735	−0.851	4.44E + 19	1.60E + 05	3.61E − 15
	62.6	0.443	0.064	1.09	0.745	1.32	0.192	1.763	0.256	−0.876	−0.127	−0.497	2.62E + 25	1.35E + 05	5.16E − 21
	73.6	0.443	0.064	2.866	3.635	4.629	0.672	5.072	0.736	−4.186	−0.608	−0.825	2.34E + 20	1.59E + 05	6.81E − 16
	119.8	0.443	0.064	2.649	1.633	3.111	0.452	3.555	0.516	−2.668	−0.387	−0.751	1.00E + 22	1.83E + 05	1.83E − 17
	128.3	0.443	0.064	1.931	0.67	2.043	0.297	2.487	0.361	−1.6	−0.232	−0.644	4.98E + 23	1.96E + 05	3.94E − 19
	187.9	0.443	0.064	1.2	0.416	1.27	0.184	1.713	0.249	−0.827	−0.12	−0.483	3.69E + 25	6.09E + 05	1.65E − 20
	137	0.443	0.064	2.968	1.03	3.141	0.456	3.585	0.52	−2.698	−0.392	−0.753	9.15E + 21	4.69E + 05	5.12E − 17
Launch (phase I)	35	3.700	0.537	2.455	4.988	5.559	0.807	9.259	1.344	−1.859	−0.27	−0.201	8.40E + 18	8.17E + 01	9.72E − 18
	33	3.700	0.537	1.265	6.303	6.428	0.933	**10.129**	1.47	−2.728	−0.396	−0.269	2.46E + 18	7.70E + 01	3.13E − 17
	96	3.700	0.537	0.874	5.769	5.835	0.847	9.535	1.384	−2.135	−0.31	−0.224	5.59E + 18	2.25E + 02	4.02E − 17
	104.7	3.700	0.537	4.114	3.469	5.382	0.781	9.082	1.318	−1.681	−0.244	−0.185	1.10E + 19	2.44E + 02	2.22E − 17
	62.6	3.700	0.537	0.342	0.234	0.415	0.06	4.115	0.597	3.286	0.477	0.798	1.57E + 27	1.46E + 02	9.33E − 26
	73.6	3.700	0.537	0.955	1.212	1.543	0.224	5.243	0.761	2.157	0.313	0.411	1.88E + 23	1.72E + 02	9.15E − 22
	119.8	3.700	0.537	3.002	1.85	3.526	0.512	7.227	1.049	0.174	0.025	0.024	3.48E + 20	2.80E + 02	8.03E − 19
	128.3	3.700	0.537	2.342	0.812	2.479	0.36	6.179	0.897	1.222	0.177	0.198	5.48E + 21	2.99E + 02	5.46E − 20
	187.9	3.700	0.537	3.888	1.349	4.115	0.597	7.815	1.134	−0.415	−0.06	−0.053	1.01E + 20	4.38E + 02	4.36E − 18
	137	3.700	0.537	2.199	0.763	2.327	0.338	6.028	0.875	1.373	0.199	0.228	8.86E + 21	3.20E + 02	3.61E − 20
	471.7	3.700	0.537	1.101	1.354	1.745	0.253	5.445	0.79	1.955	0.284	0.359	7.64E + 22	1.10E + 03	1.44E − 20
	696.8	3.700	0.537	1.264	1.117	1.687	0.245	5.387	0.782	2.013	0.292	0.374	9.81E + 22	1.63E + 03	1.66E − 20
	378.1	3.700	0.537	0.725	1.385	1.564	0.227	5.264	0.764	2.137	0.31	0.406	1.71E + 23	8.82E + 02	5.17E − 21
	950.4	3.700	0.537	0.417	0.309	0.519	0.075	4.219	0.612	3.182	0.462	0.754	3.57E + 26	2.22E + 03	6.20E − 24
	311.3	3.700	0.537	0.219	0.567	0.608	0.088	4.308	0.625	3.092	0.449	0.718	1.24E + 26	7.26E + 02	5.88E − 24

(continued)

Table 4.31 (continued)

Loading Phase	Frequency (Hz)	Static analysis		Harmonic analysis									Fatigue analysis		
		Mean Stress		Real	Imag	Amp. Stress		Max Stress		Min Stress		R	N (cycles)	n (cycles)	Damage
		MPa	Ksi	MPa	MPa	MPa	Ksi	MPa	Ksi	MPa	Ksi				
Launch (phase II)	35	3.700	0.537	2.455	4.988	5.559	0.807	9.259	1.344	−1.859	−0.27	−0.201	8.40E + 18	1.94E + 03	2.31E − 16
	33	3.700	0.537	1.265	6.303	6.428	0.933	**10.129**	1.47	−2.728	−0.396	−0.269	2.46E + 18	1.83E + 03	7.43E − 16
	96	3.700	0.537	0.704	4.65	4.703	0.683	8.403	1.22	−1.003	−0.146	−0.119	3.37E + 19	5.33E + 03	1.58E − 16
	104.7	3.700	0.537	3.067	2.586	4.012	0.582	7.712	1.119	−0.311	−0.045	−0.04	1.24E + 20	5.80E + 03	4.70E − 17
	62.6	3.700	0.537	0.342	0.234	0.415	0.06	4.115	0.597	3.286	0.477	0.798	1.57E + 27	3.47E + 03	2.22E − 24
	73.6	3.700	0.537	0.955	1.212	1.543	0.224	5.243	0.761	2.157	0.313	0.411	1.88E + 23	4.08E + 03	2.17E − 20
	119.8	3.700	0.537	2.013	1.241	2.365	0.343	6.065	0.88	1.335	0.194	0.22	7.84E + 21	6.64E + 03	8.47E − 19
	128.3	3.700	0.537	1.499	0.52	1.587	0.23	5.287	0.767	2.114	0.307	0.4	1.54E + 23	7.11E + 03	4.63E − 20
	187.9	3.700	0.537	2.136	0.741	2.261	0.328	5.961	0.865	1.439	0.209	0.241	1.10E + 22	1.04E + 04	9.44E − 19
	137	3.700	0.537	1.349	0.468	1.428	0.207	5.128	0.744	2.272	0.33	0.443	3.30E + 23	7.59E + 03	2.30E − 20
	471.7	3.700	0.537	0.433	0.532	0.686	0.1	4.386	0.637	3.014	0.437	0.687	5.47E + 25	2.61E + 04	4.77E − 22
	696.8	3.700	0.537	0.642	0.567	0.857	0.124	4.557	0.661	2.844	0.413	0.624	1.20E + 25	3.86E + 04	3.21E − 21
	378.1	3.700	0.537	0.249	0.477	0.538	0.078	4.238	0.615	3.162	0.459	0.746	2.80E + 26	2.10E + 04	7.47E − 23
	950.4	3.700	0.537	0.213	0.158	0.265	0.038	3.965	0.576	3.435	0.499	0.866	2.89E + 28	5.27E + 04	1.83E − 24
	311.3	3.700	0.537	0.111	0.289	0.309	0.045	4.01	0.582	3.391	0.492	0.846	1.06E + 28	1.73E + 04	1.63E − 24
					Maximum stress (MPa)			10.129				Cumulative fatigue damage			2.56E − 14

Table 4.32 Thermal and overall fatigue damage calculation for the critical points located in the basis unit block module

Structural module	Critical points	Thermoelastic analysis				Thermal fatigue analysis				Dynamic damage	Overall fatigue damage
		σ_{max}		σ_{min}		R	N (cycles)	n (cycles)	Thermal damage		
		MPa	Ksi	MPa	Ksi						
Basis plate	11	36.617	5.315	32.200	4.674	0.88	1.72E + 19	26,808	4.40E − 16	4.41E − 03	4.41E − 03
	12	4.313	0.626	3.545	0.514	0.82	2.13E + 27	26,808	1.83E − 24	2.68E − 01	2.68E − 01
	13	7.602	1.103	6.217	0.902	0.82	6.98E + 24	26,808	5.31E − 22	1.68E − 02	1.68E − 02
Basis walls	14	27.956	4.058	23.575	3.422	0.84	4.84E + 19	26,808	1.04E − 16	8.24E − 07	8.24E − 07
	15	19.673	2.855	15.926	2.311	0.81	4.58E + 20	26,808	7.32E − 18	1.09E − 03	1.09E − 03
	16	29.530	4.286	28.697	4.165	0.97	1.17E + 24	26,808	2.30E − 20	9.74E − 02	9.74E − 02
	17	4.006	0.590	1.855	0.269	0.46	3.75E + 24	26,808	3.17E − 24	3.85E − 01	3.85E − 01
	18	9.342	1.356	7.868	1.142	0.84	2.24E + 24	26,808	2.21E − 21	4.01E − 06	4.01E − 06
	30	125.937	18.278	94.158	13.666	0.75	9.33E + 11	26,808	1.64E − 09	2.558E − 14	1.64E − 09

References

1. ANSYS release 9.0 Documentation
2. Tedesco J, McDougal G, Ross C (1999) Structural dynamics—theory and applications. Addison Wesley Longmann, CA
3. Bruhn EF (1973) Analysis and design of flight vehicle structures. Tri-State Offset Company, USA
4. Crandall SH (1958) Random vibration. Technology Press, Wiley, New York
5. Sarafin TP, Larson WJ (eds) (1995) Spacecraft structures and mechanisms—from concept to launch. Microcosm Press and Kluwer Academic Publishers, Torrance
6. Abdelal GF, Abuelfoutouh N, Hamdy A (2008) Mechanical fatigue and spectrum analysis of small-satellite structure. Springer Int J Mech Mater Design 4(3):265–278
7. Abdelal GF, Abuelfoutouh N, Hamdy A, Atef A (2006) Thermal fatigue analysis of small-satellite structure. Springer Int J Mechand Mater Design 3(2):145–159

Chapter 5
Stochastic Finite Element and Satellite Structure Design

Abstract Throughout design development of satellite structure, stress engineer is usually challenged with randomness in applied loads and material properties. To overcome such problem, a risk-based design is applied which estimates satellite structure probability of failure under static and thermal loads. Determining probability of failure can help to update initially applied factors of safety that were used during structure preliminary design phase. These factors of safety are related to the satellite mission objective. Sensitivity-based analysis is to be implemented in the context of finite element analysis (probabilistic finite element method or stochastic finite element method (SFEM)) to determine the probability of failure for satellite structure or one of its components.

5.1 Introduction to Stochastic Finite Element Theory

The concept of risk-based design was introduced by Freudenthal [1] and was summarized by Freudenthal et al. [2]. Considering the load on structure, S, and the resistance of the structure, R are random in nature. Their randomness can be defined by their means and standard deviations and corresponding probability density functions, $f_S(s)$, $f_R(r)$. The objective of safe design in deterministic design procedures can also be achieved by selecting the design variables so that the area of overlap between the two probability density functions is as small as possible. The area of overlap depends on the relative positions of the two curves (mean value of the two variables), the dispersion of the two variables (standard deviation), and the shape of the curves (normal, lognormal, or beta distribution). Deterministic design approaches achieve this objective by shifting the positions of the curves through the use of safety factors. A more rational approach would be to

G. F. Abdelal et al., *Finite Element Analysis for Satellite Structures*,
DOI: 10.1007/978-1-4471-4637-7_5, © Springer-Verlag London 2013

compute the risk of failure by accounting for all three overlap factors and selecting the design variables so that acceptable risk of failure is achieved.

The first step in evaluation the reliability of failure of a structure is to decide on specific performance criteria and the relevant load and resistance parameters, called the basic variables X_i, and the functional relationship among them corresponding to each performance criterion. Then define limit state equation, which can be explicit or implicit function of the basic random variables. The probability of failure is calculated using first-order reliability methods (FORM) or second-order reliability methods (SORM). Most of these methods are applicable if the limit state equation is explicit. Estimating the probability of failure using these methods requires a background in probability and statistics. Simple simulation technique can be used to estimate probability of failure with minimum knowledge of probability and statistics. So, in the simulation process, each random variable is sampled several times to represent its real distribution. Considering each realization of all the random variables in the problem produces a set of numbers that indicates one realization of the problem itself. Solving the problem deterministically for each realization is known as a simulation cycle. Using many simulation cycles gives the overall probabilistic of the problem. The method commonly used for this purpose is called the Monte Carlo simulation technique.

Several computational approaches could be pursued for the reliability analysis of structures with implicit performance functions. These can be divided into three categories, (1) Monte Carlo simulation, (2) response surface approach, and (3) sensitivity-based analysis. The Monte Carlo simulation and response surface approach could be time-consuming. Thus, combination of sensitivity analysis and FORM/SORM methods for reliability calculations for functions with implicit performance functions is more efficient technique. This technique is integrated with the powerful tool of FE that realistically represents the structure problem to estimate the reliability of failure. This methodology leads to the stochastic finite element method (SFEM).

The developed approach consists of three main steps. First step is to develop a finite element program using MATLAB to analyze structure under static loading with 3D beam elements. Second step is to develop the sensitivity-based analysis program based on Halder and Mahadeven [3]. The last stream is to integrate both programs together and validate the approach on different cases already available in the literature.

5.2 First-Order Reliability Method

The failure surface (limit state), which is the boundary between safe and unsafe regions in the design parameter space can be defined as $Z = 0$. The limit state equation plays an important role in the development of structural reliability assessment. A limit state can be explicit or implicit function of the basic random variables.

$$Z = g(X_1, X_2, \ldots, X_n) \tag{5.1}$$

Failure occurs when $Z < 0$. Therefore, the probability of failure, p_f, is given by the integral,

$$p_f = \int \ldots \int_{g()<0} f_X(x_1, x_2, \ldots, x_n) dx_1\, dx_2 \ldots dx_n \tag{5.2}$$

in which $f_X(x_1, x_2, \ldots, x_n)$ is the joint density function for the basic random variables X_1, X_2, …, X_n. If the random variables are statically independent, then the joint probability density function may be replaced by the product of the individual probability density functions in the integral. Evaluation of the multiple integral is extremely complicated. One approach is to use analytical approximations of this integral that are simpler to compute, such as FORM.

A Taylor series expansion of the performance function about the mean values gives,

$$Z = g(\mu_X) + \sum_{i=1}^{n} \frac{\partial g}{\partial X_i}(X_i - \mu_{X_i}) + \frac{1}{2}\sum_{i=1}^{n}\sum_{j=1}^{n} \frac{\partial^2 g}{\partial X_i\, \partial X_j}(X_i - \mu_{X_i})\left(X_j - \mu_{X_j}\right) + \cdots \tag{5.3}$$

where the derivatives are evaluated at the mean values of the random variables (X_1, X_2,…,X_n) and μ_{Xi} is the mean value of X_i. Truncating the series at the linear terms, we obtain the first-order approximate mean and variance of Z as

$$\mu_Z \approx g\left(\mu_{X_1}, \mu_{X_2}, \ldots, \mu_{X_n}\right); \quad \sigma_Z^2 \approx \sum_{i=1}^{n}\sum_{j=1}^{n} \frac{\partial^2 g}{\partial X_i\, \partial X_j} \mathrm{Cov}\left(X_i, X_j\right) \tag{5.4}$$

where Cov(X_i, X_j) is the covariance of X_i, X_j. If the variables are uncorrelated, then the variance is

$$\sigma_Z^2 \approx \sum_{i=1}^{n} \left(\frac{\partial g}{\partial X_i}\right)^2 \mathrm{Var}(X_i) \tag{5.5}$$

The probability of failure depends on the ratio of the mean value of Z to its standard deviation. This ratio is defined as the safety index β:

$$\beta = \frac{\mu_Z}{\sigma_Z}; \quad p_f = 1 - \Phi(\beta)$$

where Φ is the CDF of the standard normal variant.

Hasofer and Lind [4] proposed the advanced first-order second moment (AFOSM) method. This method is applicable for normal variables. It first defines the reduced variables as,

$$X_i^{'} = \frac{X_i - \mu_{Xi}}{\sigma_{Xi}} \tag{5.6}$$

where $X_i^{'}$ is a random variable with zero mean and unit standard deviation. Equation (7) is used to transform the original limit state $g(X) = 0$ to the reduced limit state $g(X') = 0$. The safety index β is defined as the minimum distance from the origin of the axes in the reduced coordinate system to the limit state surface (failure surface). It can be expressed as

$$\beta = \sqrt{(x^{'*})^t (x^{'*})} \tag{5.7}$$

The minimum distance point on the limit state surface is called the design point. It is defined by the vector x^* in the original coordinate system and by vector $x^{'*}$ in the reduced coordinate system. It is obvious that the nearer x^* is to the origin, the larger is the failure probability. Thus, the minimum distance point on the limit state surface is also the most probable failure point. This point represents the worst combination of the stochastic variables and is named the design point or the most probable point (MPP) of failure. For nonlinear limit states, the computation of the MPP becomes as an optimization problem,

$$\begin{aligned} \text{Minimize } D &= \sqrt{x^{'t} x^{'}} \\ \text{Subject to the constraint } g(X) &= g\left(X^{'}\right) = 0 \end{aligned} \tag{5.8}$$

where $x^{'}$ represents the coordinates of the checking point on the limit state equation in the reduced coordinates to be estimated. Using the method of Lagrange multipliers, we can obtain the minimum distance as,

$$\beta = -\frac{\sum_{i=1}^{n} x_i^{'*} \left(\frac{\partial g}{\partial X_i^{'}}\right)^*}{\sqrt{\sum_{i=1}^{n} \left(\frac{\partial g}{\partial X_i^{'}}\right)^{2*}}} \tag{5.9}$$

where $\left(\partial g/\partial X_i^{'}\right)^*$ is the ith partial derivative evaluated at the design point with coordinates$\left(x_1^{'*}, x_1^{'*}, \ldots, x_n^{'*}\right)$. The disadvantage in the Hasofer-Lind method that is applicable only to normal variables. If not all the variables are normally distributed, it is necessary to transform the non-normal variables into equivalent normal variables. Rackwitz and Fiessler [5] estimated the parameters of the equivalent normal distribution, μ_{Xi}^N and σ_{Xi}^N, by imposing two conditions. The cumulative distribution functions and the probability density functions of the actual variables and the equivalent normal variables should be equal at the design point $\left(x_1^{'*}, x_1^{'*}, \ldots, x_n^{'*}\right)$on the failure surface

$$\Phi\left(\frac{x_i^* - \mu_{Xi}^N}{\sigma_{Xi}^N}\right) = F_{Xi}(x_i^*)$$
$$\mu_{Xi}^N = x_i^* - \Phi^{-1}\left[F_{Xi}(x_i^*)\right]\sigma_{Xi}^N \quad (5.10)$$

in which $\Phi()$ is the CDF of the standard normal variable, μ_{Xi}^N and σ_{Xi}^N are the mean and standard deviation of the equivalent normal variable at the design point. Equating the PDFs of the original variable and the equivalent normal variable at the design point results in

$$\frac{1}{\sigma_{Xi}^N}\phi\left(\frac{x_i^* - \mu_{Xi}^N}{\sigma_{Xi}^N}\right) = f_{Xi}(x_i^*)$$
$$\sigma_{Xi}^N = \frac{\phi\{\Phi^{-1}\left[F_{Xi}(x_i^*)\right]\}}{f_{Xi}(x_i^*)} \quad (5.11)$$

in which $\phi()$ and $f_{Xi}(x_i^*)$ are the PDFs of the equivalent standard normal and original non-normal random variable.

In cases of the performance function $g(X)$ being a complicated, nonlinear or implicit function, Rackwitz and Fiessler [6] have proposed using Newton–Raphson type recursive algorithm. The algorithm linearizes the performance function at each iteration point; however, instead of solving the limit state equation explicitly for β, it uses the derivatives to find the next iteration point.

FORM Method can be described as follows:

- Define appropriate performance function $g(X)$.
- Assume initial values of the design point x_i^*, i = 1, 2..., n, and compute the corresponding value of the performance function g(). The initial point can be the mean values of the random variables.
- Compute the mean and standard deviation at the design point of the equivalent normal distribution for those variables that are nonnormal.

$$\mu_{Xi}^N = x_i^* - \Phi^{-1}\left[F_{Xi}(x_i^*)\right]\sigma_{Xi}^N$$
$$\sigma_{Xi}^N = \frac{\phi\{\Phi^{-1}\left[F_{Xi}(x_i^*)\right]\}}{f_{Xi}(x_i^*)} \quad (5.12)$$
$$y_i^* = \frac{x_i^* - \mu_{Xi}^N}{\sigma_{Xi}^N}$$

- Compute the partial derivative $\partial g/\partial X_i$ evaluated at the design pointx_i^*.
- Compute the partial derivatives $\partial g/\partial Y_i$ in the equivalent standard normal space by the chain rule of differentiation as

$$\frac{\partial g}{\partial Y_i} = \frac{\partial g}{\partial X_i}\frac{\partial X_i}{\partial Y_i} = \frac{\partial g}{\partial X_i}\,\sigma_{Xi}^N \tag{5.13}$$

The components of the corresponding unit vector are the direction cosines of the performance function

$$\alpha_i = \frac{\left(\frac{\partial g}{\partial Y_i}\right)^*}{\sqrt{\sum_{i=1}^{n}\left(\frac{\partial g}{\partial Y_i}\right)^{2*}}} = \frac{\left(\frac{\partial g}{\partial X_i}\right)^* \sigma_{Xi}^N}{\sqrt{\sum_{i=1}^{n}\left(\frac{\partial g}{\partial X_i}\,\sigma_{Xi}^N\right)^{2*}}} \tag{5.14}$$

- Compute the new values for the design point in the equivalent standard normal space (x_i^*) using:

$$y_{k+1}^* = \frac{1}{\left|\nabla g\left(y_k^*\right)\right|^2}\left[\nabla g\left(y_k^*\right)^t x_k^{\prime *} - g\left(y_k^*\right)\right]\nabla g\left(y_k^*\right) \tag{5.15}$$

- Compute the distance to this new design point from the origin

$$\beta = \sqrt{\sum_{i=1}^{n}\left(y_i^*\right)^2} \tag{5.16}$$

Check convergence criterion for β.

- Compute the new values for the design point in the original space (x_i^*) as

$$x_i^* = \mu_{Xi}^N + \sigma_{Xi}^N\, y_i^* \tag{5.17}$$

Check convergence criterion for g(); that is, check that the value of g() is very close to zero.

5.3 3D Beam Finite Element Program

Finite element code is written in MATLAB programming language to simulate static load (thermal—Acceleration) for frame structure that can be modeled as 3D beam elements. The program evaluates the deformations and stresses in the 3D beam elements. Many programs exists that can solve for 3D beam elements structures, though this program is linked to ANSYS GUI, which enables user to build his model in ANSYS and then generate four files:

- 'ELIST.lis': file that contains element numbering, nodes, material set number, and real constants set number.
- 'NLISTt.lis': file that contains node numbers and locations in the [x–y–z] coordinate system.
- 'BFELI.lis': file that contains Temperature distribution on all elements.
- 'FLIST.lis': file that contains concentrated forces information (magnitude—location—direction).

Then, the program that is written in MATLAB reads these four files and solves for the deformations and stresses in all beams. This solution will be used as Step 2 in the FORM Method-2 later to determine the structure probability of failure.

The FE-3D Beam program consists of the following:

- BEAM3D main program to calculate deformations and stresses.
- Input3R input data of real constants sets, material properties sets, and boundary conditions (this part is not read from ANSYS so the user has to define node numbers that are constrain as defined in the Input3R file).
- BMCOL3 procedure to construct the element stiffness matrix.
- FORMKV procedure to construct the global stiffness matrix.
- GSTRNG procedure to determine node numbering for each element and if it is constrained or not. As during the assembling of the global stiffness matrix, the constrained degrees of freedom are excluded.

Finite Element for 3D Beam Elements List

```
clear
%%%%% DO NOT CHANGE THIS PART
IKM=2;              % No. of nodes per element.
NODOF=6;              % No. of Degree of freedom per Node.
IDOF = IKM*NODOF;
format short e
%%%%%%%%%%%%%%%%%%%%%%%%%%%%%%%%%%%%%%%%%%%%%%%%%%%%%%%%%%%%%%%%%%%%%%%%%%%%%%%%%
NXE = 3;  % No of elements.
N  = 12;  % Total no of degree of freedom excluding B.Cs ones.
NN  = 3;  % No of nodes in the mesh
NR  = 2;  % No of restrained B.C.
NODE(:,1) = [0; 5; 5; 5];   % X-coord of each node.
NODE(:,2) = [5; 5; 5; 0];   % Y-coord of each node.
NODE(:,3) = [5; 5; 0; 0];   % Z-coord of each node.
EL_table = [  1  1 2 ;
          2  2 3 ;
          3  3 4 ] ;   % EL_table = [ Element No.,  First Node No., Second Node No.]

 for i = 1:NXE
   STOREC(i,1:6) = [ NODE(EL_table(i,2),:)   NODE(EL_table(i,3),:)];
end
EA =  [4e6; 4e6; 4e6];    % Modulus of elasticity * Cross scetion Area.
```

```
EIY = [1e6; 1e6; 1e6];    % Modulus of elasticity * Inertia-Y.
EIZ = [0.3e6; 0.3e6; 0.3e6];    % Modulus of elasticity * Inertia-Z.
GJ  = [0.3e6; 0.3e6; 0.3e6];   % Modulus of rigidity * polar moment of inertia.
Alphe =[ 0; 0; 0];            % Alpha (rotation of each element @ element local x-axis)
KV2(1:N,1:N) = 0 ;
NF(1:NN,1:NODOF) = 1;
G(1:IKM*NODOF) = 0;
% Define Node B.C.
 NF(1,:) =[0,0,0,0,0,0];
 NF(4,:) =[0,0,0,0,0,0];
 %Define Loads
 NL = 1 ;  % no of loaded degree of freedom
 LOADS(1:N) = 0;
 LOADS(2) = -100;
```

Main Program Body

```
% INPUT SECTION C
clear
input3d;
% NODE FREEDOM DATA
NF = READNF(NN,NODOF,NR,NF) ;
% GLOBAL STIFFNESS MATRIX ASSEMBLY
for IP=1:NXE
   KM = BMCOL3(EA,EIY,EIZ,GJ,Alphe,IP,STOREC);
   G = GSTRNG(EL_table,IP,NODOF,NF) ;
   STOREG(IP,:) = G;
   [KV2] = FORMKV(KV2,KM,IKM,G,N,NR,IDOF);
end
 Disp2 = inv(KV2)*LOADS';
%RETRIEVE ELEMENT END FORCES AND MOMENTS

for IP=1:NXE
   [KM] = BMCOL3(EA,EIY,EIZ,GJ,Alphe,IP,STOREC);
   for i=1:IDOF
      if (STOREG(IP,i)==0)
         Edisp(i)=0;
      else
         Edisp(i) = Disp2(STOREG(IP,i));
      end
   end
   Efor = KM*Edisp';
   Eforce(IP,1:IDOF) = Efor';
end
```

Subroutine to Calculate Element Stiffness Matrix

```
function [KMF] = BMCOL3(EA1,EIY1,EIZ1,GJ1,Alphe1,IP,COORD)
%THIS SUBROUTINE FORMS THE STIFFNESS MATRIX OF AN INCLINED 2-D BEAM-
COLUMN ELEMENT

EA = EA1(IP); EIY = EIY1(IP); EIZ = EIZ1(IP); GJ = GJ1(IP); alph = Alphe1(IP)*(pi/180);
X1=COORD(IP,1); Y1=COORD(IP,2); Z1 = COORD(IP,3);
X2=COORD(IP,4); Y2=COORD(IP,5); Z2 = COORD(IP,6);

ELL=sqrt( (Y2-Y1)^2 + (X2-X1)^2 + (Z2-Z1)^2 )
CX = (X2-X1)/ELL;  CY = (Y2-Y1)/ELL;  CZ = (Z2-Z1)/ELL;
CXZ = sqrt( CX^2 + CZ^2 );

KM(1:12,1:12) = 0;

KM(1,1) = EA/ELL;  KM(1,7) = -EA/ELL;
KM(2,2) = 12*EIZ/(ELL^3);  KM(2,6) = 6*EIZ/(ELL^2);  KM(2,8) = -12*EIZ/(ELL^3);   KM(2,12) =
6*EIZ/(ELL^2);
KM(3,3) = 12*EIY/(ELL^3);  KM(3,5) = -6*EIY/(ELL^2);  KM(3,9) = -12*EIY/(ELL^3);  KM(3,11)
= -6*EIY/(ELL^2);


KM(4,4) = GJ/ELL;  KM(4,10) = -GJ/ELL;
KM(5,3) = KM(3,5);  KM(5,5) = 4*EIY/(ELL);  KM(5,9) = 6*EIY/(ELL^2);  KM(5,11) =
2*EIY/(ELL);
KM(6,2) = KM(2,6);  KM(6,6) = 4*EIZ/(ELL);  KM(6,8) = -6*EIZ/(ELL^2);   KM(6,12) =
2*EIZ/(ELL);


KM(7,1) = KM(1,7);  KM(7,7) = EA/ELL;
KM(8,2) = KM(2,8);  KM(8,6) = KM(6,8);  KM(8,8) = 12*EIZ/(ELL^3);   KM(8,12) = -
6*EIZ/(ELL^2);
KM(9,3) = KM(3,9);  KM(9,5) = KM(5,9);  KM(9,9) = 12*EIY/(ELL^3);  KM(9,11) =
6*EIY/(ELL^2);
KM(10,4) = KM(4,10);  KM(10,10) = GJ/ELL;
KM(11,3) = KM(3,11);  KM(11,5) = KM(5,11);  KM(11,9) = KM(9,11);  KM(11,11) = 4*EIY/(ELL);
KM(12,2) = KM(2,12);  KM(12,6) = KM(6,12);  KM(12,8) = KM(8,12);   KM(12,12) = 4*EIZ/(ELL);
if CXZ==0
   Alph = [ 0            CY          0;
        -CY*cos(alph)  0  sin(alph);
        CY*sin(alph)   0  cos(alph) ];
else

   Alph = [ CX                              CY                              CZ;
        (-CX*CY*cos(alph)-CZ*sin(alph))/CXZ         CXZ*cos(alph)        (-
CY*CZ*cos(alph)+CX*sin(alph))/CXZ;
        (CX*CY*sin(alph)-CZ*cos(alph))/CXZ          -CXZ*sin(alph)
(CY*CZ*sin(alph)+CX*cos(alph))/CXZ   ] ;
```

```
end

GAMM (1:12,1:12) = 0;
GAMM(1:3,1:3) = Alph;
GAMM(4:6,4:6) = Alph;
GAMM(7:9,7:9) = Alph;
GAMM(10:12,10:12) = Alph;
KM1 = GAMM'*KM;
KMF = KM1*GAMM;
```

Subroutine to Calculate Global Stiffness Matrix

```
function [BK2] = FORMKV(BK2,KM,IKM,G,N,NR,IDOF)
% C C THIS SUBROUTINE FORMS THE GLOBAL STIFFNESS MATRIX C STORING THE
UPPER TRIANGLE AS A VECTOR BK(N*(IW+l))
for I =1:IDOF
  if (G(I) == 0) ,continue ,end
  for J =1:IDOF
    if (G(J) == 0) ,continue ,end
    BK2(G(I),G(J)) = BK2(G(I),G(J)) + KM(I,J);
  end
end
```

Subroutine to extract ANSYS file to MATLAB

```
echo off
%          ext56uxuyuz.m                    extraction of ux, uy and uz dof's only
%          10-13-00
           clear all;
%          bring up a window with the file names of defined extension to choose from
  filename=uigetfile('*.lis','Select ELEMENT Input file from list - ELIST')
           fid = fopen(filename,'rt');  %% Open in read mode as text.
           if fid == -1
                     error('Error opening the file')
           end
  ic=1;
           while 1
                     nextline = fgetl(fid);
                                          %          read a line
       %             end of the file
    if nextline == -1, break, end
                     if ext56chk(nextline, ' ELEM') & ext56chk(nextline, 'MAT')
                     emp = 0;
                               while 1
                     nextline = fgetl(fid);
                     c = sscanf(nextline,' %f');
```

```
                    if isempty(c),
        emp = emp+1;
        if emp > 1, break, end;
        continue;
      end
                    array(ic,:) = c';
      ic = ic + 1;
                    end
  end
end
      fclose(fid);
filename=uigetfile('*.lis','Select NODES Input file from list - NLIST')
      fid = fopen(filename,'rt');  %% Open in read mode as text.

      if fid == -1
                    error('Error opening the file')
end
ic = 1;
      while 1
                    nextline = fgetl(fid);
                                  %         read a line
                    if nextline == -1, break, end;
                    %         end of the file
                    if ext56chk(nextline, ' X') & ext56chk(nextline, ' Y')
      emp=0;
                           while 1
                    nextline = fgetl(fid);
                    c2 = sscanf(nextline,' %f');

                    if isempty(c2),
        emp = emp+1;
        if emp > 0, break, end;
        continue;
      end
       array2(ic,:) = c2';
      ic = ic + 1;
                    end
  end
end
      fclose(fid);
NE = size(array(:,1));
filename=uigetfile('*.lis','Select TEMP Input file from list - BFLIST')
      fid = fopen(filename,'rt');  %% Open in read mode as text.

TEMP = [];
```

```
        if fid == -1

                error('Error opening the file')
        end
ipe=1;
        while 1
                nextline = fgetl(fid);
                                        %       read a line
                if ~isstr(nextline), break, end;
                %       end of the file

                if ext56chk(nextline, ' ELEMENT=') & ext56chk(nextline, ' TEMPERATURES')
                disp(nextline);
      EL(ipe) = sscanf(nextline, ' ELEMENT=       %f ');

      nextline = fgetl(fid);
      c3 = sscanf(nextline,' %f');
      TEMP(ipe,:) = [EL(ipe) c3(1,1)];
              ipe = ipe+1;
       %if ip > NE, break,end;
    end
 end
        fclose(fid);

 filename=uigetfile('*.lis','Select NODES FORCES Input file from list - FLIST')
        fid = fopen(filename,'rt');  %% Open in read mode as text.
FORC_E = [];
        if fid == -1
                error('Error opening the file')
 end
        while 1
                nextline = fgetl(fid);
                                        %       read a line
                if ~isstr(nextline), break, end
                %       end of the file
                if ext56chk(nextline, ' NODE') & ext56chk(nextline, ' IMAG')


        ic2=1;
                        while 1
                nextline = fgetl(fid);
        fc = sscanf(nextline,'      %i %s %f')
        if isempty(fc), break, end;

                FORC_E(ic2,1:6) = [fc(1) fc(2) fc(3) fc(4) fc(5) fc(6)];
       ic2 = ic2 + 1;
```

```
                    end
        end
    end

            fclose(fid);
```

Subroutine to Input Parameters (mean and covariance)

```
%%% Verify with ANSYS file 'beam4.db' with no Temp. -  'beam3.db' with Temp
%%%%%%%%%%%%%%%%%%%%
extract;
%pack;
format short e
%%%%%%% DO NOT CHANGE THIS PART
%%%%%%%%%%%%%%%%%%%%%%%%%%%%%%%%%%%%5
IKM =2;  % No. of nodes in each Element
NODOF=6; % No. of Degree of Freedom at each node.
IDOF= IKM*NODOF; % No. of Degree of Freedom for each element.

% X - Axis = is Horizontal and from left to right.
% Y - Axis = is Vertical and from down to up.
% Theta_Z  = is counter clockwise.
%%%%%%%%%%%%%%%%%%%%%%%%%%%%%%%%%%%%%%%%%%%%%%%%%
%%%%%%%%%%%%%%%%%%%

%%%%%%% Define FE Mesh %%%%%%%%%%%%%%%%%%%%%%%%%%%%%
a11 = size(array(:,1)); NXE = a11(1);  % No of elements.
a22 = size(array2(:,1)); NN = a22(1);  % No of nodes in the mesh
NF(1:NN,1:NODOF) = -1;
G(1:IKM*NODOF) = 0;
%%%%%%%%%%%%%%%%%%  Enter Node (x1,y1) Location & Element Node Numbering

NODE(:,1:4) = array2(:,1:4);  % [ Node No., X,  Y,  Z] location of each node.

%%%%%% EL_table = [Element No.,  First Node No., Second Node No., Real_Constants No, Material_Constants No. ]
EL_table(:,1) = array(:,1);  EL_table(:,2:3) = array(:,7:8);  EL_table(:,4) = array(:,4);  EL_table(:,5) = array(:,2);

%%%%%%%%%%%%%%%%%%%%%  Define Real constants of the cross section properties of the 2-D BEAMS %%%%%%%%%%%%%%%%%%%%%%%%%%%%%%%%%%%
% %%  You can define any No. of Real_Constants Sets according to your problem
%%%%%%%%%%%%%%%%%%%%%%%%%%%%%%%%%%%%%%%%%%%%%%%%%
%%
% Enter Dist. Type of each variable %%%%%%%%%%%%%%%%%%%%%%
 % 0 = Normal.
 % 1 = Lognormal.
```

```
% 2 = Type-I.
% Real_Constants[Set No.] = [Ty_mean,Ty_COV, Tz_mean,Tz_COV, Dist_ty, Dist_tz, Alphe]
NREAL = 6;   % No. of real constants sets
Real_Constants(1,1:7) = [ 25 0.05  15 0.05  1 1  0];
Real_Constants(2,1:7) = [ 20 0.05  12 0.05  1 1  0];
Real_Constants(3,1:7) = [ 15 0.05  10 0.05  1 1  0];
Real_Constants(4,1:7) = [ 12 0.05  6 0.05  1 1  0];
Real_Constants(5,1:7) = [ 11 0.05  5 0.05  1 1  0];
Real_Constants(6,1:7) = [ 10 0.05  5 0.05  1 1  0];
% You can define any No. of Real_Constants Sets

%%%%%%%%%%%%%%% Define Sets of Material Properties
%%%%%%%%%%%%%%%%%%%%%%%%%%%%%%%%%%%%%%%%%%%%
%%%%%%%%%%%%%%%%%%%%%%%%%%%%
% Material_Properties[Set. no.] = [E_mean, E_COV, Fy_mean, Fy_COV, Density_mean, Densi-
ty_cov, Thermal-exp_mean, Thermal-exp_cov, Dist_E, Dist_Fy,
%                              Dist_density, Dist_thermal_exp]
NM = 1;    % No of Material Properties Sets.
Material_Properties(1,1:12) = [ 72e6 0.06  150e3 0.11  2.63e-6 0  23e-6 0  1 1 1 1];
T_ref = 20;
%Material_Properties(1,1:9) =  [ 29e3 0.06  39.6 0.11  0 0  0 0 0];
%Material_Properties(2,1:9) = [ 39e3 0     30      0 0 0  0 0 0];

%%%%%%%%%%%% DO NOT CHANGE THIS PART %%%%%%%%
%%%%%%%%%%%%%%%%%%%%%%%%%%%%%%%%%%%%%%%%%%%%
%%%%%%%%%%%%%%
for i = 1:NXE
     STOREC(i,1:NODOF) = [ NODE(EL_table(i,2),2:4)    NODE(EL_table(i,3),2:4)];
end
%%%%%%%%%%%%%%%%%%%%%%%%%%%%%%%%%%%%%%%%%%%%
%%%%%%%%%%%%%%%%%%%%%%%%%%%%%%%%%%%%%%%%%%

% Define Node B.C. , 0 = fixed , 1 = free
 NF(78,:) =[0,0,0,0,0,0];
 NF(79,:) =[0,0,0,0,0,0];
 NF(80,:) =[0,0,0,0,0,0];
 NF(81,:) =[0,0,0,0,0,0];
 NR  = 4;  % No of restrained B.C.
 N = (NN - NR)*6; % Total no of degree of freedom excluding B.Cs ones.

 %%%%%%% Define Loads Sets
%%%%%%%%%%%%%%%%%%%%%%%%%%%%%%%%%%%%%%%%%%%%
%%%%%%%%%%%%%%%%%%%%%%%%%%%%%

% Loads_Set[Set No.] = [Key, Value_mean, Value_COV, Distribution Type, Dir-X, Dir_Y, Dir_Z,
Node No.(if %concentrated load) OR No. of Element(NE)
```

```
% (Thermal Load), Element No. OR Element No.1 (Thermal Load) , Angle between Load and the
Beam OR %Element No.2 (thermal load), Element No. 3, .... % ......., Elemnt no. NE ];
%[Dir-X, Dir-Y, Dir-Z] = for distributed load, define the plane of working load, i.e. x-y plane = [ 1  1
0];  i.e. %x-z plane = [0  1  1]; x-y-z are
%the element coord. system
%[Dir-X, Dir-Y, Dir-Z] = for concentrated load are the Global Coord. system
% Key = 0  - Concentrated Load.
% Key = 1  - Distributed Load.
% Key = 2  - Thermal Load.
% Key = 3  - Acceleration.
% Enter Dist. Type of each variable %%%%%%%%%%%%%%%%%%%%%%%%
 % 0 = Normal.
 % 1 = Lognormal.
 % 2 = Type-I.
NLS = 1;       % No. of Load Sets.
Loads_Set(1,1:10) = [ 3  10e4 0.1 2    0 1 0   0  0  0];

if ~isempty(FORC_E)
   a22 = size(FORC_E(:,1)); n3 = a22(1);
   for i3 = 1:n3
      NLS = NLS + 1;
      Loads_Set(NLS,1) = 0;
      Loads_Set(NLS,2) = FORC_E(i3,4);
      Loads_Set(NLS,3) = 0.1;           % Coefficient of Variation
      Loads_Set(NLS,8) = FORC_E(i3,1);
      if FORC_E(i3,3) == 88
         Loads_Set(NLS,5:7) = [1 0 0];
      elseif FORC_E(i3,3) == 89
         Loads_Set(NLS,5:7) = [0 1 0];
      elseif FORC_E(i3,3) == 90
         Loads_Set(NLS,5:7) = [0 0 1];
      end

   end
end

if ~isempty(TEMP)

         [TEMP2,I2] = sort(TEMP(:,2)); ik=1;
         while 1
      NLS = NLS + 1; ik2=1;
      while 1
         Loads_Set(NLS,1) = 2; % Key for Thermal Load
         Loads_Set(NLS,2) = TEMP2(ik); Loads_Set(NLS,3) = 0.05;
         Loads_Set(NLS,9+(ik2-1)) = I2(ik);
         Loads_Set(NLS,8) = ik2;
```

```
        ik2 = ik2+1;
        if ik < (ipe-1)
          if TEMP2(ik+1) ~= TEMP2(ik)
            ik = ik+1;
            break;
          end
        end
         ik = ik + 1;
        if ik > (ipe-1) , break, end;
      end
    if ik > (ipe-1), break, end;
          end

end
%Loads_Set(2,1:11) = [ 2  50  0.1 2   0 0 0  3 1 2 3 ];
%Loads_Set(3,1:11) = [ 2  40  0.1 2   0 0 0  3 4 5 6 ];
%Loads_Set(4,1:11) = [ 2  -10 0.1 2   0 0 0  3 7 8 9 ];
%Loads_Set(2,1:10) = [ 3  20e4 0.1 2   0 1 0  0 3 0];

%%%%%%%%%%%%%%%%%%%% DO NOT change this part
%%%%%%%%%%%%%%%%%%%%%%%%%

 DVn = 0;

   for iL=1:NLS     % No of Load sets
    if Loads_Set(iL,3) ~= 0
       DVn = DVn + 1; dist(DVn) = Loads_Set(iL,4);
       Xs(1,DVn) = Loads_Set(iL,2);   % Mean value
       Xs(2,DVn) = Loads_Set(iL,3);   % COV value
    end
   end

   for ir = 1:NREAL
     if Real_Constants(ir,2)~= 0
       DVn = DVn + 1; dist(DVn) = Real_Constants(ir,5);
       Xs(1,DVn) = Real_Constants(ir,1);   % Mean value
       Xs(2,DVn) = Real_Constants(ir,2);   % COV value
     end
     if Real_Constants(ir,4)~= 0
       DVn = DVn + 1; dist(DVn) = Real_Constants(ir,6);
       Xs(1,DVn) = Real_Constants(ir,3);   % Mean value
       Xs(2,DVn) = Real_Constants(ir,4);   % COV value
     end
   end

   for im = 1:NM
```

```
        if Material_Properties(im,2)~= 0
          DVn = DVn + 1; dist(DVn) = Material_Properties(im,9);
          Xs(1,DVn) = Material_Properties(im,1);   % Mean value
          Xs(2,DVn) = Material_Properties(im,2);   % COV value
        end
        if Material_Properties(im,4)~= 0
          DVn = DVn + 1; dist(DVn) = Material_Properties(im,10);
          Xs(1,DVn) = Material_Properties(im,3);   % Mean value
          Xs(2,DVn) = Material_Properties(im,4);   % COV value
        end
        if Material_Properties(im,6)~= 0
          DVn = DVn + 1; dist(DVn) = Material_Properties(im,11);
          Xs(1,DVn) = Material_Properties(im,5);   % Mean value
          Xs(2,DVn) = Material_Properties(im,6);   % COV value
        end
        if Material_Properties(im,8)~= 0
          DVn = DVn + 1; dist(DVn) = Material_Properties(im,12);
          Xs(1,DVn) = Material_Properties(im,7);   % Mean value
          Xs(2,DVn) = Material_Properties(im,8);   % COV value
        end
    end
```

Subroutine to Calculate Displacement and Forces in Frame Structure

```
function[Disp2,Eforce,STOREG,NF]=
Beam3D(IKM,IDOF,NODOF,NXE,N,NN,NR,E,Iy,Iz,A,Alphr,STOREC,NF,EL_table,LOADS,T_ref,
Loads_Set,Alpha_T,NLS)

% NODE FREEDOM DATA
T_t(1:NXE) = T_ref;
NF = READNF(NODOF,NXE,EL_table,NF);

%%%%% IW = subtract minimum DOF from maximum DOF for each element, then take the largest
number among all elements.

for i= 1:NXE
  G = GSTRNG(EL_table,i,NODOF,NF);
  STOREG(i,:) = G;
  if ( min(STOREG(i,:)) == 0), continue, end;
  %IW_array(i) =  max(STOREG(i,:)) - min(STOREG(i,:));
end

%IW = max(IW_array);
%IW
%KV2(1:N*(IW+1))=0;
KV2(1:N,1:N) = 0;
% GLOBAL STIFFNESS MATRIX ASSEMBLY
for IP=1:NXE
```

```
  [KM,KMe,Gam] = BMCOL3(E,A,Iy,Iz,Alphr,IP,STOREC);  % Ke is the KM-without rotation
  G = STOREG(IP,:);
  [KV2] = FORMKV(KV2,KM,IKM,G,N,NR,IDOF);
 % [KV2] = FORMKV2(KV2,KM,IKM,G,N,NR,IDOF);
end

%[KV2] = BANRED(KV2,N,IW);
%[Disp2] = BACSUB(KV2,LOADS,N,IW);

Disp2 = inv(KV2)*LOADS';

%RETRIEVE ELEMENT END FORCES AND MOMENTS

for IP=1:NXE

  [KM,KMe,Gam] = BMCOL3(E,A,Iy,Iz,Alphr,IP,STOREC);   % Calcluate Orginal KM-element

  for i=1:IDOF
    if (STOREG(IP,i)==0)
      Edisp(i)=0;
    else
      Edisp(i) = Disp2(STOREG(IP,i));  % Displacement of element nodes in Global-Coord
    end
  end

  X1=STOREC(IP,1); Y1=STOREC(IP,2); Z1 = STOREC(IP,3);
  X2=STOREC(IP,4); Y2=STOREC(IP,5); Z2 = STOREC(IP,6); alph = Alphr(IP)*(pi/180);

  ELL=sqrt( (Y2-Y1)^2 + (X2-X1)^2 + (Z2-Z1)^2 );
  CX = (X2-X1)/ELL;  CY = (Y2-Y1)/ELL;  CZ = (Z2-Z1)/ELL;
  CXZ = sqrt( CX^2 + CZ^2 );

  if CXZ==0

  Alph = [ 0          CY          0;
      -CY*cos(alph)  0  sin(alph);
      CY*sin(alph)  0  cos(alph) ];
  else

  Alph = [ CX                              CY                                   CZ;
      (-CX*CY*cos(alph)-CZ*sin(alph))/CXZ         CXZ*cos(alph)         (-
CY*CZ*cos(alph)+CX*sin(alph))/CXZ;
      (CX*CY*sin(alph)-CZ*cos(alph))/CXZ         -CXZ*sin(alph)
(CY*CZ*sin(alph)+CX*cos(alph))/CXZ  ] ;
  end
```

```
        Pm(1:12,1:12) = 0;
        Pm(1:3,1:3) = Alph;
        Pm(4:6,4:6) = Alph;
        Pm(7:9,7:9) = Alph;
        Pm(10:12,10:12) = Alph;

%%%% Define each element Temperature from Load Sets
 for IL=1:NLS
    ch = Loads_Set(IL,1);
    if ch == 2
        NE = Loads_Set(IL,8);
        for IT = 1:NE
         IE = Loads_Set(IL,9+(IT-1));
         T_t(IE) = Loads_Set(IL,2);
        end
    end
 end

    LTe = E(IP)*A(IP)*Alpha_T(IP)*(T_t(IP)-T_ref)*[-1;0;0;0;0;0;1;0;0;0;0;0];

    Edispe = Pm*Edisp';
    Efor = KMe*Edispe - LTe;
    Eforce(IP,1:IDOF) = Efor';
 end
```

5.3.1 *Sensitivity-Based Approach with Classic Perturbation (SFEA)*

In this method, the sensitivity of the structure response to the input variables is computed and used in the FORM method. The basic concept of the FORM method, the search for the design point, requires only the value and gradient of the performance function at each iteration. The value of the performance function is available from deterministic structural analysis. The gradient is computed using sensitivity analysis. The sensitivity-based reliability analysis approach is more efficient than Monte Carlo approach, as the latest could be time-consuming. While the response surface approach that requires the construction of an approximate close-form expression of the performance function, may need quite large number of deterministic analysis for problems with a large number of random variables. The reliability estimate using the response surface approach is as accurate as the closed-form approximation to the performance function. If the implicit performance function is highly nonlinear and the close-form approximation is too approximate, the reliability estimate may also be too approximate. The combination of sensitivity analysis and FORM for the reliability analysis with implicit performance functions does not suffer from the drawbacks of Monte Carlo simulation or response surface approach. It uses the information about the actual value and the actual gradient of the performance function at each iteration of the search for the design point and uses an optimization scheme to converge to the minimum distance point.

In case the performance function Z can be defined as $Z = g(X_1, X_2, \ldots, X_n)$, the forward difference approach can be used to compute the derivatives $\partial Z/\partial X_1, \partial Z/\partial X_2, \ldots, \partial Z/\partial X_n$ at the point $\left(X_1^0, X_2^0, \ldots, X_n^0\right)$, as follows:

1. First Compute $Z_0 = g\left(X_1^0, X_2^0, \ldots, X_n^0\right)$.
2. Change the value of X_1 to $X_1^0 + \Delta X_1$, where ΔX_1 is a small number (perturbation value). All other variables stay at the same value. Compute the new value of Z as $Z_1 = g\left(X_1^0 + \Delta X_1, X_2^0, \ldots, X_n^0\right)$ and the change in its value $\Delta Z = Z_1 - Z_0$.
3. Compute the approximate derivative of Z with respect to X_1 as $\Delta Z/\Delta X_1$.
4. Repeat steps 2 and 3 for each variable X_2 to X_{n}. It is common to use perturbation values in proportion size of one-tenth of the standard deviation for each variable.

The numerical values of the derivatives computed above are valid only at the mean values of the random variables. During the iterations of the FORM method, the derivatives need to be recalculated at each iteration.

The steps of SFEM-based reliability analysis are as follows. FORM method requires the value of the performance function $G(Y)$, and its gradient $\nabla G(Y)$:

1. Using the parameters of the structure, assemble the global stiffness matrix K and the global nodal load vector F.
2. Solve for the displacement, U, using the finite element equation,

$$KU = F \tag{5.18}$$

3. Compute the vector of desired response quantities S (e.g., stress) from the computes displacement using a transformation of the form

$$S = Q^t U + S_0 \tag{5.19}$$

where Q^t is a transformation matrix relating U and S, and S_0 is the reference vector for $U = 0$.

4. Compute the performance function

$$g(X) = g\{R(X), S(X)\} \tag{5.20}$$

where R is the vector of resistance variable, S is the vector of response quantities occurring in the performance function, and X is the vector of the original random variables.

5. Transform the original random variable X to equivalent reduced normal variables Y. Compute $\nabla G(Y)$ in order to implement the FORM algorithm.

5.4 Program Sensitivity-Based Analysis for Implicit Performance Functions for 2-D Beam Elements

In the case where '$g(x)$' (performance function) is Implicit function, the following methods can be applied to evaluate probability of failure:

- Monte Carlo Simulation.
- Response Surface Approach.
- Sensitivity-Based Approach with Finite difference.

Steps for static analysis of 2-D beam structures:

1. Assemble global stiffness matrix.
2. Solve for the displacement U.
3. Compute response quantities S.
4. Compute performance function $g(x)$.
5. Transform random variables (X) to reduced random variables (Y).
6. Evaluate $g(Y)$ and $\nabla g(Y)$.
7. Apply FORM method to search for y*.

Matlab program is written to evaluate probability of failure for a structure that consists from 2D beam elements (frame structure). The program consists of main body which is called FORM2 and the following procedures:

- INPUT2R—input parameters (Structure configuration, material properties, and stochastic properties).
- FORM2 main body of the program that applies the FORM method. This program uses partial derivatives to calculate derivative of the performance function with respect to the random variables.
- BMCOL2—Evaluate element K-matrix for 2D beam element.
- BMCOL2KdA—Evaluate differentiation of K-matrix with respect to Area.
- BMCOL2KdE—Evaluate differentiation of K-matrix with respect to E (modulus of elasticity).
- BMCOL2KdI—Evaluate differentiation of K-matrix with respect to I (moment of inertia).
- FORMKV—Construct Global K-matrix.
- FORMLV—Construct GLOBAL Load vector.
- GSTRNG—Determines degrees of freedom numbers for each element.

- GUMBLEPDF—PDF of Gumbel probability distribution.
- GUMBLECDF—CDF of Gumbel probability distribution.
- LOADS1—Evaluate Load vector for each load case.
- PERFORMANCE_FUNCTION—Evaluate performance function for each set of random variables.
- DPERFORMANCE_FUNCTION—Evaluate differential of performance function with respect to to each set of random variables.
- Smatrix—Response evaluation.

FORM Method-2

```
% Step-1: Define Random variables parameters [mean - std - distribution
% type & FE Mesh
 clear all
 format short e

 Input2R

% Step-2: Define initial design point xo
  xm0(1:DVn) = Xs(1,1:DVn); cov0(1:DVn) = Xs(2,1:DVn);
  for i1=1:DVn
    std0(i1) = cov0(i1) * xm0(i1);
  end
  xstar(1:DVn) = xm0(1:DVn);
  Beta_old=0;

 for k=1:100

% Calculate response parameters and g-value for the current random variables.
    for in=1:NXE
      No_r = EL_table(in,4);   % Set No. for element Real Constants
      No_m = EL_table(in,5);   % Set No. for element Material Properties
      E(in) = Material_Properties(No_m,1);
      I(in) = Real_Constants(No_r,3);
      A(in) = Real_Constants(No_r,1);
      Z(in) = Real_Constants(No_r,5);
      Fy(in) = Material_Properties(No_m,3);
    end

    [LOADS,DLOADS] =
LOADS1(IDOF,NN,NR,N,NXE,NLS,DVn,NODOF,EL_table,NF,Loads_Set,STOREC);
    [Disp2,Eforce,STOREG] =
Beam2D(IKM,IDOF,NODOF,NXE,N,NN,NR,E,I,A,STOREC,NF,EL_table,KV2,LOADS);
    S = Smatrix(Eforce,EL_table,Rel_E_N);
    [gvalue] = performance_function(S,xstar,Rel_E_N,STOREC,E,I,A,Z,Fy);
```

```
% Step-3: Define mean and std of the equivalent normal dist. And evaluate
% the equivalent normal variables
  [xm,std] = meanstd(DVn,xstar,xm0,std0,dist);
  Bmatrix(1:DVn,1:DVn) = 0;
  for i2=1:DVn
     Bmatrix(i2,i2) = 1/std(i2);
  end

  for i=1:DVn
     xstar_dash(i) = (xstar(i) - xm(i))/std(i);
  end

% Step-4: Compute partial derivative at design point x_star

  [dgdx] = Dperfor-
mance_function(IKM,IDOF,NODOF,NXE,N,NR,STOREG,S,xstar,Disp2,DLOADS,Rel_E_N,EL_tabl
e,STOREC,E,I,A,Z,Fy,NLS,NREAL,NM,Loads_Set,Real_Constants,Material_Properties);

  for i3=1:DVn
     Xv = xstar;
     dx = std0(i3)/10;
     Xv(i3) = xstar(i3) + dx;
     [Loads_Set,Real_Constants,Material_Properties] = Con-
vert(NLS,NREAL,NM,Xv,Loads_Set,Real_Constants,Material_Properties);
     for in=1:NXE
      No_r = EL_table(in,4);   % Set No. for element Real Constants
      No_m = EL_table(in,5);   % Set No. for element Material Properties
      A(in) = Real_Constants(No_r,1);
      I(in) = Real_Constants(No_r,3);
      Z(in) = Real_Constants(No_r,5);
      E(in) = Material_Properties(No_m,1);
      Fy(in) = Material_Properties(No_m,3);
      Density(in) = Material_Properties(No_m,5);
     end

     [LOADS2,DLOADS] =
LOADS1(IDOF,NN,NR,N,NXE,NLS,DVn,NODOF,EL_table,NF,Loads_Set,STOREC);
     [Disp22,Eforce,STOREG] =
Beam2D(IKM,IDOF,NODOF,NXE,N,NN,NR,E,I,A,STOREC,NF,EL_table,KV2,LOADS2);
     S2 = Smatrix(Eforce,EL_table,Rel_E_N);
     [gvalue2] = performance_function(S2,xstar,Rel_E_N,STOREC,E,I,A,Z,Fy);
     dgdx2(i3) = (gvalue2-gvalue)/dx;
     [Loads_Set,Real_Constants,Material_Properties] = Con-
vert(NLS,NREAL,NM,xstar,Loads_Set,Real_Constants,Material_Properties);
  end
```

```
 dgdx
 dgdx2
% Step-5: Compute partial derivative in the equivalent normal space

 dgdxstar_dash = [inv(Bmatrix)] * dgdx;
 alpha1 = dgdxstar_dash/norm(dgdxstar_dash)

% Step-6: Evaluate coordinates of the new design point in the equivalent
% normal space using eqn 3.48
%if norm(dgdx2)==0
%    xstar_dash_new(1:DVn) = 0;
%else
   xstar_dash_new = (1/norm(dgdxstar_dash))^2 * ( DOT(dgdxstar_dash,xstar_dash) - gvalue
)*dgdxstar_dash;
%end

% Step-7: Evaluate Beta - distance to the new design values

Beta = norm(xstar_dash_new);
delta_Beta = abs(Beta - Beta_old);
Beta_old = Beta;

% Step-8: Evaluate new design values in the orginal space
 for i=1:DVn
    xstar(i) = xm(i) + std(i)*xstar_dash_new(i);
 end

 [Loads_Set,Real_Constants,Material_Properties] = Con-
vert(NLS,NREAL,NM,xstar,Loads_Set,Real_Constants,Material_Properties)

            xstar_dash_k(:,k) = xstar_dash';
            xstar_dash_new_k(:,k) = xstar_dash_new;
            xm_k(:,k) = xm';
            std_k(:,k) = std';
            dgdx_k(:,k) = dgdx;
            dgdxstar_dash_k(:,k) = dgdxstar_dash;
            gvalue_k(k) = gvalue;
            Beta_k(k) = Beta;
            xstar_k(:,k) = xstar';

            if (abs(gvalue) < 0.0001) & (delta_Beta < 0.001)
      break
   end
end
```

INPUT2R – List

```
%%%%%% DO NOT CHANGE THIS PART
IKM =2; % No. of nodes in each Element
IDOF=6; % No. of Degree of Freedom for each element.
NODOF=3; % No. of Degree of Freedom at each node.

% X - Axis = is Horizontal and from left to right.
% Y - Axis = is Vertical and from down to up.
% Theta_Z = is counter clockwise.
%%%%%%%%%%%%%%%%%%%%%%%%%%%%%%%%%%%%%%%%%%%%%%%%%%%%%%%%%%%%%%%%%%%%%%%%%%%%%%%%%%%%
%%%%%% Define FE Mesh %%%%%%%%%%%%%%%%%%%%%%%%%%%%%%%%%%%%%%%%%%%%%%%%%%%%%%%%%%%%%%%%%%%%%
NXE = 3; % No of elements.
N  = 6; % Total no of degree of freedom excluding B.Cs ones.
NN  = 4; % No of nodes in the mesh
NR  = 2; % No of restrained B.C.

%%%%%%%%%%%%%%%%%%%%%%%%%%%%%%%%%%% Enter Node (x1,y1) Location & Element Node Numbering
NODE(:,1) = 12*[0; 0; 30; 30]; %X-Coord. of each node.
NODE(:,2) = 12*[0; 12; 12; 0]; %Y-Coord. of each node.
% EL_table = [Element No., First Node No., Second Node No., Real_Constants No, Materi-
al_Constants No. ]
EL_table = [ 1 1 2 1 1;
         2 2 3 1 1;
         3 3 4 1 1];

%%%%%%%%%%%%%%%%%%%%%%%%%%%%%%%%%%%%%% Define Real constants of the cross section properties of
the 2-D BEAMS %%%%%%%%%%%%%%%%%%%%%%%%%%%%%%%%%%%%%%%%%%%%%%%%%%%%%%%%%%
% %% You can define any No. of Real_Constants Sets according to your problem % Enter Dist. Type
of each variable %%%%%%%%%%%%%%%%%%%%%%%%%%%%%%%%%%%%%%%%%%%%%%%%
 % 0 = Normal.
 % 1 = Lognormal.
 % 2 = Type-I.
 % Real_Constants[Set No.] = [Area_mean, A_COV, I_mean, I_COV, Z_mean, Z_COV, Dist_A,
Dist_I, Dist_Z]
NREAL = 1;   % No. of real constants sets
Real_Constants(1,1:9) = [ 4.41 0.05  68.9 0.05  16 0.05  0 0 0];
Real_Constants(2,1:9) = [ 5   0.07  70  0.06  20 0.04  0 0 0];
 % You can define any No. of Real_Constants Sets

%%%%%%%%%%%%%%%%%%%%%%%%%%%%%% Define Sets of Material Properties
% Material_Properties = [E_mean, E_COV, Fy_mean, Fy_COV, Density_mean, Density_COV]
% Material_Properties[Set. no.] = [E_mean, E_COV, Fy_mean, Fy_COV, Density_mean, Densi-
ty_cov, Dist_E, Dist_Fy, Dist_density]
NM =1;    % No of Material Properties Sets.
Material_Properties(1,1:9) = [ 29e3 0.06 39.6 0.11 0 0 0 0 0];
Material_Properties(2,1:9) = [ 39e3 0   30    0 0 0  0 0 0];
```

```
%%%%%%%%%%%% DO NOT CHANGE THIS PART %%%%%%%%
for i = 1:NXE
   STOREC(i,1:4) = [ NODE(EL_table(i,2),:)   NODE(EL_table(i,3),:)];
end

KV2(1:N,1:N) = 0 ;
NF(1:NN,1:NODOF) = 1;
G(1:IKM*NODOF) = 0;
%%%%%%%%%%%%%%%%%%%%%%%%%%%%%%%%%%%%%%%%%%%%
% Define Node B.C. , 0 = fixed , 1 = free
 NF(1,:) =[0,0,0];
 NF(4,:) =[0,0,0];

%%%%%%% Define Loads Sets
% Loads_Set[Set No.] = [Key, Value_mean, Value_COV, Distribution Type, Dir-X, Dir_Y, Dir_Z,
Node No.(if concentrated load), Element No. , Angle between Load and the Beam ];
% Key = 0  - Concentrated Load.
% Key = 1  - Distributed Load.
% Enter Dist. Type of each variable %%%%%%%%%%%%%%%%%%%%%%%%
 % 0 = Normal.
 % 1 = Lognormal.
 % 2 = Type-I.
NLS = 3;      % No. of Load Sets.
Loads_Set(1,1:10) = [ 1 0.44/12 0.1  0    0  0  0    0  2 90];
Loads_Set(2,1:10) = [ 1 0.05/12 0.25 2    0  0  0    0  2 90];
Loads_Set(3,1:10) = [ 0 0.41    0.1  2    1  0  0    2  1  0];

% Enter Element No. with its Node no. that Reliability of Failure is needed to be calculated.

  Rel_E_N = [ 3  3];  % [Element No. ,  Node No.]

%%%%%%%%%%%%%%%%%%%%% DO NOT change this part
  DVn = 0;

     for iL=1:NLS      % No of Load sets
      if Loads_Set(iL,3) ~= 0
         DVn = DVn + 1; dist(DVn) = Loads_Set(iL,4);
         Xs(1,DVn) = Loads_Set(iL,2);    % Mean value
         Xs(2,DVn) = Loads_Set(iL,3);    % COV value
      end
     end

     for ir = 1:NREAL
        if Real_Constants(ir,2)~= 0
          DVn = DVn + 1; dist(DVn) = Real_Constants(ir,7);
          Xs(1,DVn) = Real_Constants(ir,1);    % Mean value
```

```
      Xs(2,DVn) = Real_Constants(ir,2);    % COV value
    end
    if Real_Constants(ir,4)~= 0
      DVn = DVn + 1; dist(DVn) = Real_Constants(ir,8);
      Xs(1,DVn) = Real_Constants(ir,3);    % Mean value
      Xs(2,DVn) = Real_Constants(ir,4);    % COV value
    end
    if Real_Constants(ir,6)~= 0
      DVn = DVn + 1; dist(DVn) = Real_Constants(ir,9);
      Xs(1,DVn) = Real_Constants(ir,5);    % Mean value
      Xs(2,DVn) = Real_Constants(ir,6);    % COV value
    end
  end

  for im = 1:NM
    if Material_Properties(im,2)~= 0
      DVn = DVn + 1; dist(DVn) = Material_Properties(im,7);
      Xs(1,DVn) = Material_Properties(im,1);    % Mean value
      Xs(2,DVn) = Material_Properties(im,2);    % COV value
    end
    if Material_Properties(im,4)~= 0
      DVn = DVn + 1; dist(DVn) = Material_Properties(im,8);
      Xs(1,DVn) = Material_Properties(im,3);    % Mean value
      Xs(2,DVn) = Material_Properties(im,4);    % COV value
    end
    if Material_Properties(im,6)~= 0
      DVn = DVn + 1; dist(DVn) = Material_Properties(im,9);
      Xs(1,DVn) = Material_Properties(im,5);    % Mean value
      Xs(2,DVn) = Material_Properties(im,6);    % COV value
    end
  end
```

BEAM2D – List

```
%function [Disp2,Eforce] =
Beam2D(IKM,IDOF,NODOF,NXE,N,NN,NR,EI,EA,STOREC,NF,EL_table,KV2,LOADS);
function [Disp2,Eforce,STOREG] =
Beam2D(IKM,IDOF,NODOF,NXE,N,NN,NR,E,I,A,STOREC,NF,EL_table,KV2,LOADS)

% NODE FREEDOM DATA
NF = READNF(NN,NODOF,NR,NF);

% GLOBAL STIFFNESS MATRIX ASSEMBLY
for IP=1:NXE
```

```
  %[KM] = BMCOL2(EI,EA,IP,STOREC)
  [KM,KMe,Gam] = BMCOL2(E,I,A,IP,STOREC);  % Ke is the KM-without rotation
  G = GSTRNG(EL_table,IP,NODOF,NF);
  STOREG(IP,:) = G;
  [KV2] = FORMKV(KV2,KM,IKM,G,N,NR,IDOF);

end
STOREG;
Disp2 = inv(KV2)*LOADS';
Disp2;

%RETRIEVE ELEMENT END FORCES AND MOMENTS

for IP=1:NXE

  [KM,KMe,Gam] = BMCOL2(E,I,A,IP,STOREC);   % Calcluate Orginal KM-element without rota-
tion

  for i=1:IDOF
    if (STOREG(IP,i)==0)
      Edisp(i)=0;
    else
      Edisp(i) = Disp2(STOREG(IP,i));
    end
  end
  G1 = STOREG(IP,:); Gmatrix(1:IDOF,1:N) = 0;
  for i2=1:IDOF
   if G1(i2)~=0, Gmatrix(i2,G1(i2)) = 1;, end;   % Gmatrix to transform from Global Displacement in
Global-Coord to Global Displ in Element-Coord
  end

  Edisp;
  Edisp2 = Gmatrix*Disp2;

  X1=STOREC(IP,1); Y1=STOREC(IP,2);
  X2=STOREC(IP,3); Y2=STOREC(IP,4);
  L=sqrt( (Y2-Y1)^2+(X2-X1)^2 ) ;
  C1=(X2-X1)/L ;
  S1=(Y2-Y1)/L ;

  gam = [[C1, S1, 0]; [-S1, C1, 0];[0, 0, 1]];
  Pm(1:6,1:6) = 0; Pm(1:3,1:3)=gam; Pm(4:6,4:6) = gam;

  Edispe = Pm*Edisp';
  Efor = KMe*Edispe;
  Eforce(IP,1:IDOF) = Efor';
end
```

DPEROFORMANCE _FUNCTION

```
function [DgDX] = Dperfor-
mance_function(IKM,IDOF,NODOF,NXE,N,NR,STOREG,S,X,Disp,DLOADS,Rel_E_N,EL_table,C
OORD,E2,I2,A2,Z2,Fy2,NLS,NREAL,NM,Loads_Set,Real_Constants,Material_Properties)
% R(1) = Pn, R(2) = Mn
% S(1) = Pu, S(2) = Mu
phi = 1/1.25;
S;
X;
%%%%%%%%%% Determine Properties of the Reliability element
IP = Rel_E_N(1,1);
X1=COORD(IP,1); Y1=COORD(IP,2);
X2=COORD(IP,3); Y2=COORD(IP,4);
L =sqrt( (Y2-Y1)^2+(X2-X1)^2 );
%%%%%%%%%%%%%%%%%%%%%%%%%%%%%%%%%%%%%A1 = A2(IP); I1 =
I2(IP); Z1= Z2(IP); E1 = E2(IP); Fy = Fy2(IP);
Pu = abs(S(1)); Mu = abs(S(2));

Keff = 1;
r1=(I1/A1)^0.5;
Lamd = ( Keff*L/(r1*pi) )*(Fy/E1)^0.5

if (S(1) > 0)    % it means the force is in the positive direction of the element = compression
  if (Lamd <= 1.5)

    Fcr = (0.658^(Lamd^2)) * Fy ;
    Pn = A1*Fcr;

    DPnDE = 0.41855*A1*0.658^(Keff^2*L^2*(Fy/E1)^0.10e1*(I1/A1)^(-
0.10e1)/pi^2)*Keff^2*L^2*(I1/A1)^(-0.10e1)/pi^2*Fy^2/E1^2;
    DPnDA = 0.658^(Keff^2*L^2*(Fy/E1)^0.10e1*(I1/A1)^(-0.10e1)/pi^2)*Fy -
0.4185503477e0/A1*0.658e0^(Keff^2*L^2*(Fy/E1)^0.10e1*(I1/A1)^(-
0.10e1)/pi^2)*Keff^2*L^2*(Fy/E1)^0.10e1*(I1/A1)^(-0.20e1)/pi^2*I1*Fy;
    DPnDI = 0.4185503477e0*0.658e0^(Keff^2*L^2*(Fy/E1)^0.10e1*(I1/A1)^(-
0.10e1)/pi^2)*Keff^2*L^2*(Fy/E1)^0.10e1*(I1/A1)^(-0.20e1)/pi^2*Fy;
    DPnDFy = -0.4185503477e0*A1*0.658e0^(Keff^2*L^2*(Fy/E1)^0.10e1*(I1/A1)^(-
0.10e1)/pi^2)*Keff^2*L^2*(I1/A1)^(-
0.10e1)/pi^2/E1*Fy+A1*0.658e0^(Keff^2*L^2*(Fy/E1)^0.10e1*(I1/A1)^(-0.10e1)/pi^2);
  else

    Fcr = (0.877/(Lamd^2))* Fy
    Pn = A1*Fcr

    DPnDE = 0.8770e0*A1/Keff^2/L^2*(Fy/E1)^(-0.20e1)*(I1/A1)^0.10e1*pi^2*Fy^2/E1^2;
```

```
   DPnDA = 0.877e0/Keff^2/L^2*(Fy/E1)^(-0.10e1)*(I1/A1)^0.10e1*pi^2*Fy -
0.8770e0/A1/Keff^2/L^2*(Fy/E1)^(-0.10e1)*pi^2*Fy*I1;
   DPnDI = 0.8770e0/Keff^2/L^2*(Fy/E1)^(-0.10e1)*pi^2*Fy;
   DPnDFy = -0.8770e0*A1/Keff^2/L^2*(Fy/E1)^(-0.20e1)*(I1/A1)^0.10e1*pi^2*Fy/E1 +
0.877e0*A1/Keff^2/L^2*(Fy/E1)^(-0.10e1)*(I1/A1)^0.10e1*pi^2;

   end
  else

   Pn = A1*Fy;
   DPnDE =0;
   DPnDA = Fy;
   DPnDFy = A1;
   DPnDI = 0;

end

Mn = Z1*Fy;
Pn
DPnDX1 = 0; DPnDX2 = 0; DPnDX3 = 0; DPnDZ = 0;
DMnDX1 = 0; DMnDX2 = 0; DMnDX3 = 0;
DMnDA = 0; DMnDI = 0; DMnDZ = Fy; DMnDE = 0; DMnDFy = Z1;
%Jr = [ DPnDX1 DPnDX2 DPnDX3 DPnDA DPnDI DPnDZ DPnDE DPnDFy;
% DMnDX1 DMnDX2 DMnDX3 DMnDA DMnDI DMnDZ DMnDE DMnDFy]' % Transpose
to make it (8x2) matrix.

DVn1 = 0;
for iL=1:NLS     % No of Load sets
     if Loads_Set(iL,3) ~= 0
        DVn1=DVn1+1;
        Jr(DVn1,1) = DPnDX1;  Jr(DVn1,2) = DMnDX1;
     end
    end
 for ir = 1:NREAL
       if Real_Constants(ir,2)~= 0     % Write KG w.r.t. Area of Real_const=ir
          DVn1 = DVn1 + 1;
          Jr(DVn1,1) = DPnDA;  Jr(DVn1,2) = DMnDA;
       end
       if Real_Constants(ir,4)~= 0     % Write KG w.r.t. Moment of Interia of Real_const=ir
          DVn1 = DVn1 + 1;
          Jr(DVn1,1) = DPnDI;  Jr(DVn1,2) = DMnDI;
       end
       if Real_Constants(ir,6)~= 0     % Write KG w.r.t. Z of Real_const=ir
          DVn1 = DVn1 +1;
          Jr(DVn1,1) = DPnDZ;  Jr(DVn1,2) = DMnDZ;
       end
```

```
    end

    for im = 1:NM
      if Material_Properties(im,2)~= 0        % Write KG w.r.t. E of Real_const=ir
        DVn1 = DVn1 + 1;
        Jr(DVn1,1) = DPnDE;  Jr(DVn1,2) = DMnDE;
      end
      if Material_Properties(im,4)~= 0        %Write KG w.r.t. Fy of Real_const=ir
        DVn1 = DVn1 + 1;
        Jr(DVn1,1) = DPnDFy;  Jr(DVn1,2) = DMnDFy;
      end
      if Material_Properties(im,6)~= 0        %Write KG w.r.t. Density of Real_const=ir
        DVn1 = DVn1 + 1;
        Jr(DVn1,1) = 0;  Jr(DVn1,2) = 0;
      end
    end
if (Pu/(phi*Pn)) >= 0.2
                DgDPn = Pu/Pn^2;
                DgDMn = (8/9)*Mu/(Mn^2);

                Dgr = [DgDPn; DgDMn];          % (2x1) matrix.

                DgDPu = -1/Pn;
                DgDMu = -8/(9*Mn);

                Dgs = [DgDPu; DgDMu];          % (2x1) matrix.
else

    DgDPn = Pu/(2*(Pn)^2);
                DgDMn = Mu/(Mn^2);

                Dgr = [DgDPn ; DgDMn];

                DgDPu = -1/(2*Pn);
                DgDMu = -1/Mn;

                Dgs = [DgDPu; DgDMu];
end

%%%%%%%%%%%%%% Evaluate Qt Matrix & its Dervative
KMG(1:N,1:N) = 0; KdAG = KMG; KdIG = KMG; KdEG = KMG; Qt(1:2,1:N) = 0;
 [KM,KMe,Gam]  = BMCOL2(E2,I2,A2,IP,COORD);
[KdA,KdAe] = BMCOL2KdA(E2,I2,A2,IP,COORD);
[KdI,KdIe] = BMCOL2KdI(E2,I2,A2,IP,COORD);
[KdE,KdEe] = BMCOL2KdE(E2,I2,A2,IP,COORD);
```

```
NODE(1:2) = EL_table(IP,2:3);
G1 = STOREG(IP,:); Gmatrix(1:IDOF,1:IDOF) = 0;
for i2=1:IDOF
  if G1(i2)~=0, Gmatrix(i2,G1(i2)) = 1;, end;
end

FORCE_tra = KMe*Gam*Gmatrix*Disp;
 Q1 = KMe*Gam*Gmatrix;
Q2 = KdAe*Gam*Gmatrix;
Q3 = KdIe*Gam*Gmatrix;
Q4 = KdEe*Gam*Gmatrix;
 DVn1=0;
    for iL=1:NLS      % No of Load sets
     if Loads_Set(iL,3) ~= 0
        DVn1=DVn1+1;
        DQtDX(DVn1,1:2,1:N) = 0;
     end
    end

    for ir = 1:NREAL
      if Real_Constants(ir,2)~= 0     % Write DQt w.r.t. Area of Real_const=ir
        DVn1 = DVn1 + 1;
        if Rel_E_N(1,2) == NODE(1)
          DQtDX(DVn1,1,1:N) = Q2(1,1:N); DQtDX(DVn1,2,1:N) = Q2(3,1:N);
        else
          DQtDX(DVn1,1,1:N) = Q2(4,1:N); DQtDX(DVn1,2,1:N) = Q2(6,1:N);
        end
      end
      if Real_Constants(ir,4)~= 0     % Write DQt w.r.t. Moment of Interia of Real_const=ir
        DVn1 = DVn1 + 1;
        if Rel_E_N(1,2) == NODE(1)
          DQtDX(DVn1,1,1:N) = Q3(1,1:N); DQtDX(DVn1,2,1:N) = Q3(3,1:N);
        else
          DQtDX(DVn1,1,1:N) = Q3(4,1:N); DQtDX(DVn1,2,1:N) = Q3(6,1:N);
        end
      end
      if Real_Constants(ir,6)~= 0
        DVn1 = DVn1 +1;
        DQtDX(DVn1,1:2,1:N) = 0;
      end
    end

    for im = 1:NM
      if Material_Properties(im,2)~= 0
        DVn1 = DVn1 + 1;
        if Rel_E_N(1,2) == NODE(1)
```

```
            DQtDX(DVn1,1,1:N) = Q4(1,1:N); DQtDX(DVn1,2,1:N) = Q4(3,1:N);
          else
            DQtDX(DVn1,1,1:N) = Q4(4,1:N); DQtDX(DVn1,2,1:N) = Q4(6,1:N);
          end
        end
        if Material_Properties(im,4)~= 0
          DVn1 = DVn1 + 1;
          DQtDX(DVn1,1:2,1:N) = 0;
        end
        if Material_Properties(im,6)~= 0
          DVn1 = DVn1 + 1;
          DQtDX(DVn1,1:2,1:N) = 0;
        end
      end

   if Rel_E_N(1,2) == NODE(1)    % First nodi is required, so we choose 1st & 3rd rows of KM only
and the rest is zero
      Qt(1,1:N) = Q1(1,:);  Qt(2,1:N) = Q1(3,:);
   else     % second node is required, so we choose 2nd & 4th rows of KM only and the rest is zero
      Qt(1,1:N) = Q1(4,:);  Qt(2,1:N) = Q1(6,:);
   end

%%%%%%%%% Dervative of the LOADS Vector

%%%%%%%%%%%%%%%%%%%%%%%%%% Evaluate Dervative of S wrt X   KMG(1:N,1:N) =
0; KdAG = KMG; KdIG = KMG; KdEG = KMG;
for i1=1:NXE

   [KM,KMe,Gam]  = BMCOL2(E2,I2,A2,i1,COORD);
   [KdA,KdAe] = BMCOL2KdA(E2,I2,A2,i1,COORD);
   [KdI,KdIe] = BMCOL2KdI(E2,I2,A2,i1,COORD);
   [KdE,KdEe] = BMCOL2KdE(E2,I2,A2,i1,COORD);
   G = STOREG(i1,:);

   [KMG]  = FORMKV(KMG,KM,IKM,G,N,NR,IDOF);
   [KdAG] = FORMKV(KdAG,KdA,IKM,G,N,NR,IDOF);
   [KdIG] = FORMKV(KdIG,KdI,IKM,G,N,NR,IDOF);
   [KdEG] = FORMKV(KdEG,KdE,IKM,G,N,NR,IDOF);
end

%%%%%% Dervative of the K-Global Matrix
DVn1=0;
      for iL=1:NLS     % No of Load sets
       if Loads_Set(iL,3) ~= 0
          DVn1=DVn1+1;
          DKDX(DVn1,1:N,1:N) = 0;
```

```
     end
    end

    for ir = 1:NREAL
      if Real_Constants(ir,2)~= 0    % Write KG w.r.t. Area of Real_const=ir
        DVn1 = DVn1 + 1;
        DKDX(DVn1,1:N,1:N) = KdAG;
      end
      if Real_Constants(ir,4)~= 0    % Write KG w.r.t. Moment of Interia of Real_const=ir
        DVn1 = DVn1 + 1;
        DKDX(DVn1,1:N,1:N) = KdIG;
      end
      if Real_Constants(ir,6)~= 0    % Write KG w.r.t. Z of Real_const=ir
        DVn1 = DVn1 +1;
        DKDX(DVn1,1:N,1:N) = 0;
      end
    end

    for im = 1:NM
      if Material_Properties(im,2)~= 0        % Write KG w.r.t. E of Real_const=ir
        DVn1 = DVn1 + 1;
        DKDX(DVn1,1:N,1:N) = KdEG;
      end
      if Material_Properties(im,4)~= 0        %Write KG w.r.t. Fy of Real_const=ir
        DVn1 = DVn1 + 1;
        DKDX(DVn1,1:N,1:N) = 0;
      end
      if Material_Properties(im,6)~= 0        %Write KG w.r.t. Density of Real_const=ir
        DVn1 = DVn1 + 1;
        DKDX(DVn1,1:N,1:N) = 0;
      end
    end

  QtKinv = Qt*inv(KMG);

for i3=1:DVn1
   DSDX = 0;
   DQt(:,:) = DQtDX(i3,:,:); DKG(:,:) = DKDX(i3,:,:)
   DSDX = DQt*Disp + QtKinv* ( DLOADS(:,i3) - DKG*Disp)
   Js(i3,:) = DSDX';
end

DgDX = Jr*Dgr + Js*Dgs;
```

BMCOL2 – List

```
function [KM,K,Pm] = BMCOL2(E1,I1,A1,IP,COORD)
%function [KM] = BMCOL2(EI1,EA1,IP,COORD)
%THIS SUBROUTINE FORMS THE STIFFNESS MATRIX OF AN INCLINED 2-D BEAM-
COLUMN ELEMENT

E = E1(IP); A=A1(IP); I = I1(IP);
%EI = EI1(IP); EA = EA1(IP);
X1=COORD(IP,1); Y1=COORD(IP,2);
X2=COORD(IP,3); Y2=COORD(IP,4);
L=sqrt( (Y2-Y1)^2+(X2-X1)^2 );
C1=(X2-X1)/L ;
S1=(Y2-Y1)/L ;

gam = [[C1, S1, 0]; [-S1, C1, 0];[0, 0, 1]];

Pm(1:6,1:6) = 0; Pm(1:3,1:3)=gam; Pm(4:6,4:6) = gam;

K=[ [(E*A)/L,0,0,(-E*A)/L,0,0];
 [0,(12*E*I)/(L^3),(6*E*I)/(L^2),0,(-12*E*I)/(L^3),(6*E*I)/(L^2)];
 [0,(6*E*I)/(L^2),(4*E*I)/(L),0,(-6*E*I)/(L^2),(2*E*I)/(L)];
 [(-E*A)/(L),0,0,(E*A)/(L),0,0];
 [0,(-12*E*I)/(L^3),(-6*E*I)/(L^2),0,(12*E*I)/(L^3),(-6*E*I)/(L^2)];
 [0,(6*E*I)/(L^2),(2*E*I)/(L),0,(-6*E*I)/(L^2),(4*E*I)/(L)] ];

KM = Pm'*K*Pm;
```

5.5 Program Sensitivity-Based Analysis for Implicit Performance Functions for 3D Beam Elements

The MATLAB program that was written for the 2D beam elements was modified to fit 3D frame structures. The modifications are as follows:

- BMCOL3 Evaluate element K-matrix for 3D beam element.
- MMCOL3 Evaluate element Mass-matrix for 3D beam element.
- Input3R Includes the files that link the program to ANSYS, load input data, real constants, and material properties. Types of loading are:
  - Concentrated load at nodes.
  - Thermal load (Element temperature).
  - Acceleration loading.
- FORM3 [11] main body of the program that applies the FORM method. The program for the 2D Beam element used partial derivatives to calculate

derivative of the performance function with respect to the random variables, where for the 3D frame structures, finite difference is used instead due to memory restriction, as the partial derivative requires derivative of the stiffness matrix of each element with respect to the random variables.

FORM Method-3

```
% Step-1: Define Random variables parameters [mean - std - distribution
% type & FE Mesh
 clear all
 format short e
  Input3R
% Step-2: Define initial design point xo
  xm0(1:DVn) = Xs(1,1:DVn); cov0(1:DVn) = Xs(2,1:DVn);
  for i1=1:DVn
   std0(i1) = cov0(i1) * xm0(i1);
  end
  xstar(1:DVn) = xm0(1:DVn);
  Beta_old=0;

 for k=1:100

% Calculate response parameters and g-value for the current random variables.

   [No_r,No_m,Ty,Tz,A,Iy,Iz,Alphe,E,Fy,Density,Alpha_T] = Proper-
ties(NXE,EL_table,Real_Constants,Material_Properties);

   for j1=1:1e4   % Loop to increase value of Load to make 'g( ) < 0.7' for convergence reason. done
for the first loop only.
       [LOADS,DLOADS] =
LOADS1(IDOF,NN,NR,N,NXE,NLS,DVn,NODOF,EL_table,NF,Loads_Set,Alphe,STOREC,E,A,Alp
ha_T,T_ref,Density);
       [Disp2,Eforce,STOREG,NF] =
Beam3D(IKM,IDOF,NODOF,NXE,N,NN,NR,E,Iy,Iz,A,Alphe,STOREC,NF,EL_table,LOADS,T_ref,
Loads_Set,Alpha_T,NLS);
       [S,Rel_E_N,Stress,S_DIR] = Smatrix(Eforce,EL_table,NXE,A,Iy,Iz,Ty,Tz);
       [gvalue] = performance_function(S,xstar,Rel_E_N,STOREC,Fy);
       gvalue

       if (abs(gvalue) < 0.75) | (k > 1) , break, end;
       DVnL=0;
       for iL=1:NLS      % No of Load sets
           if Loads_Set(iL,3) ~= 0
              DVnL = DVnL + 1;
              Loads_Set(DVnL,2) = 1.25*Loads_Set(DVnL,2);    % Mean value
```

```
            xm0(DVnL) = Loads_Set(DVnL,2); std0(DVnL) = cov0(DVnL) * xm0(DVnL);
            xstar(DVnL) = xm0(DVnL);
          end
      end
  end

% Step-3: Define mean and std of the equivalent normal dist. And evaluate
% the equivalent normal variables

  [xm,std] = meanstd(DVn,xstar,xm0,std0,dist);
  Bmatrix(1:DVn,1:DVn) = 0;
  for i2=1:DVn
     Bmatrix(i2,i2) = 1/std(i2);
  end
  for i=1:DVn
     xstar_dash(i) = (xstar(i) - xm(i))/std(i);
  end
 % Step-4: Compute partial derivative at design point x_star

%[dgdx]=Dperformance_function(IKM,IDOF,NODOF,NXE,N,NR,STOREG,S,xstar,Disp2,DLOADS
,Rel_E_N,EL_table,STOREC,E,Iy,Iz,A,Zy,Zz,Fy,Alphe,NLS,NREAL,NM,Loads_Set,Real_Constants,
Material_Properties);
  for i3=1:DVn
     Xv = xstar;
     dx = std0(i3)/10;
     Xv(i3) = xstar(i3) + dx;
     [Loads_Set,Real_Constants,Material_Properties] = Con-
vert(NLS,NREAL,NM,Xv,Loads_Set,Real_Constants,Material_Properties);
     [No_r,No_m,Ty,Tz,A,Iy,Iz,Alphe,E,Fy,Density,Alpha_T] = Proper-
ties(NXE,EL_table,Real_Constants,Material_Properties);
     [LOADS,DLOADS] =
LOADS1(IDOF,NN,NR,N,NXE,NLS,DVn,NODOF,EL_table,NF,Loads_Set,Alphe,STOREC,E,A,Alp
ha_T,T_ref,Density);
     [Disp2,Eforce,STOREG] =
Beam3D(IKM,IDOF,NODOF,NXE,N,NN,NR,E,Iy,Iz,A,Alphe,STOREC,NF,EL_table,LOADS,T_ref,
Loads_Set,Alpha_T,NLS);
     [S2,Rel_E_N] = Smatrix(Eforce,EL_table,NXE,A,Iy,Iz,Ty,Tz);
     [gvalue2] = performance_function(S2,xstar,Rel_E_N,STOREC,Fy);
     dgdx(i3) = (gvalue2-gvalue)/dx;
     [Loads_Set,Real_Constants,Material_Properties] = Con-
vert(NLS,NREAL,NM,xstar,Loads_Set,Real_Constants,Material_Properties);

  end
  [NLS]=sensitivity(NLS,NREAL,NM,Xs,Loads_Set,Real_Constants,Material_Properties,dgdx)
```

```
% Step-5: Compute partial derivative in the equivalent normal space
  dgdxstar_dash = [inv(Bmatrix)] * dgdx';
  alpha1 = dgdxstar_dash/norm(dgdxstar_dash)

% Step-6: Evaluate coordinates of the new design point in the equivalent
% normal space using eqn 3.48

 xstar_dash_new = (1/norm(dgdxstar_dash))^2 * ( DOT(dgdxstar_dash,xstar_dash) - gvalue
)*dgdxstar_dash;

% Step-7: Evaluate Beta - distance to the new design values
Beta = norm(xstar_dash_new);
delta_Beta = abs(Beta - Beta_old);
Beta_old = Beta;

% Step-8: Evaluate new design values in the orginal space
  for i=1:DVn
     xstar(i) = xm(i) + std(i)*xstar_dash_new(i);
  end
 [Loads_Set,Real_Constants,Material_Properties] = Con-
vert(NLS,NREAL,NM,xstar,Loads_Set,Real_Constants,Material_Properties);
             xstar_dash_k(:,k) = xstar_dash';
             xstar_dash_new_k(:,k) = xstar_dash_new;
             xm_k(:,k) = xm';
             std_k(:,k) = std';
             dgdx_k(:,k) = dgdx';
             dgdxstar_dash_k(:,k) = dgdxstar_dash;
             gvalue_k(k) = gvalue;
             Beta_k(k) = Beta;
             xstar_k(:,k) = xstar';

             if (abs(gvalue) < 0.0001) & (delta_Beta < 0.001)
       break
   end
end
```

Subroutine to Assemble Load Vector

```
function [LV2] = FORMLV(LV2,LM,G,N,IDOF)
% C C THIS SUBROUTINE FORMS THE GLOBAL STIFFNESS MATRIX

   for J =1:IDOF
     if (G(J) == 0) ,continue ,end
     LV2(G(J)) = LV2(G(J)) + LM(J);
   end
```

Subroutine to Calculate Performance function

```
function [gv] = performance_function(S,X,Rel_E_N,COORD,Fy)
IP = Rel_E_N(1);
X1=COORD(IP,1); Y1=COORD(IP,2); Z1 = COORD(IP,3);
X2=COORD(IP,4); Y2=COORD(IP,5); Z2 = COORD(IP,6);
L=sqrt( (Y2-Y1)^2 + (X2-X1)^2 + (Z2-Z1)^2 );
gv = 1 - (S / Fy(IP));
```

Subroutine to Calculate extract Statistical Parameters from Input File

```
function [Loads_Set,Real_Constants,Material_Properties] = Con-
vert(NLS,NREAL,NM,Xs,Loads_Set,Real_Constants,Material_Properties)
    DVn=0;
    for iL=1:NLS     % No of Load sets
     if Loads_Set(iL,3) ~= 0
       DVn = DVn + 1;
       Loads_Set(iL,2)=Xs(DVn);    % Mean value
    end
    end

    for ir = 1:NREAL
      if Real_Constants(ir,2)~= 0
        DVn = DVn + 1;
        Real_Constants(ir,1)=Xs(DVn);    % Mean value

      end
      if Real_Constants(ir,4)~= 0
        DVn = DVn + 1;
        Real_Constants(ir,3)=Xs(DVn);    % Mean value

      end
    end

    for im = 1:NM
      if Material_Properties(im,2)~= 0
        DVn = DVn + 1;
        Material_Properties(im,1)=Xs(DVn);    % Mean value

      end
      if Material_Properties(im,4)~= 0
        DVn = DVn + 1;
        Material_Properties(im,3)=Xs(DVn);    % Mean value

      end
      if Material_Properties(im,6)~= 0
```

```
        DVn = DVn + 1;
        Material_Properties(im,5)=Xs(DVn);    % Mean value

      end
      if Material_Properties(im,8)~= 0
        DVn = DVn + 1;
        Material_Properties(im,7)=Xs(DVn);    % Mean value

      end
    end
```

Subroutine to Calculate sensitivity of performance function to random variables

```
function [NLS] = sensitivi-
ty(NLS,NREAL,NM,Xs,Loads_Set,Real_Constants,Material_Properties,dgdx)

DVn = 0;
for iL=1:NLS      % No of Load sets
     if Loads_Set(iL,3) ~= 0
       DVn = DVn + 1;
       fprintf(1,'Sensitivity of performance function w.r.t Load Set %i = %f \n', iL,dgdx(DVn))
     end
    end

    for ir = 1:NREAL
      if Real_Constants(ir,2)~= 0
        DVn = DVn + 1;
        fprintf(1,'Sensitivity of performance function w.r.t Ty of real constant set %i = %f \n',
ir,dgdx(DVn))
      end
      if Real_Constants(ir,4)~= 0
        DVn = DVn + 1;
        fprintf(1,'Sensitivity of performance function w.r.t Tz of real constant set %i = %f \n',
ir,dgdx(DVn))
      end
   end

    for im = 1:NM
      if Material_Properties(im,2)~= 0
        DVn = DVn + 1;
        fprintf(1,'Sensitivity of performance function w.r.t E of Material set %i = %f \n',
im,dgdx(DVn))
      end
      if Material_Properties(im,4)~= 0
        DVn = DVn + 1;
        fprintf(1,'Sensitivity of performance function w.r.t Fy of Material set %i = %f \n',
im,dgdx(DVn))
```

```
    end
    if Material_Properties(im,6)~= 0
      DVn = DVn + 1;
      fprintf(1,'Sensitivity of performance function w.r.t Density of Material set %i = %f \n',
im,dgdx(DVn))
    end
    if Material_Properties(im,8)~= 0
      DVn = DVn + 1;
      fprintf(1,'Sensitivity of performance function w.r.t Thermal_exp_coefficient of Material set %i
= %f \n', im,dgdx(DVn))
    end
  end
```

Subroutine to convert between random states to loads, material, and real constants

```
function [Loads_Set,Real_Constants,Material_Properties] = Convert(NLS,NREAL,NM,Xs,Loads_Set,F
Constants,Material_Properties)

  DVn=0;
  for iL=1:NLS     % No of Load sets
   if Loads_Set(iL,3) ~= 0
     DVn = DVn + 1;
     Loads_Set(iL,2)=Xs(DVn);   % Mean value

   end
  end
  for ir = 1:NREAL
    if Real_Constants(ir,2)~= 0
      DVn = DVn + 1;
      Real_Constants(ir,1)=Xs(DVn);    % Mean value

    end
    if Real_Constants(ir,4)~= 0
      DVn = DVn + 1;
      Real_Constants(ir,3)=Xs(DVn);    % Mean value

    end
  end
  for im = 1:NM
    if Material_Properties(im,2)~= 0
      DVn = DVn + 1;
      Material_Properties(im,1)=Xs(DVn);    % Mean value

    end
    if Material_Properties(im,4)~= 0
      DVn = DVn + 1;
      Material_Properties(im,3)=Xs(DVn);    % Mean value
```

```
  end
  if Material_Properties(im,6)~= 0
    DVn = DVn + 1;
    Material_Properties(im,5)=Xs(DVn);   % Mean value

  end
  if Material_Properties(im,8)~= 0
    DVn = DVn + 1;
    Material_Properties(im,7)=Xs(DVn);   % Mean value

  end
end
```

5.6 Test Case-1

The satellite structure used in the test case is a $1.03 \times 0.8 \times 1.0255\ \mathrm{m}^3$ with four solar panels attached to the satellite by means of rotation mechanism (Fig. 5.1). Four heat shields are installed on the satellite structure to prevent internal instruments from direct environmental heat loads. The highest power consuming components will be placed away from hot heat shields subjected to solar radiation. The main elements of the satellite structure are shown in Fig. 5.2.

Base plate of satellite structure is studied in this example. The Base plate will be simplified to 3D Beam elements, while the plate elements effect will be ignored. The structure material of the satellite will be assumed to behave in the elastic region. Material properties are listed in Table 5.1.

Loads on satellite structure are 10 g acceleration perpendicular to the plain of the base plate and are divided on eight nodes (as the primary structure components are assembled through eight studs). Satellite mechanical design is discussed in Chap. 4 using deterministic finite element analysis [7]. Deterministic temperature profile during the maximum load case during orbit was evaluated in Sect. 4.14 (Table 4.27) [8, 9].

Four load sets are defined, Load Set 1 (mean = −5.3), Load Set 2 (mean = −1.8), Load Set 3 (mean = 5.4), and Load Set 4 (mean = 7.2). Frame of base plate is meshed with 3D Beam elements and is shown in Fig. 5.3 [10].

Probability of failure is evaluated using the following performance function

$$g(x_1, x_2, \ldots, x_n) = 1 - \frac{\sigma}{\sigma_y}$$

$$\sigma_y(\text{mean}) = 150\ \text{MPa}$$

Frame of base plate is meshed with 3D Beam elements with two types of sets of real constants. Each set represents the following: thickness mean in the y-dir (t_y),

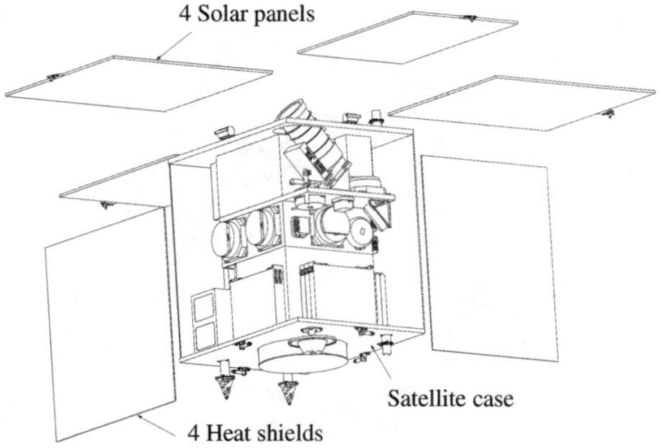

Fig. 5.1 Layout of the satellite under study [11]

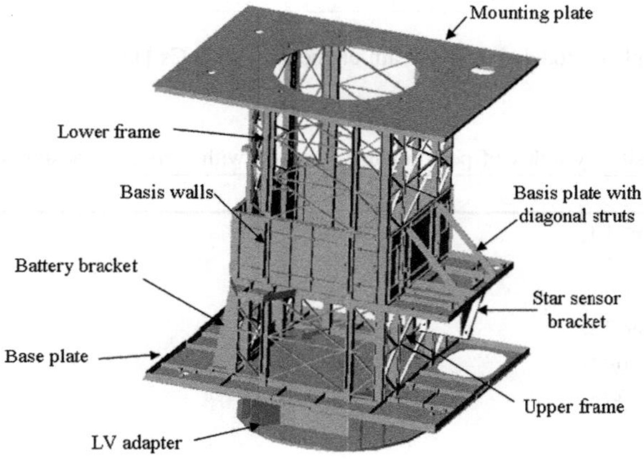

Fig. 5.2 3-D model of primary satellite structure [11]

Table 5.1 Mechanical and thermal material properties

Mechanical material properties	Modulus of elasticity, E = 69 (10^6) kg.mm/s^2/mm^2 Density, $\rho = 2.71$ (10^{-6}) kg.mm/s^2/mm^2
Thermal material properties	Thermal conductivity, K = 117 W/m.°C Coefficient of thermal expansion, $\alpha = 24.7 \times 10^{-6}$ K^{-1}

covariance of t_y, thickness mean in the z-dir (t_z), covariance of t_z, distribution type (Normla (0) or Lognormal (1)), and beam rotation angle around its center.

- Real_Constants(1,1:7) = [5 0.05 10 0.05 1 1 0];
- Real_Constants(2,1:7) = [7 0.05 14 0.05 1 1 0];

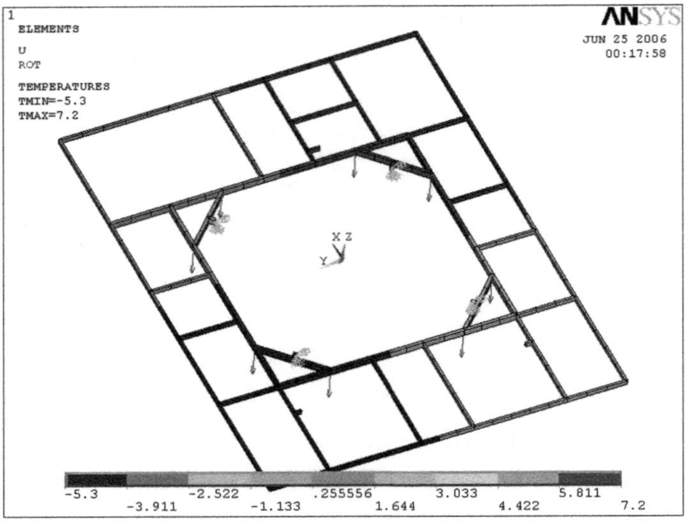

Fig. 5.3 Base plate modeled as 3D beam elements with B.Cs [11]

Table 5.2 Sensitivity index of performance function with respect to design space parameters

Variable	Sensitivity index
Real-constant set-1 t_y	0.117683
Real-constant set-1 tz	0.110029
Real-constant set-2 t_y	0.012153
Real-constant set-2 tz	0.017118
Modulus of Elasticity	−0.0000001
Material yield stress	0.000005
Load set-1	0.002447
Load set-1	0.000341
Load set-1	−0.000092
Load set-1	−0.003401

A set of material properties and acceleration load are defined as: modulus of elasticity mean, modulus of elasticity covariance, material yield stress (σ_y), σ_y covariance, material density mean, material density covariance, thermal expansion mean, thermal expansion covariance, modulus of elasticity distribution type, σ_y distribution type, density distribution type, thermal expansion distribution type.

– Material_Properties(1,1:12) = [72e6 0.06 150e3 0.11 2.63e-6 0 23e-6 0 1 1 1 1].

Simulations results are shown in Tables 5.2 and 5.3. The critical value of probability of failure is 0.5, so the satellite structure has safe design. The ratio of the calculated probability of failure to the its critical value proves that the factor of safety used during the deterministic FE analysis can be reduced.

Table 5.3 Reliability analysis of the satellite structure

No. of Iteration	Beta	Performance function	
1	2.9316	2.977 (10^{-1})	The probability of failure = 1−normcdf(2.4189) = 7.7838 (10^{-3})
2	2.4555	−1.2469 (10^{-1})	
3	2.4184	−8.3486 (10^{-3})	
4	2.4189	3.8566 (10^{-5})	

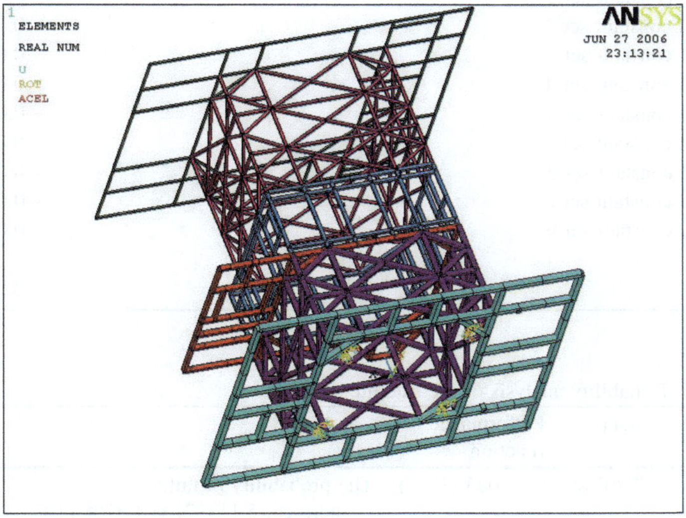

Fig. 5.4 Satellite structure modeled as 3-D beam elements with 6 real constants [11]

5.7 Test Case-2

The same satellite structure used in Test Case-1 is used here. The reliability analysis is performed on the whole satellite structure (Fig. 5.4) and not only on the base plate. Probability of failure is evaluated using the following performance function

$$g(x_1, x_2, \ldots, x_n) = 1 - \frac{\sigma}{\sigma_y}$$

$$\sigma_y = 150 \text{ MPa}$$

Similar to Test Case-1, the number of real constants used = 6;

- Real_Constants(1,1:7) = [25 0.05 15 0.05 1 1 0]; Base Plate
- Real_Constants(2,1:7) = [20 0.05 12 0.05 1 1 0]; Lower Frame
- Real_Constants(3,1:7) = [15 0.05 10 0.05 1 1 0]; Basis Plate

Table 5.4 Sensitivity index of performance function with respect to design space parameters

Variable	Sensitivity index
Load set 1	−0.000002
Load set 2	0.021617
Load set 3	−0.000379
Load set 4	−0.000078
Load set 5	0.003533
Ty of real constant set 1	−0.004883
Tz of real constant set 1	0.017350
Ty of real constant set 2	0.002037
Tz of real constant set 2	0.004685
Ty of real constant set 3	−0.001636
Tz of real constant set 3	−0.002339
Ty of real constant set 4	−0.002149
Tz of real constant set 4	−0.004768
Ty of real constant set 5	−0.005229
Tz of real constant set 5	−0.011481
Ty of real constant set 6	−0.003135
Tz of real constant set 6	−0.006266
E	0
Fy	0.000003

Table 5.5 Reliability analysis of the satellite structure

No. of Iteration	Beta	Performance function	
1	7.7051e	4.9003 (10^{-1})	The probability of failure = 1–normcdf(5.11367) = 1.5798 (10^{-7})
2	7.2552e	−1.4444 (10^{-1})	
3	5.6596e	−5.5430 (10^{-1})	
4	5.1767e	1.3402 (10^{-1})	
5	5.1368e	−1.0334e (10^{-2})	
6	5.11367e	2.8054e (10^{-5})	

- Real_Constants(4,1:7) = [12 0.05 6 0.05 1 1 0]; Basis Wall
- Real_Constants(5,1:7) = [11 0.05 5 0.05 1 1 0]; Upper Frame
- Real_Constants(6,1:7) = [10 0.05 5 0.05 1 1 0]; Mounting Plate
- Material_Properties(1,1:12) = [72e6 0.06 150e3 0.11 2.63e-6 0 23e-6 0 1 1 1 1];

Reference Temperature used = 20 C°.

Simulations results are shown in Tables 5.4 and 5.5. The critical value of probability of failure is 0.5, so the satellite structure has safe design. The ratio of the calculated probability of failure to the its critical value proves that the factor of safety used during the deterministic FE analysis can be reduced.

5.8 Conclusion

SFEM is used to evaluate sensitivity index of design space parameters with respect to the selected performance function and probability of failure of satellite structure under static and thermal loads. It is important for the design engineer to determine the structure response sensitivity to the random design variables (geometrical, material properties, loads). If the uncertainty in one of the design space variables has significant effect on the selected performance function, then it is important to reduce its uncertainty by collecting more data using testing or increase the value of safety factor. This will improve the design reliability and could help to reduce cost. Reducing cost can be achieved by assigning different safety factors for the design space parameters based on their uncertainty and sensitivity index. Sensitivity analysis can help the designer to eliminate the variables that have lower sensitivity index from the design optimization process. The response sensitivities are computed in this paper using the finite difference approach, by perturbing each design variable and computing the change in the performance function. The proposed approach is applied after simplifying the structure to a group of 3D beam elements and ignoring the plate effect.

References

1. Freudenthal AM (1956) Safety and probability of structural failure. ASCE Trans 121:1337–1397
2. Freudenthal AM, Garrelts JM, and Shinozuka M0 (1966) The analysis of structural safety. J Struct Eng ASCE 92(ST1):267–324
3. Halder A, Mahadeven S (2000) Reliability assessment using stochastic finite element analysis. John Wiley and Sons, New York
4. Hasofer AM, Lind NC (1974) Exact and invariant second moment code format, J Eng Mech Divisions ASCE 100(EM1):111–121
5. Rackwitz R, Fiessler B (1976) Note on discrete safety checking when using non-normal stochastic models for basic variables. Load Project Working Session, Cambridge
6. Rackwitz R, Fiessler B (1978) Structural reliability under combined random load sequences. Comput Struct 9(5):484–494
7. Gasser F (2008) Abdelal, nader abuelfoutouh, and ahmed hamdy: mechanical fatigue and spectrum analysis of small-satellite structure. Springer Int J Mech Mater in Des 4(3):265–278
8. Gasser F (2006) Abdelal, nader abuelfoutouh, ahmed hamdy, and ayman atef: thermal fatigue analysis of small-satellite structure. Springer Int J Mech Mater in Des 3(2):145–159
9. Tsai JR (2004) Overview of satellite thermal analytical model. J Spacecraft Rockets 41:120–125
10. ANSYS professional, software package, ver. 10, ANSYS, Inc., southpointe, 275 Technology drive canonsburg, PA 15317
11. Abdelal GF, Cooper JE, Robotham AJ (2011) Reliability assessment of 3D space frame structures applying stochastic finite element analysis. Springer Int J Mech Mater in Des 1–9. doi:10.1007/s10999-011-9168-0

Chapter 6
Qualification Testing Phase of Satellite Structure

Abstract The development of the satellite structure necessitates a good understanding of the static and dynamic characteristics of the satellite to establish design and test loads. Static strength verification is accomplished by a static-centrifugal test. Static tests objective is verification of strength of the satellite structure in accordance with accepted types of loading and determination of load-carrying ability of the structure. Dynamic tests objectives are the estimation of the vibration-transfer coefficients to the mounting seats of the satellite's components, modal frequencies of the satellite structure components and satellite structure as a whole, effect of the impact loads on satellite's equipments mounting accuracy, and verify the electrical equipments function due to vibrations. The full list of qualification tests is discussed in this chapter. Then, finite element method is used to investigate cracks that were initiated during air transportation vibration test of satellite structure strength mockup.

Qualification tests are performed according to test programs, developed for different categories of verification requirements and in accordance with selected verification approach and composition of models and mock-ups. Full-scale mockup simulates the flight model of the satellite in size and design characteristics (except strength, mass, inertial, and magnetic characteristics). Full-scale mockup is used to validate the design configuration of the satellite, to check onboard electric harness, to inspect satellite overall dimensions, and to check-up mechanical interfaces of the satellite structural components. Strength mockup simulates the flight model in size, strength, and mass-inertial characteristics. Mockups of onboard equipment and devices, which are equivalent to their flight models in mass and in moments of inertia, are installed on the strength mockup. Static and dynamic tests are performed on the strength mockup of the satellite and are discussed in this chapter. The engineering model of the satellite structure is flight representative in form, fit, and function. Electrical tests are performed on the

G. F. Abdelal et al., *Finite Element Analysis for Satellite Structures*,
DOI: 10.1007/978-1-4471-4637-7_6, © Springer-Verlag London 2013

engineering model of the satellite to check the function of the electrical interfaces of the pyrotechnic pin-pullers, intended for locking of solar arrays, to check of electrical interfaces of electrical equipments after releasing the solar arrays once in-orbit, and modification of design documentation of the electrical tests.

The development of the satellite structure necessitates a good understanding of the static and dynamic characteristics of the satellite to establish design and test loads also considering coupled loads dynamic analysis with the launcher [1]. Static strength verification is accomplished by a static-centrifugal test. Static tests objective is the verification of strength of the satellite structure in accordance with accepted types of loading and determination of load-carrying ability of the structure. Dynamic tests objectives are determination of the vibration transfer coefficients to the mounting seats of the satellite‘s components, modal frequencies of the satellite structure components and satellite structure as a whole, effect of the impact loads on satellite's equipments mounting accuracy, and verify the electrical equipments function due to vibrations.

The focus of this book is the application of finite element in design, testing, and manufacturing of satellite structures. It is meant to help mechanical designers and stress analysts and not test engineers. The topic of vibration testing of spacecraft structures cannot be discussed in one chapter, and there are many references that discuss this topic in more details. So, only a survey of the qualification tests are discussed without going in details that are important for test engineers. Then, finite element method is used to investigate cracks that were initiated during air transportation vibration test.

6.1 Static Test Specification

The static test objective is to verify the strength of satellite structure under the loads that act on the satellite flight model during operation [2]. These loads are applied on the satellite strength mockup by inertial forces created by the centrifugal stand. Technological brackets are designed and used to fix the satellite strength mockup on the centrifugal stand. The following tasks have to be achieved,

- The strength of satellite's structural components during its transportation by automobile, railway, and air transport and during orbital injection;
- Verify the mounting accuracy of the instruments' seats that require precise referencing to the satellite's coordinate axes after loading conditions;
- Verify the function of solar array locking elements and deployment mechanisms.

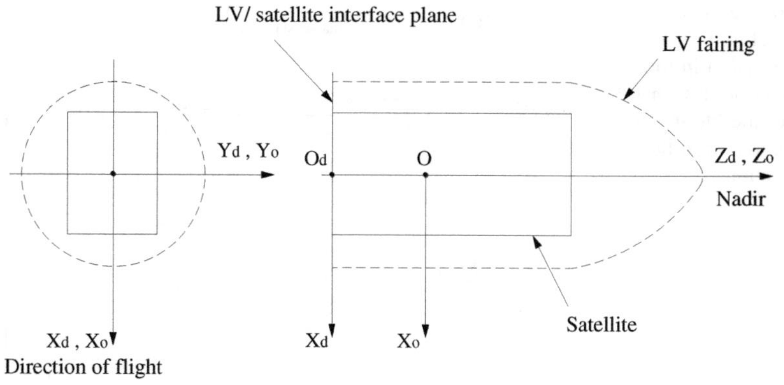

Fig. 6.1 Coordinate systems

6.1.1 Test Object

The structure of Small Sat includes the following modules as described in detail in Sects. 2.9 and 3.5:

- Base plate;
- Mounting plate;
- Basis-unit case with star sensor bracket;
- Upper frame;
- Lower frame;
- Rotation mechanisms of the solar arrays;
- Locking and releasing mechanisms;
- Fastening and mounting elements.

The strength mockup is manufactured in compliance with the flight model in respect to geometrical dimensions, materials, manufacturing technology, and instruments' seats mounting accuracy. The strength mockups of the instruments and frame modules match the original components' material, mass, and center of mass position. The coordinate systems (Sect. 2.8.6) are shown in Fig. 6.1, while the general view of the satellite is shown in Fig. 6.2.

6.1.2 Test Objectives

The main objective of the static test is to verify the strength of the primary structure elements and functionality of the secondary structure elements during satellite operation. Verification of strength of primary structure elements involves the following:

Fig. 6.2 **a** General view of the satellite in stowed configuration in the Y-direction. **b** General view of the satellite in stowed configuration in the Z-direction

(a)

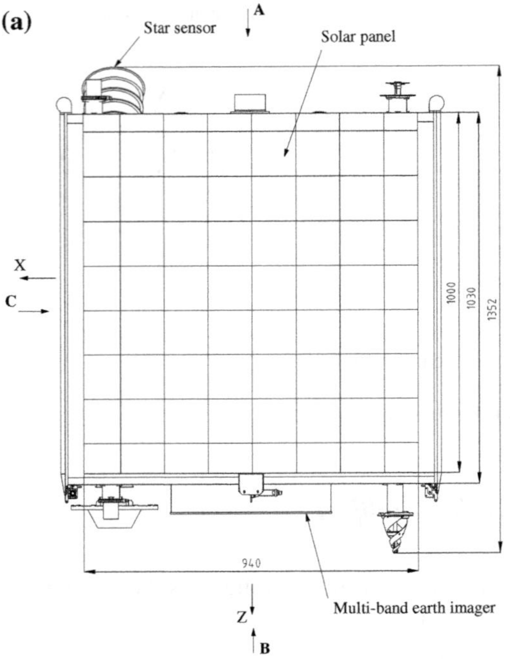

(b)

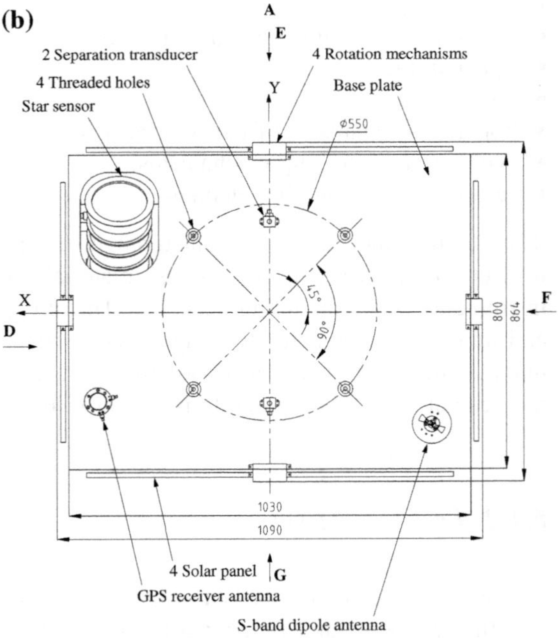

1. The integrity of structure elements with no major deformations.
2. Verifying mounting accuracy of ADCS equipment as described in the following table. Mounting accuracy is performed after the first phase of static testing.

Verifying the functionality of the secondary structure is performed on the solar array locking and releasing mechanisms after the second phase of static test.

6.1.3 Test Scheme

The satellite strength mockup is subjected to static strength tests for the below mentioned design loading cases. During the tests, the design loads are simulated in the considered cases. The design cases for Egyptsat-1 satellite are as follows:

- Case "T_{1a}"—satellite loading during transportation by automobile transport in a standard container. The loads in this case exceed the loads of other transportation cases. The operational *g*-loads are:

$$n_x^o = \pm 2; \quad n_y^o = 1 \pm 2; \quad n_z^o = \pm 1.25;$$

- Case "A"—satellite loading with maximal longitudinal g-load during orbital injection. The operational g-loads are:

$$n_x^o = 0.1 \pm 0.5; \quad n_y^o = 7 \pm 0.5; \quad n_z^o = 0.$$

In "T_{1a}" case, X-axis is directed along the motion of a transport means, Y-axis is directed downward, and Z-axis completes the coordinate system to right-handed. In "A" case, Y-axis is directed downward and coincides with the satellite's longitudinal axis.

The safety factors are adopted to be as follows:

- During autonomous ground operation for static component of g-load SF = 1.5 and for dynamic component SF = 2.0;
- For orbital injection SF = 1.3.

The test is performed on two phases. First phase is a simulation of the operation loads without applying safety factors. The purpose of this simulation is to study the effect of transportation and launching loads on mounting accuracy of ADCS equipment. Second phase is performed by applying safety factors to test functionality of the solar arrays locking and releasing mechanism under more severe conditions. Each phase consists of two stages. First stage is to simulate transportation conditions, while second phase is to simulate launching ones. Figures 6.3 and 6.4 illustrate the test setup for stage 1 and 2, respectively.

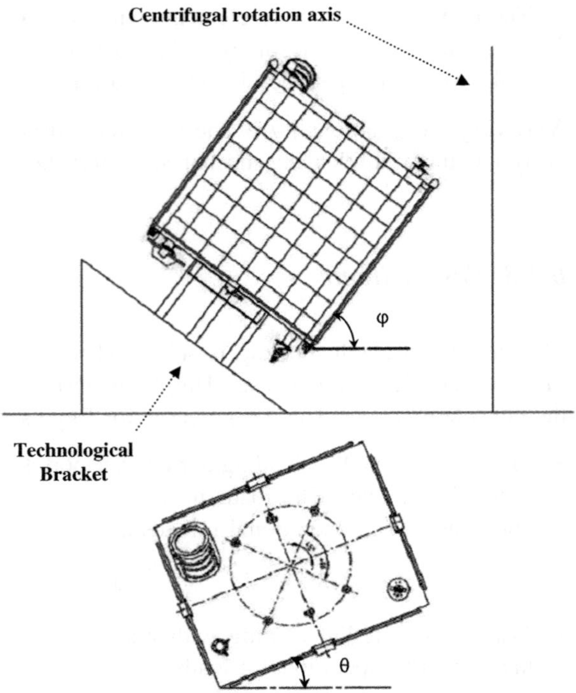

Fig. 6.3 Stage-1 of phase 1 and phase 2 of static strength test setup

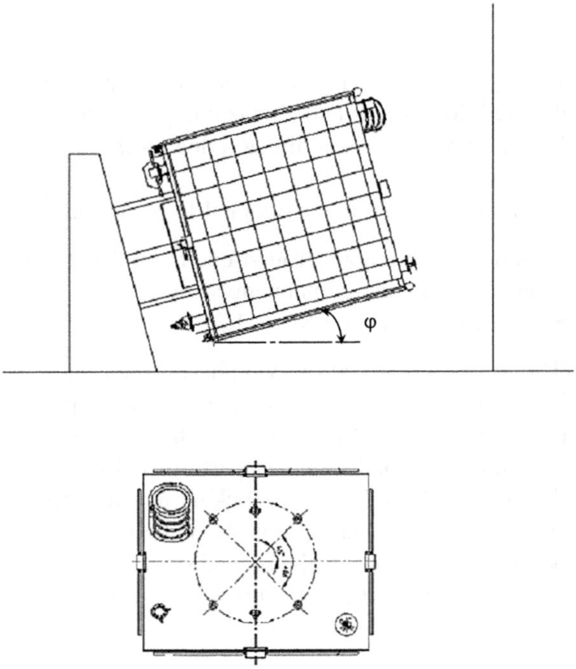

Fig. 6.4 Stage-2 of phase 1 and phase 2 of static strength test setup

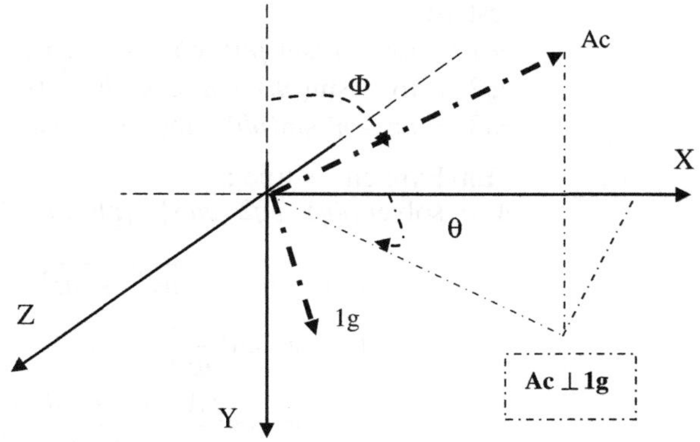

Fig. 6.5 Relationship between satellite axes, centrifugal acceleration, and gravity

6.1.4 Test Conditions Calculations

The centrifugal stand rotates with constant angular velocity (ω), which causes centrifugal acceleration equal to ($A_c = r^2 * \omega$). From Mechanical Loads Specification document, accelerations due to transportation or launching are defined in the Satellite Design Coordinates axes as (n_x, n_y, n_z). So, the problem can be defined as follows:

1. Input parameters n_x, n_y, n_z.
2. We need to determine Ac, Φ, and θ. Where,
   a. Φ = Angle between satellite Y-axis and horizontal level of centrifugal stand.
   b. θ = Angle between satellite X-axis and projection of A_c vector on the satellite X–Z plane.
   c. The effect of gravity has to be cancelled from (n_x, n_y, n_z) components and then added separately from A_c applied by centrifugal stand (Fig. 6.5).

$$Ax = Ac * \sin(\varphi) * \cos(\theta) \quad 1gx = \sin(90 - \varphi) * \cos(\theta)$$
$$Ay = -Ac * \cos(\varphi) \quad 1gy = \cos(90 - \varphi)$$
$$Az = Ac * \sin(\varphi) * \sin(\theta) \quad 1gz = \sin(90 - \varphi) * \sin(\theta)$$
$$nx = Ac * \sin(\varphi) * \cos(\theta) + \sin(90 - \varphi) * \cos(\theta)$$
$$ny = -Ac * \cos(\varphi) + \cos(90 - \varphi)$$
$$nz = Ac * \sin(\varphi) * \sin(\theta) + \sin(90 - \varphi) * \sin(\theta)$$

Using MAPLE symphonic solver to calculate Ac, φ, and θ

$$
\begin{aligned}
&\text{restart;}\\
&eq1 := nx - n^* \sin(ph)^* \cos(th) - \cos(ph)^* \cos(th) = 0;\\
&eq2 := ny - \sin(ph) + n^* \cos(ph) = 0;\\
&eq3 := nz - n^* \sin(ph)^* \sin(th) - \cos(ph)^* \sin(th) = 0;\\
&\_\text{EnvExplicit} := \text{true} :\\
&A := \text{solve}(\{eq1, eq2, eq3\}, \{ph, th, n\});
\end{aligned}
$$

$$
\begin{aligned}
n &= \sqrt{nx^2 + ny^2 + nz^2 - 1}\\
\theta &= \arctan\left(\frac{nz}{nx}\right)\\
\varphi &= \arccos\left(\frac{Ac^* ny^* \cos(\theta) - nx}{\cos(\theta)^* (n^2 + 1)}\right)
\end{aligned}
\tag{6.1}
$$

6.2 Dynamic Test Specification

Shaker vibration tests cover dynamic response and support subsequent functional tests, e.g. deployment of solar arrays [2]. As a result, the unavailability of shaker vibration tests will have distinct implications on functional and satellite alignment tests, particularly in cases where such tests cannot be easily performed at component and subsystem level. Consequently, alternative test verification of the satellite system to the classical shaker vibration test will vary significantly from project to project.

6.2.1 Test Object

The structure of strength mockup is described in Sect. 6.1.1.

6.2.2 Test Objectives

The purpose of the dynamic tests is to verify vibration strength of satellite structure. The following tasks have to be achieved:

1. To evaluate vibration-transfer coefficients for attachment points of electrical components mounting seats.
2. To verify vibration strength of the satellite structure for road transportation, rail transportation, air transportation (from manufacturing works to Processing Facilities), and launch.
3. To verify locking and releasing devices for solar arrays after vibration loading.

4. To verify mounting accuracy of the instruments mounting seats, which require precise installation relative to the satellite axes, after dynamic loading.

The loads applied to the satellite strength mockup within the dynamic tests are forced by the following ways:

1. Vibration loads are forced by shakers.
2. Shock environment is forced by actuating nominal pyrotechnic devices powered from special boards, which simulate components of onboard electrical harness.

Scope of the dynamic tests is performed as follows:

1. Tests for case of road, rail, and air transportation.
2. Tests for case of launch using LV.
3. Tests for cases of Stage separation and Fairing jettison.
4. Tests for case of satellite separation.
5. Tests for case of release and deployment of solar arrays.

Visual inspection of the satellite strength mock-up is done after each test. Checks of stability of mounting seats for instruments according to instruction and inspection for defects of the Satellite Structure are performed.

6.2.3 Test Scheme

1. Tests start with environmental tests,
   - Vibration tests of transportation using rail, motor, and air modes of transport;
   - Determination of Transmissibility coefficients and natural frequencies of satellite.
   - Vibration strength tests for design case of injection of the satellite in orbit using launch vehicle;
   - Fairing drop and separation of LV third stage tests;
   - Satellite separation from LV tests;
2. Functional Tests:
   - Solar array panels unlocking and releasing tests.
   - Structure integrity test (optical alignment) is carried out before and after each test to make sure that the optical and sensors alignment are within the accepted tolerance.

Strength mockup is fixed on vibration stand and control oscillator that controls vibration loads. This control oscillator is located in a different room and controls vibrations by averaging the output signal of four accelerometers, which are attached to the technological bracket of vibration stand itself. Data from attached accelerometers are recorded using data acquisition system, which is also located in a different room. Test schedule is listed in Table 6.1.

Table 6.1 Dynamic tests schedule

Type of operation	Number of days required
Vibration due to transportation (Z-dir) [auto—train—aircraft]	7
Vibration due to transportation (Y-dir) [auto—train—aircraft]	7
Vibration due to transportation (X-dir) [auto—train—aircraft]	7
Modal and vibration due to launching (X-dir)	7
Modal vibration due to launching (Y-dir)	7
Modal and vibration due to launching (Z-dir)	7
Shock test (2nd stage separation and fairing jettison)	4
Separation of satellite strength mockup.	7
Unfolding of solar panels	2
Check mounting accuracy	7

During vibration tests, acceleration is recorded at attachment points of satellite components, and at four attachment points of the container (during tests for case of transportation) or at three points of attachment of the adapter to the mechanical support equipment (during tests for case of launch). The recorded data is used to calculate the vibration-transfer coefficients at the mounting seats of the satellite components (ratio of the acceleration at mounting seats of satellite components to the average acceleration at mounting seats of the container during the tests for case of transportation or at the mounting seats of the adapter during the tests for case of launch). Shock strength in the cases of Stage 2/Upper Stage separation and Payload Fairing jettison is simulated. During tests for cases of satellite separation and solar array deployment, the pyrotechnic devices are actuated, then accelerometers measure vibration acceleration at different locations on the satellite, and on the solar array rotation mechanisms.

6.2.4 Environmental Tests

The purpose of the dynamic environmental tests is to verify the satellite structure strength under vibration loads while transporting satellite using rail and car. These vibrations are random vibrations, which are defined in terms of PSD (g^2/Hz) as shown in Table A.2 for rail transport and Table A.4 for car transport. First, the vibration strength tests for the case of transportation using motor transport are performed, then followed by the case of transportation by railway. Strength mock-up is mounted on technological facilities (Fig. 6.6), then covered by a container. During testing by shaker the random vibration is applied alternately in three mutually perpendicular directions:

- vertical (Y-axis),
- in moving direction (X-axis),
- in direction perpendicular to transportation direction (Z-axis).

Fig. 6.6 Environmental test setup

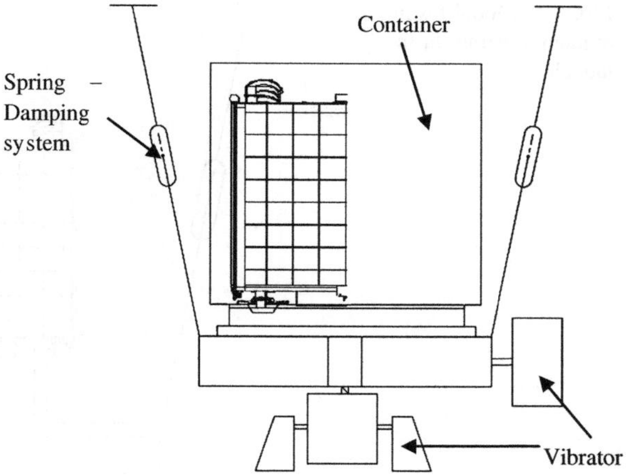

Next, verification of satellite structure components strength under vibration loads during transportation by air is performed, as defined in Table A.7. The air transport vibration test setup is similar to the previous rail and road transportation, except the random vibration is applied alternately in three mutually perpendicular directions:

- vertical (Y-axis),
- in moving direction (X-axis),
- in direction perpendicular to moving direction (Z-axis).

Next, modal testing is performed to estimate the coefficient of vibration transfer and estimate satellite structure natural frequencies, to verify finite element strength model and modify it if necessary. The test setup for the modal testing is similar to the transportation testing except the satellite is not stored inside a container, and it is fixed on a LV adaptor, as shown in Fig. 6.7. Usually, the finite element modal analysis does not match the experimental one from the first iteration, due to two factors. First one is the material models used in finite element analysis that does not model the real material behavior accurately. The second factor is stresses and deformations that are induced in the structural components during manufacturing processes. So, actually, the stress engineer will modify some of the finite element modeling assumptions, such as numerical contact conditions, or material constants to match the experimental modal results. There are two approaches to reduce the uncertainty in these. The first approach is to use stochastic finite element analysis as described in Chap. 5, but some investigation is required to apply it on modal analysis. The second approach, which is more realistic but more expensive, is the MultiScale material modeling and simulates how manufacturing process is affecting material behavior. MultiScale material modeling involves integrating the material behavior on two separate levels, local and global ones. The local level is used to model the microstructure of a material and to study the effect of its variability, while the global one models geometry of components and loading conditions [8]. Manufacturing simulation using continuum material modeling is

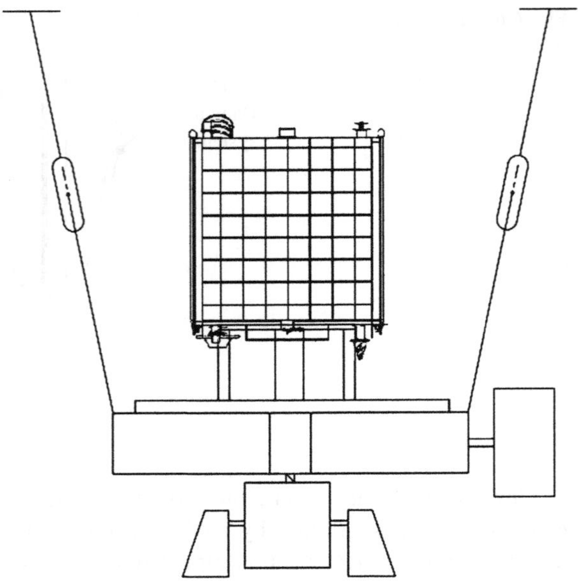

Fig. 6.7 Modal testing and dynamic testing during launch

discussed in the next chapter. Numerical strength analysis can be improved by including the residual stresses and deformations that are induced in the structure parts due to the manufacturing process. Though only the continuum approach is used here instead of the MultiScale, yet it will be an interesting research topic to compare how accurate and efficient it is to apply these two approaches. Yet, if the material modeling is based on modified constants till the numerical model matches the experimental results, then bringing the MultiScale level down to the nanolevel may be required.

Following the modal testing, the strength mockup is tested under the effect of vibrations during launch. These vibration levels are defined in Tables A.9, A.10, and A.12. Test setup is similar to the modal testing one.

6.2.5 Shock Tests

Two types of shock testing were performed during the qualification phase of the satellite design project. First shock testing is to evaluate satellite structure mechanical design under the effect of shock loads due to separation of third stage of the launcher. The shock spectrum is defined for the Dnepr launcher [3] in Table 6.2. The third stage of the launcher is separated from the main body through the use of pyrotechnic devices that are activated using electric signals. Details of the shock tests and their data analysis are out of the scope of this book. It has to be noted that no shock numerical analysis was performed using finite element, and satellite structure shock strength has to be verified using only experimental setup.

Table 6.2 Shock spectrum loads during third stage LV separation [2]

Frequency band (Hz)	30–50	50–100	100–200	200–500	500–1,000	1,000–2,000	2,000–5,000
Shock spectrum, g (at figure of merit $Q = 10$)	5–10	10–25	25–100	100–350	350–1,000	1,000	1,000

The main reason is again lacking the appropriate material model, and this can be an interesting graduation project topic.

Second shock test is the verification of the satellite structure strength for the load case of separation from the launch vehicle adaptor. Similar to the first shock test, the source for the shock loads in this case is the pyrotechnic locks that connect the satellite base plate to the launch vehicle adaptor. The test setup is shown in Fig. 6.8. The satellite is connected to the LV adaptor through the use of pyrotechnic locks and Zero-g kit is used to simulate the effect of no gravity in space, once the satellite is released in-orbit. The Zero-g kit consists of support frame that is connected to the satellite, a group of pulleys, cables, and virtual weight that is equivalent to the satellite weight. The shock spectrum is recorded at different locations on the satellite to verify the function of the electronic modules. So, what happens if there is a problem with the shock spectrum at one or more locations?

Meaning, the shock spectrum does not meet the allowed vibrations at some locations on the satellite components. In this case, usually washers and shock absorbers have to be added at those locations to reduce shock effect. These additions will solve the satellite mechanical design problem but may cause an issue to its thermal design. As adding washers to the mechanical design may lead to a different thermal profile that has an effect on one or more of the electric components. This will require more design iterations and optimization that may lead to longer time and increased cost. Improving the finite element models by using advanced material modeling can lead to early predictions to these issues and reduce time and cost.

6.2.6 Functional Testing

Functional tests are meant to verify the locking and releasing mechanisms of the solar panels are carrying out their purposes, and the deviations in the mounting accuracy of the satellite equipments seats are within the specified tolerances, after the strength mockup passed all other dynamic tests.

Test setup for the first functional test is similar to the satellite separation from LV one, except that the Zero-g kit is used on the solar panel instead of the whole satellite. Pyrotechnic devices are used for the release of the solar panels, then the rotation mechanisms that connect the solar panels to the satellite structure create a moment to rotate the solar panels and lock them at the required function angle. The tasks of this test are:

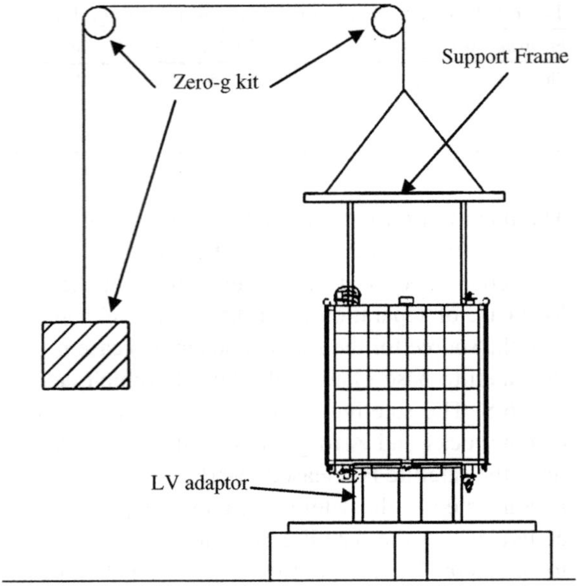

Fig. 6.8 Test setup for shock load under separation of Satellite from LV

- To verify the release of the solar panels.
- To verify the locking mechanism.
- To verify the strength of the rotation mechanism.

Following the first functional test, the satellite is disassembled to three main subassemblies; base-plate subassembly, basis-plate subassembly, and mounting plate subassembly. Each of these subassemblies is fixed on a technological support and the mounting accuracy deviations at the equipments' mounting seats are measured using optical equipments. Mounting accuracy test setup is out of this book's scope.

6.3 Test Case

Mounting accuracy deviations of satellite payload seats, due to mechanical and thermal loads, are one of the requirements to achieve the satellite mission with acceptable performance. Mounting seats deviations of the Multi-Band-Earth-Imager (MBEI) are caused by cracks in the plate of the basis unit and bracket for attachment of MBEI. Cracks were detected during inspection of the satellite strength mockup after vibration testing for air transportation phase. Most likely the cracks were due to mechanical fatigue damage as strength mockup structure was subjected to extended vibration loads during test. Total vibration duration during testing is about 56 h. In order to study the cracking reasons, finite element modeling of the structural parts of the basis unit including MBEI bracket and instrument MBEI is subjected to harmonic response to simulate vibration loading for the case of air transportation. Numerical results are compared with the experimental ones, and mechanical design of the basis-plate unit is modified.

Rate gyro kits and reaction wheels are located inside the basis unit case. Star sensor, interface units of the rate gyro kits, magnetorquers, and magnetometer of ADCS, MBEI and middle infrared earth imager (MIREI) are located outside the basis-unit case. Star sensor, MBEI, and MIREI are installed on basis unit so that optical axis of the star sensor during oriented flight of the satellite in orbital coordinate system is directed to zenith, and optical axes of MBEI and MIREI are directed to nadir. Figure 6.9 shows a general view of the basis-unit case.

During air transportation vibration tests, accelerations are measured at attachment points of satellite components, at four attachment points of the container (during tests for case of transportation) or at three points of attachment of the adapter to the mechanical support equipment (during tests for case of launch). Measurement data are processed and vibration-transfer coefficients at attachment points of satellite components (ratio of acceleration at attachment points of satellite components to the average acceleration at attachment points of the container during the tests for case of transportation and at attachment points of the adapter during the tests for case of launch) are estimated. Following the air transportation tests the strength mockup is inspected and the following were observed:

- The function of the rotation mechanisms of the solar arrays was verified.
- Deviation of mounting seats for instrument MBEI relative to Y-axis exceeded the allowable value.
- Cracks in the plate of the basis unit and bracket for MBEI attachment were detected during inspection. Cracks locations are shown in Fig. 6.9.

Two cracks are located on the basis plate, and their locations are symmetrical with respect to the Y–Z plane (location 1 in Fig. 6.9). The third crack location is on the MBEI bracket, where the bracket interfaces with the basis-plate (location 2 in Fig. 6.9). Mounting accuracy deviation of the MBEI seat is caused by cracks in the plate of the basis unit and bracket for attachment of MBEI. Two possible sources of these cracks were as folows: one is the fatigue damage due to prolonged vibration loads during various loading cases and second is the induced stresses and deformation during manufacturing process. In order to investigate the cracks causes, a low cycle fatigue analysis is performed on the basis-unit subassembly for the case of air transportation. Simulation of the low cycle fatigue of the basis-unit subassembly under air transportation vibration loads requires three steps: first step is to build the finite element model in terms of geometry, meshing, and boundary conditions; second step is to determine the inelastic parameters of material used in terms of plasticity parameters; and last step is to run simulations and compare results with the experimental ones.

One of the most difficult problems in structural analysis is idealizing of the structure. Assumptions have to be made to idealize a structure; these assumptions can be the strength or the weakness of the analysis. They may simplify a problem

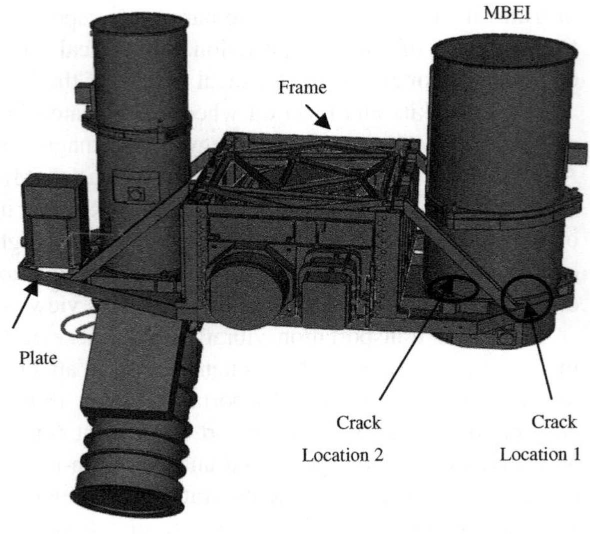

Fig. 6.9 Basis-unit case layout [4]

so that its solution may be estimated in a couple of hours while it would otherwise take days, or they may give wrong answers. Finite element modeling should provide the following objectives: provide a model that will allow obtaining the necessary information at the desired accuracy (e.g., strains, stresses, mode shapes, displacements, and stiffness), and simplify the problem to the greatest extent while still satisfying the objective.

Modeling the whole satellite structure is not an efficient way to investigate the crack problem, in terms of running time. But to build a valid finite element model for dynamic analysis of the basis unit, it has to be integrated with a virtual (simplified model) of the rest of the satellite structure. The FE model will focus on the basis-unit subassembly that has the cracks. Model is shown in Fig. 6.10. Substructure method is used to study the fatigue of satellite structure under vibrations during air transportation. Boundary conditions are: fix in three directions the three cylindrical surfaces where the satellite is connected to the adaptor.

Aluminum alloy AMg6 is the main material for the basis-unit case. It has a density of approximately 2,630 kg/m^3, which is roughly one-third the density of steel with slightly inferior strength. AMg6 alloy is used to manufacture and weld constructions working at temperatures from −196 to +150 °C. Mechanical elastic properties of alloy AMg6 according to the standards are given in [4]. Titanium alloy is used to manufacture the MBEI casing, operating within a temperature range from −120 to +300 °C. Titanium alloy is also selected because of its very low creep deformation, which minimizes loss of stud tightness during the satellite's service life. Mechanical properties of titanium alloy VT-16 are given in Table 6.3.

A linear kinematic hardening model or a nonlinear isotropic/kinematic hardening can be used to simulate the behavior of materials that are subjected to cyclic loading. The evolution law in these models consists of a kinematic hardening

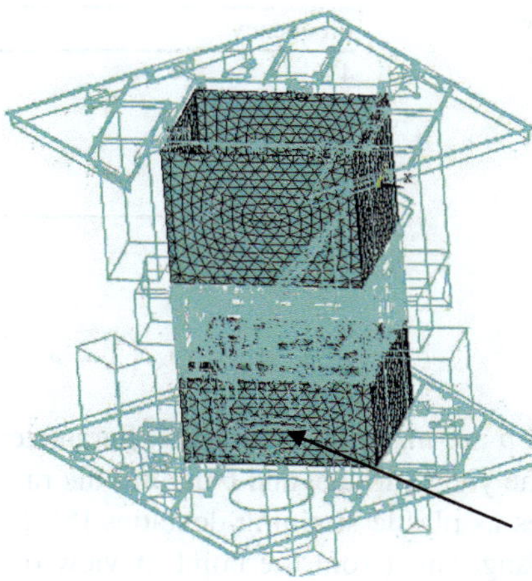

Fig. 6.10 Low fatigue FE model [4]

Table 6.3 Mechanical properties of alloys AMg6 and titanium

Material	ρ (kg/m^3)	σ_u (MPa)	σ_y	E
AMg6	2,630	310	156	72×10^3
Titanium	4,430	923	903	110×10^3

component (which describes the translation of the yield surface in stress space) and, for the nonlinear isotropic/kinematic hardening model, of an isotropic component (which describes the change of the elastic range). The nonlinear isotropic/kinematic hardening model provides more accurate predictions. Ratcheting and relaxation of the mean stress are accounted for only by the nonlinear isotropic/kinematic model. The kinematic hardening component is defined to be an additive combination of a purely kinematic term (linear Ziegler hardening law) and a relaxation term (the *recall* term), which introduces the nonlinearity. When temperature and field variable dependencies are omitted, the hardening law is

$$\dot{\alpha} = C \frac{1}{\sigma|_o} (\sigma - \sigma_y) \dot{\bar{\varepsilon}}^{pl} - \gamma \alpha \dot{\bar{\varepsilon}}^{pl} \tag{6.2}$$

where α is the translation of the yield surface in stress space through the back-stress, $\dot{\bar{\varepsilon}}^{pl}$ is the equivalent plastic strain rate, C is the initial kinematic hardening modulus, and γ determines the rate at which the kinematic hardening modulus decreases with increasing plastic deformation. In this model the equivalent stress defines the size of the yield surface, $\sigma|_o$ defining the size of the yield surface at zero plastic strain. C and γ are material parameters that must be calibrated from cyclic test data [5]. [$C = 11{,}800$ MPa, $\gamma = 103$].

The isotropic hardening behavior of the model defines the evolution of the yield surface size, $\sigma|_o$, as a function of the equivalent plastic strain, $\bar{\varepsilon}^{pl}$, using the simple exponential law

Table 6.4 Natural frequencies of the satellite [4]

Frequency	Damping
43	0.083
50	0.081
80	0.07
100	0.076

$$\sigma^0 = \sigma|_o + Q_\infty\left(1 - e^{-b\bar{\varepsilon}^{\mathrm{pl}}}\right) \tag{6.3}$$

where Q_∞ and b are material parameters. Parameter Q_∞ is the maximum change in the size of the yield surface, and b defines the rate at which the size of the yield surface changes as plastic straining develops [5]. [156 MPa, b = 5.5].

The most dangerous (from the point of view of vibration strength) vibrations are with a frequency not more than 100 Hz. Therefore, in calculations the vibrations are taken into account only in the frequency band of 100 Hz. At first, modal analysis was carried out. Table 6.4 shows the natural frequencies—the result of modal analysis and damping ratio—the value used in harmonic analysis. For all loading cases the satellite structural analysis is performed with a consideration of the action of sinusoidal vibrations at each resonance frequency f_i. In the analysis a damping value is given, which equals [6],

$$\xi = 1/(10 + 0.05f) \tag{6.4}$$

Harmonic analysis is performed to calculate peak stresses in the finite element model of the basis-unit subassembly and the virtual satellite structure due to harmonic loads [6]. The output results from the harmonic analysis will be used in the fatigue analysis. The harmonic tension stresses are responsible for dynamic fatigue failure. Harmonic loads do not act individually on structural elements, but they operate simultaneously with quasi-static loads. Quasi-static loads comprise both static and dynamic loads, and are applied at a frequency sufficiently below the first natural frequency of the structure. Therefore, the quasi-static loads are independent of time or vary slowly, so that the dynamic response of the structure due to the dynamic component is not significant. The maximum stresses affecting the structure are a combination of the equivalent amplitude stresses due to fully reversed harmonic loads and the equivalent stresses from the static components of the quasi-static loads. Since the finite element model of Small Sat is linear, superposition is applicable to calculate the maximum stresses in each structural module.

When the cyclic load level varies during the fatigue process, a cumulative damage model is often hypothesized. The generalization of this approach is called Miner's Law and can be written

$$\sum \frac{n_j}{N_j} = D \tag{6.5}$$

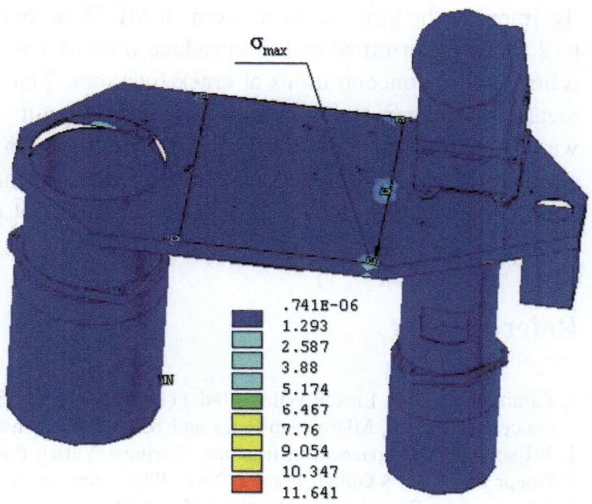

Fig. 6.11 Stress distribution for the case of basis unit [4]

Calculate the number of cycles n_j corresponding to each loading phase and selected resonance frequency. It is calculated from:

$$n_j \quad = \quad \sum f_j \, t_j \tag{6.6}$$

where f_j is the resonance frequency and t_j is the corresponding time duration of applied load in each phase. The following empirical stress equation provides an appropriate approximation for the fatigue behavior data of wrought products made of 6061-T6 aluminum alloy at room temperature [7]:

$$\log N_j = a_1 - a_2 \log[\sigma_{\max}(1 - R)^n - a_3] \tag{6.7}$$

$\sigma_{\max}$: Maximum value of stress in the stress cycle (ksi)
R: stress ratio; $R = \sigma_{\min}/\sigma_{\max}$
a_1, a_2, a_3, and n are empirical constants having the following values:
$a_1 = 20.68$, $a_2 = 9.84$, $a_3 = 0$, $n = 0.63$.

Figure 6.11 shows the FE results for the maximum mean stress and for the basis-unit case under quasi-static load = 1 g.

Fatigue damage calculations show that basis plate is safe and should not suffer any cracks. Performing fatigue analysis for the rest of the transportation cases does not cause enough damage that may cause crack initiation (<0.04). In connection with the cracks that occurred in the structure during tests in the case of air transportation, the following decisions were made:

- To improve damaged structural elements;
- To revaluate strength of improves structure;
- To repeat tests for the case of air transportation.

To improve the bracket for instrument MBEI an increase of its thickness from 1.5 to 2 mm is determined and to introduce fillet of 4-mm radius. These modifications relieve stress concentrations at crack locations. Plate of basis unit is improved by means of straps riveted to ribs of plate of basis unit. The strength of modified case was estimated by means of analysis. According to analysis after improvement, stress in crack location reduced several times as much (3.6 times as much) and their value became smaller than endurance limit of plate's material.

References

1. Sarafin Thomas P, Larson Wiley J (eds) (1995) Spacecraft Structures and mechanisms—From concept to launch. Microcosm Press and Kluwer Academic Publishers, Torrance, CA
2. Wijker J (2008) Spacecraft Structures. Springer-Verlag Publisher. Berlin, Germany
3. Dnepr SLS User's Guide, Issue 2, Nov. 2001. http://www.kosmotras.ru
4. Abdelal GF, Bakr Elhady A, Gad AH (2009) Experimental and numerical techniques to investigate fracture of a micro-satellite structure, March 2009, IEEE aerospace conference, Big Sky, MT USA
5. Francois M (2001) A plasticity model with yield surface distortion for non proportional loading. Int J Plast 17:703–717
6. Abdelal GF, Abuelfoutouh N, Hamdy A (2008) International Journal of Mechanics and Materials in Design. Springer 4(3):1569–1713
7. Daniel Schiff. Dynamic Analysis and Failure Modes of Simple Structures. John Wiley & Sons, Inc. NY, 1990
8. Soutis C, Beaumont P (2005) Multiscale modelling of composite material systems. Woodhead Publishing Limited, Abington Hall, Abington Cambridge, England

Chapter 7
Manufacturing Simulation Using Finite Element

Abstract Using nonlinear finite element method to simulate the manufacturing process can help to optimize its design parameters and produce better parts. The main topics discussed in this chapter are riveting, shot peening, and material removal (end mill). Riveting is used for assembly, shot peening is used for metal forming or to improve fatigue, while material removal is used for shaping or fixing imperfect part. The simulation steps using ABAQUS software of each of these manufacturing processes are discussed in detail including listing of ABAQUS input files. Projects are suggested at the end of each section for the reader to use it for practice or graduation projects.

7.1 Introduction

During my time as a deputy group leader with the Egyptian space program and as a senior mechanical engineer with Bluewater Energy—Netherlands, I noticed the lack of manufacturing simulation on the level of stresses and deformations. It is a reasonable assumption for the Energy industry, due to the size of the offshore structures and the high factor of safety used during the design process. The aerospace industry has the following issues:

- Geometrical tolerances that are very sensitive topic and affect assembly process. Spacecraft structures usually collect different subsystems to achieve specific objectives. Each of the subassembly has its own installation requirements, in addition to the mechanical and the thermal ones. If the mechanical part dimensions are not satisfying the allowed tolerances, it leads to either a deformed assembly or may lead to a total failure of the assembly process. This will increase the cost of the assembly process due to extra manufacturing processes required to adjust each part dimensions. For example, material removal

G. F. Abdelal et al., *Finite Element Analysis for Satellite Structures*,
DOI: 10.1007/978-1-4471-4637-7_7, © Springer-Verlag London 2013

of the extra dimensions, which will cost extra funds to perform such operation, but what is the solution in the case of curing of composite parts?. Water jets cutting is used to adjust composite parts dimensions, but not simulated in this chapter.
- Maintenance is very expensive, especially for satellite applications. Hidden cracks that are initiated during manufacturing and may lead to structure failure later during transportation or operation (Test case in Chap. 6).

Using nonlinear finite element method to simulate the manufacturing process can help to optimize its design parameters and produce better parts. The main topics discussed in this chapter are riveting, shot peening, and material removal. Riveting is used to assembly, shot peening is used in metal forming or to improve fatigue, while material removal is used for shaping or fixing imperfect part.

7.2 Electromagnetic Riveting Simulation

The LVER (low voltage electromagnetic riveter) machine uses an electrical energy stored in capacitors, which is discharged through a magnet coil, accelerating an armature that carries rivet die. A similar configuration is applied to both ends of the rivet, to deform it and fix the assembly together. The electromagnetic rivet upset process time is around 1–6 ms, where very fast plastic deformations lead to significant localized heating of the rivet material and high strain rates around $10^3\ s^{-1}$. The expansion of the rivet inside a fastener hole will create compressive residual stress around the hole. Rivet joint quality is affected by many parameters: sheet thickness, rivet diameter, rivet pitch, and squeeze force. However, the squeeze force has the most significant role Muller [1]. Muller showed that using a high squeeze riveting force can increase joint fatigue life up to 3 times. The fastening process has been investigated experimentally and numerically by other researchers [2–8]. Their work is focused mainly on how the geometrical and manufacturing parameters of the process (squeeze force, rivet type, and plate material) affect the induced residual stresses around the joint and the fatigue performance of a single rivet specimen, but none investigated head_die design effect on the residual stresses or the global panel deformations due to large number of rivets used.

7.2.1 Finite Element Model

In this section, the riveting process is simulated by applying explicit finite element analysis (ABAQUS) that includes coupled thermal stress equations, nonlinear material properties, contacts, and large deformations. This model is used to model the riveting process and predict stresses, deformations, and temperature as a

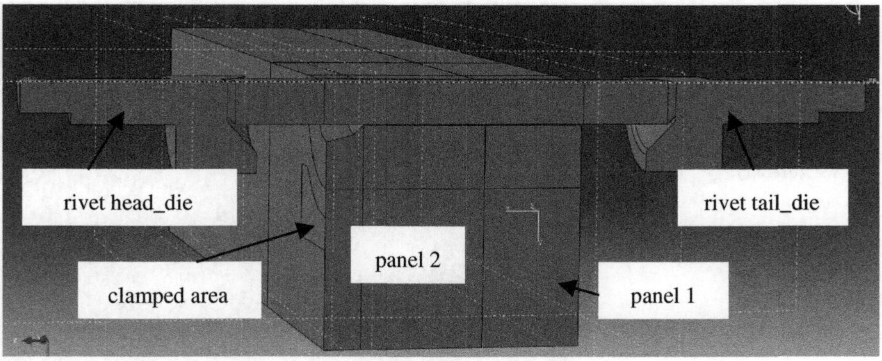

Fig. 7.1 Nonlinear finite element model for simulating single rivet insertion

Fig. 7.2 Cross section of different die designs

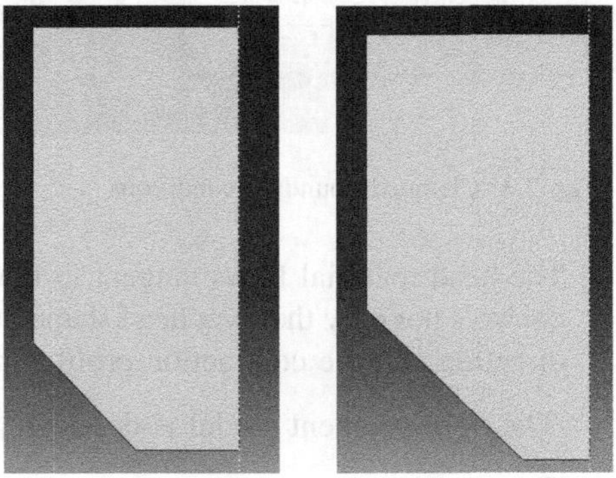

function of time. In this model, all the riveting process parameters are included in the simulation, such as head_die design, squeezed force, and assembly interface. The CAD model for the explicit finite element simulation is shown in Fig. 7.1.

The cad model consists from two panels to be assembled, rivet, head_die, tail_die, and clamped areas by the LVER. The following has to be noted,

- Rivet height is one of the main process design parameters, as rivets affect expansion in the radial direction which determines the level of residual stresses that affect fatigue life of assembly later.
- Rivet cad model is divided into three parts; head part, insert part, and tail one. The height of these parts also plays role in the final residual stresses after forming the rivet.
- Head_die and tail_die designs have effect on the final residual stresses, as they control the way the rivet is forming under the squeezed force. Figure 7.2 shows different dies cross sections. Due to the high strain rates during rivet forming, the deformed head yields as its temperature rises close to material melting point.

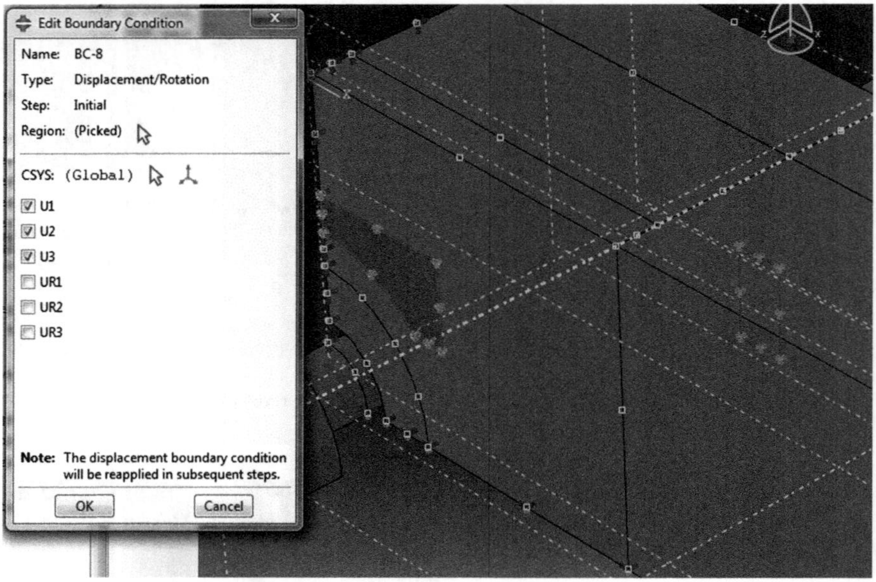

Fig. 7.3 Clamped boundary conditions

The head material flows tangent to the die surface as it deforms, and this flow controls not only the rivet head shape but also the expansion profile on the radial direction and the contraction profile on the rivet axial direction.

The finite element model is described in the following steps:

- The clamping areas are constrained in x, y, and z directions Fig. 7.3.
- The tail die and the head die are modeled as rigid bodies. Each rigid body has a reference point through which the impact riveting force is applied. The applied impact force versus time profile has a triangular shape over a time of 1–6 ms, which reflects the duration of this process. Regularly, the impact force is applied once to deform the rivet, but sometimes due to the rivet size and in an effort to minimize panel damage, the impact force is applied multiple times within a period of time that may last up to 10 s Fig. 7.4.
- Symmetry boundary conditions were considered and only a quarter of the small panel assembly was modeled. Only the symmetry around X–Z plane is shown of Fig. 7.5.
- Contacts are defined between all the assembly components (rivet—dies—panel—stiffener).,
  - Finite sliding (penalty formulation) is used to define the contact between the die and the rivet:

    For the Tangential behavior, Penalty formulation is used for friction with standard coefficient of (0.47—Steel and Aluminum). For the normal behavior, hard contact is defined with allow separation after contact.

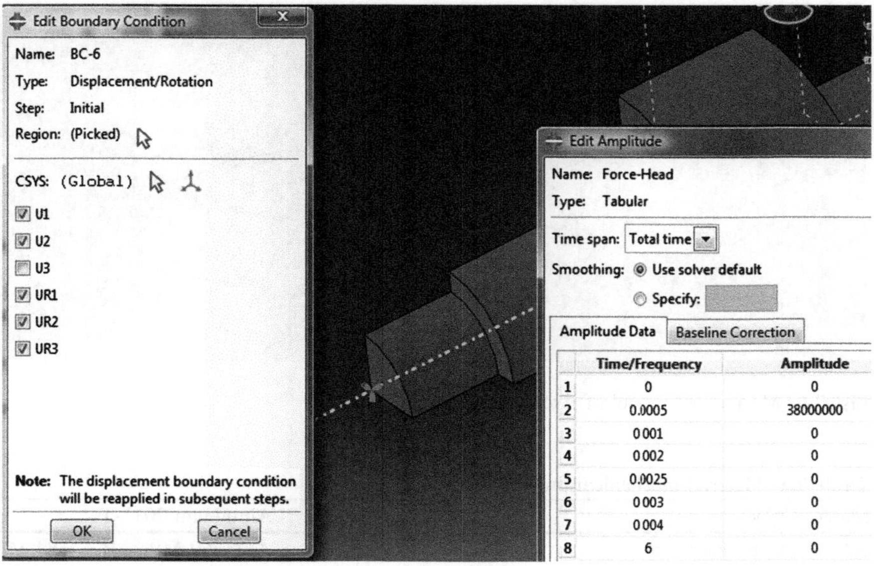

Fig. 7.4 Rivet die boundary conditions and applied impact force

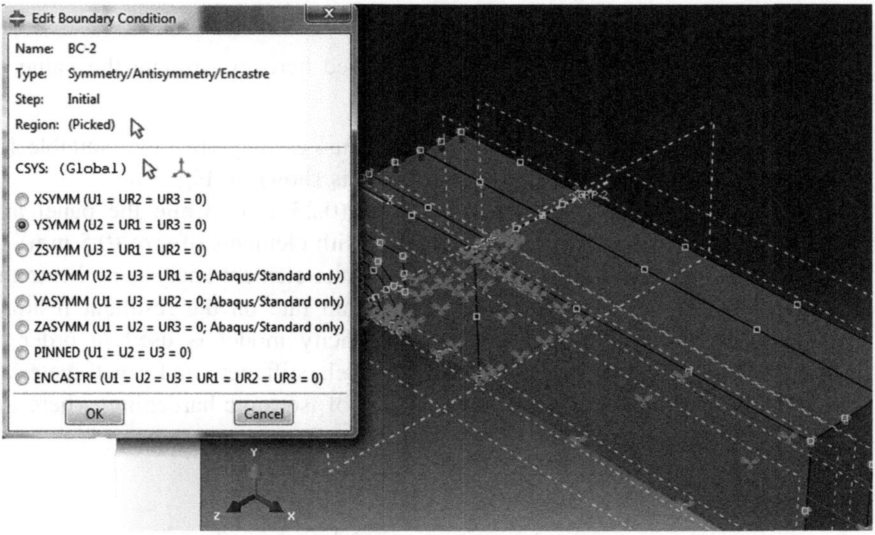

Fig. 7.5 Finite element model symmetry condition at x–z plane

– Finite sliding (penalty formulation) is used to define the contact between the rivet and the panel:
  For Tangential behavior, Frictionless formulation is used. For the Normal behavior, hard contact is used that do not allow separation. For Thermal

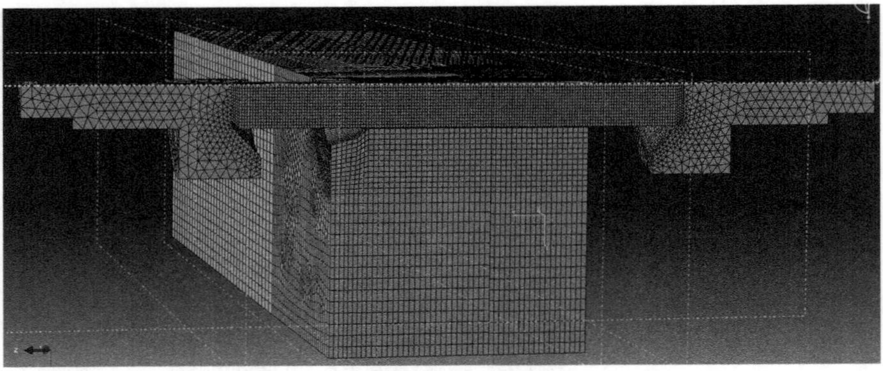

Fig. 7.6 Meshed FE model of riveting process

Table 7.1 Material mechanical properties of panel

Material type	Aluminum 2017-T4
Johnson–Cook parameters	A = 369 MPa—B = 684 MPa
	n = 0.73—m = 1.7
	C = 0.0083—ε^{o} = 1
	T_{melt} = 775 K

conductance, the work of Yuncu [9] is used here to estimate the value of contact conductance.

- The model is meshed with hexagonal elements, as they are more suitable for material forming simulations; meshed model is shown in Fig. 7.6.
- The rivet is meshed with elements of size (0.25 mm), while the panel and stiffener volumes near the rivet are meshed with elements of size (0.5 mm).
- The material models used for the nonlinear finite element analysis replicate the effect of the heat generated due to high strain rate on the resultant residual stresses. As a result, a Johnson–Cook plasticity model is used in order to describe the nonlinear behavior of the panel, stiffener, and rivet material. Johnson–Cook hardening is a particular type of isotropic hardening, where the static yield stress is assumed to be of the form

$$\bar{\sigma} = \left[A + B\left(\bar{\varepsilon}^{pl}\right)^{n}\right]\left[1 + C\,\ln\left(\frac{\dot{\bar{\varepsilon}}^{pl}}{\dot{\varepsilon}_0}\right)\right]\left(1 - \left(\frac{T}{T_{melt}}\right)^{m}\right) \tag{7.1}$$

where $\bar{\varepsilon}^{pl}$ is the equivalent plastic strain and A, B, C, $\dot{\varepsilon}_0$, n, and m are material parameters measured at or below the transition temperature (T_{trans} = 25 °C). The Johnson–Cook material parameters of the panels are listed in Table 7.1. Inelastic heat fraction coefficient has to be defined to determine the amount of strain energy converted to heat energy due to fast formation. The default value is 0.9.

The Material, Boundary conditions, and Step definitions in ABAQUS input file are:

```
**
** MATERIALS
**
*Material, name = Die_steel
*Density
0.003893,
*Elastic
2.1e + 08, 0.3
*Material, name = Panel_Bot_2024-T351
*Conductivity
121000.,
*Density
2.78e-06,
*Elastic
7.31e + 07, 0.33
*Expansion
2.47e-05,
*Inelastic Heat Fraction
0.9,
*Plastic, hardening = JOHNSON–COOK
369000., 684000., 0.34, 1.7, 638., 25.
*Rate Dependent, type = JOHNSON–COOK
0.0083,1.
*Specific Heat
9.2e + 08,
*Material, name = Rivet_2017-T4
*Conductivity
134000.,
*Density
2.78e-06,
*Elastic
7.24e + 07, 0.33
*Expansion
2.36e-05,
*Inelastic Heat Fraction
0.9,
*Plastic, hardening = JOHNSON–COOK
276000., 353000., 0.504, 0.97, 644., 25.
*Rate Dependent, type = JOHNSON–COOK
0.019,1.
*Specific Heat
```

```
8.8e + 08,
*Material, name = Rivet_2117-T4
*Conductivity
154000.,
*Density
2.75e-06,
*Elastic
7.1e + 07, 0.33
*Expansion
2.56e-05,
*Inelastic Heat Fraction
0.9,
*Plastic, hardening = JOHNSON–COOK
165000., 353000., 0.504, 1.7, 514., 25.
*Rate Dependent, type = JOHNSON–COOK
0.019,1.
*Specific Heat
8.8e + 08,
*Material, name = Rivet_7050-T7451
*Conductivity
157000.,
*Density
2.83e-06,
*Elastic
7.17e + 07, 0.33
*Expansion
2.54e-05,
*Inelastic Heat Fraction
0.9,
*Plastic, hardening = JOHNSON–COOK
435700., 2.53462e + 06, 0.504, 0.97, 488., 25.
*Rate Dependent, type = JOHNSON–COOK
0.019,1.
*Specific Heat
8.8e + 08,
*Material, name = "Rivet_Titanium 6A1-4 V"
*Conductivity
6700.,
*Density
4.43e-06,
*Elastic
1.138e + 08, 0.33
*Expansion
8.6e-06,
*Inelastic Heat Fraction
```

```
0.9,
*Plastic, hardening = JOHNSON–COOK
862000., 331000., 0.34, 0.8, 1640., 25.
*Rate Dependent, type = JOHNSON–COOK
0.012,1.
*Specific Heat
5.263e + 08,
*Material, name = Stringer_2026-T3511
*Conductivity
134000.,
*Density
2.78e-06,
*Elastic
7.4e + 07, 0.33
*Expansion
2.47e-05,
*Inelastic Heat Fraction
0.9,
*Plastic, hardening = JOHNSON–COOK
384000., 700000., 0.34, 1.7, 638., 25.
*Rate Dependent, type = JOHNSON–COOK
0.0083,1.
*Specific Heat
8.75e + 08,
**
** INTERACTION PROPERTIES
**
*Surface Interaction, name = Head_Rivet
*Friction
0.47,
*Surface Behavior, pressure-overclosure = HARD
*Surface Interaction, name = Rivet_Panel
*Friction
0.,
*Surface Behavior, no separation, pressure-overclosure = HARD
*Gap Conductance
100000.,0.
0.,2.
*Surface Interaction, name = Rivet_Stringer
*Friction
0.,
*Surface Behavior, no separation, pressure-overclosure = HARD
*Gap Conductance
100000.,0.
0.,2.
```

```
*Surface Interaction, name = Tail_Rivet
*Friction
0.47,
*Surface Behavior, pressure-overclosure = HARD
**
** BOUNDARY CONDITIONS
**
** Name: BC-1 Type: Symmetry/Antisymmetry/Encastre
*Boundary
_PickedSet99, PINNED
** Name: BC-2 Type: Symmetry/Antisymmetry/Encastre
*Boundary
_PickedSet160, YSYMM
** Name: BC-3 Type: Symmetry/Antisymmetry/Encastre
*Boundary
_PickedSet168, XSYMM
** Name: BC-6 Type: Displacement/Rotation
*Boundary
_PickedSet178, 1, 1
_PickedSet178, 2, 2
_PickedSet178, 4, 4
_PickedSet178, 5, 5
_PickedSet178, 6, 6
** Name: BC-7 Type: Displacement/Rotation
*Boundary
_PickedSet109, 1, 1
_PickedSet109, 2, 2
_PickedSet109, 4, 4
_PickedSet109, 5, 5
_PickedSet109, 6, 6
** Name: BC-8 Type: Displacement/Rotation
*Boundary
_PickedSet166, 1, 1
_PickedSet166, 2, 2
_PickedSet166, 3, 3
**
** PREDEFINED FIELDS
**
** Name: Predefined Field-1 Type: Temperature
*Initial Conditions, type = TEMPERATURE
_PickedSet159, 25.
**
**
** STEP: Step-1
**
```

```
*Step, name = Step-1
*Dynamic Temperature-displacement, Explicit
, 0.003
*Bulk Viscosity
0.06, 1.2
** Mass Scaling: Semi-Automatic
** Scale_Region
*Fixed Mass Scaling, elset = Scale_Region, factor =5000.
*Adaptive Mesh Controls, name = Ada-1, geometric enhancement = YES,
curvature refinement =2.
0., 0., 1.
**
** LOADS
**
** Name: Load-1 Type: Concentrated force
*Cload, amplitude = Force-Head
_PickedSet145, 3, -1.
** Name: Load-2 Type: Concentrated force
*Cload, amplitude = Force-Tail
_PickedSet146, 3, 1.
**
** INTERACTIONS
**
** Interaction: HeadRivet
*Contact Pair, interaction = Head_Rivet, mechanical constraint = PENALTY,
cpset = HeadRivet
Head_Die, Rivet_Head
** Interaction: PanelRivet
*Contact Pair, interaction = Rivet_Panel, mechanical constraint = KINE-
MATIC, cpset = PanelRivet
Panel_Rivet, Rivet_Body
** Interaction: StringerRivet
*Contact Pair, interaction = Rivet_Stringer, mechanical constraint = KINE-
MATIC, cpset = StringerRivet
Stringer_Rivet, Rivet_Body
** Interaction: TailRivet
*Contact Pair, interaction = Tail_Rivet, mechanical constraint = PENALTY,
cpset = TailRivet
Tail_Rivet, Rivet_Tail
**
** OUTPUT REQUESTS
**
*Restart, write, overlay, number interval = 1, time marks = NO
**
** FIELD OUTPUT: F-Output-1
```

```
**
*Output, field
*Node Output
A, RF, U, V
*Element Output, directions = YES
DMICRT, E, EVF, LE, NE, PE, PEEQ, PEEQVAVG, PEVAVG, S, STATUS,
SVAVG, TEMP, TEMPMAVG
*Contact Output
CSTRESS,
**
** HISTORY OUTPUT: H-Output-2
**
*Output, history
*Energy Output, elset = Set-Rivet-Body
ALLAE, ALLCD, ALLCW, ALLDMD, ALLFD, ALLHF, ALLIE, ALLIHE,
ALLKE, ALLMW, ALLPD, ALLPW, ALLSE, ALLVD, ALLWK, ETOTAL
**
** HISTORY OUTPUT: H-Output-3
**
*Energy Output, elset = Set-Rivet-Head
ALLAE, ALLCD, ALLCW, ALLDMD, ALLFD, ALLHF, ALLIE, ALLIHE,
ALLKE, ALLMW, ALLPD, ALLPW, ALLSE, ALLVD, ALLWK, ETOTAL
**
** HISTORY OUTPUT: H-Output-4
**
*Energy Output, elset = Set-Rivet-Tail
ALLAE, ALLCD, ALLCW, ALLDMD, ALLFD, ALLHF, ALLIE, ALLIHE,
ALLKE, ALLMW, ALLPD, ALLPW, ALLSE, ALLVD, ALLWK, ETOTAL
**
** HISTORY OUTPUT: H-Output-1
**
*Output, history, variable = PRESELECT
*End Step
```

7.2.2 Test Case 1

The numerical results using the method described in the previous section are compared to a set of experimental results that were conducted by Withers [10]. The single rivet sample is 5/16″ diameter, panels have a length of 200 mm and width of 50 mm, and their thicknesses are 15 and 8.5 mm [10]. The applied impact force is incrementally increased and decreased following a triangular form over a time of 1 ms, which reflects the duration of this process. The amplitude of the impact force applied on the head and tail die had a maximum value of 380 and 400 kN, respectively, at

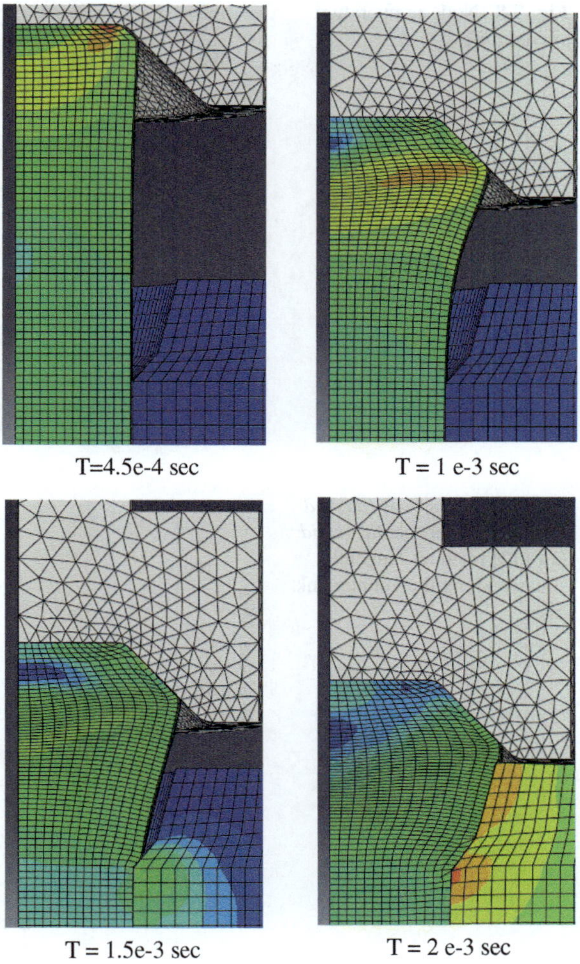

Fig. 7.7 Rivet formation under impact force

0.5 ms. These values will make some ask questions about how did we come with these values?. The answer is simply, we came with these values by trial and error until we found the best match with the experimental results. Measuring these values is a very difficult task and instead, the voltage of the LVER is used as an input to deform the rivet. The rivet expansion limits in the radial direction is of the order of 0.3 mm. The deformed process is shown in Fig. 7.7 at selected times. As the die squeeze the rivet and due to the high strain rate, 90 % of the strain energy is converted to heat energy. This rise in temperature leads to lower yield stress and the rivet head is deformed softly under squeeze pressure and follow the die curvature. Excluding the temperature effect on the material yield will produce different numerical results and should not be ignored. Temperature profile at selected node adjacent to the die versus time is shown in Fig. 7.8.

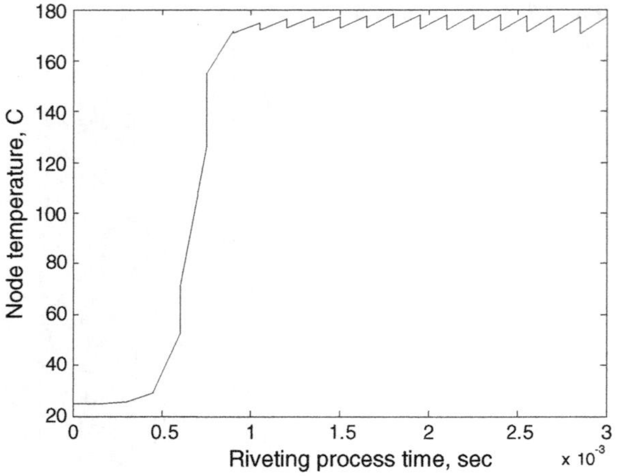

Fig. 7.8 Node temperature versus time

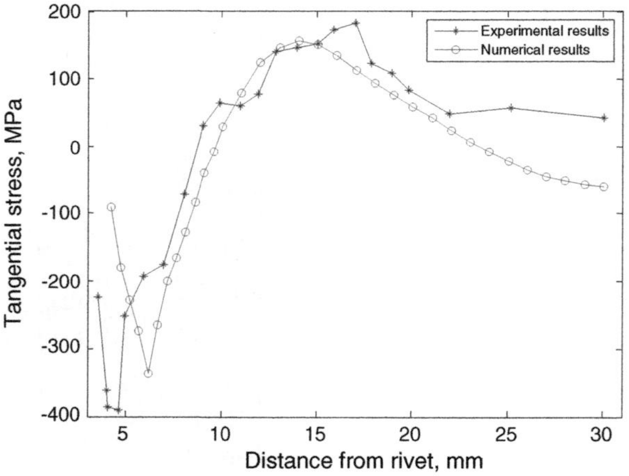

Fig. 7.9 Numerical (*dashed line*) and experimental (*solid line*) [2] residual tangential stresses along the countersink line

The residual tangential and radial stresses (σ_{xx}, σ_{yy}) along the countersink line, that were predicted from the nonlinear finite element model, were compared against the experimental results of the physical prototype, as seen in Figs. 7.9 and 7.10.

7.2.3 Test Case 2

The main objective from riveting simulation is to estimate total deformations on panels and their fatigue life due to residual stresses induced during the assembly. Panels size to be assembled determines the required number of rivets, which can be in the range of hundreds. Manufacturing simulation of such number of rivets using explicit finite element can be expensive. A surrogate model for estimating the induced residual stresses due to the riveting process was created and compared

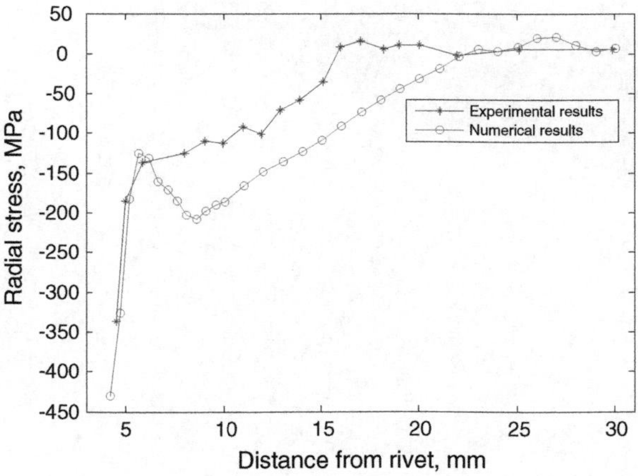

Fig. 7.10 Numerical (*dashed line*) and experimental (*solid line*) [9] residual radial stresses along the countersink line

against the experimental measurements conducted by Withers [10] on a single rivet sample. The modeling scheme used is based on the work of Jachimowicz [10], where orthotropic thermal expansion of the rivet (axial-radial-tangential) was considered and temperature boundary conditions on the body of the rivet were applied in an effort to simulate the expansion and contraction of the rivet. The temperature boundary conditions were specified from the work of Repetto [11]. In this model, heat transfer between the rivet and the panel was not taken into account.

The finite element model of the single rivet specimen is shown in Fig. 7.11. Brick solid elements were used for the discretization of the model, where the panel, stiffener, and rivet were modeled as one part and different material type for each region was defined.

The objective of the single rivet surrogate model is to estimate the stress field due to the rivet insertion by varying the applied temperature boundary conditions on the rivet volume and the material thermal expansion coefficients of the rivet body. Selecting the temperature to be 270 °C and the expansion coefficients (axial-radial-tangential) to be −0.0002, 0.0001, 0.0001 1/ °C, respectively, the developed residual tangential and radial stresses are shown in Figs. 7.12 and 7.13. Direct comparison of the numerical results and experimental data presented by Fox and Withers [10] provided a level of confidence about this surrogate model approach, although there is some discrepancy between results closer to the rivet center. Future work can be done to relate rivet forming process parameters to the surrogate model ones. This can be done by using surface approximation approach, such as Kriging.

Three projects can be suggested at this stage:

- Project 1: Model a coupon that has at least 10 rivets and compare results of explicit FE to the surrogate FE described in Test case 2.

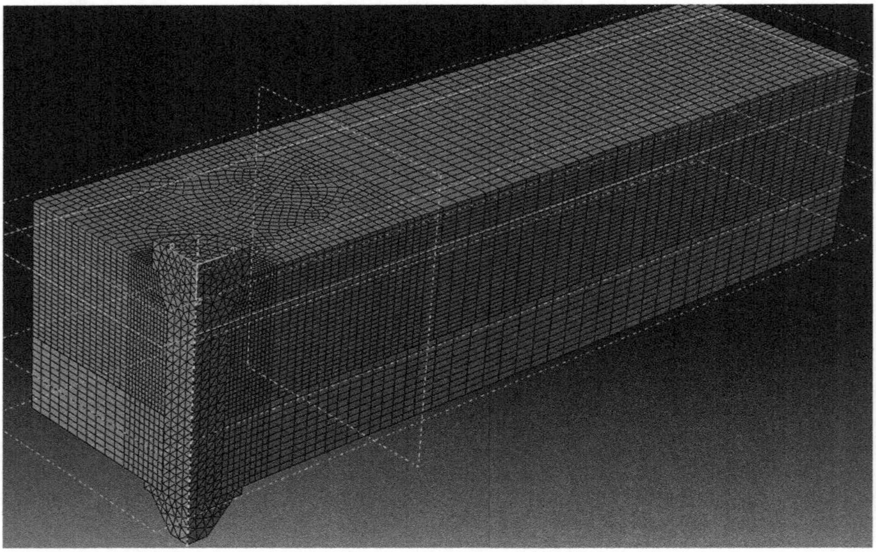

Fig. 7.11 FE surrogate of a single rivet forming process

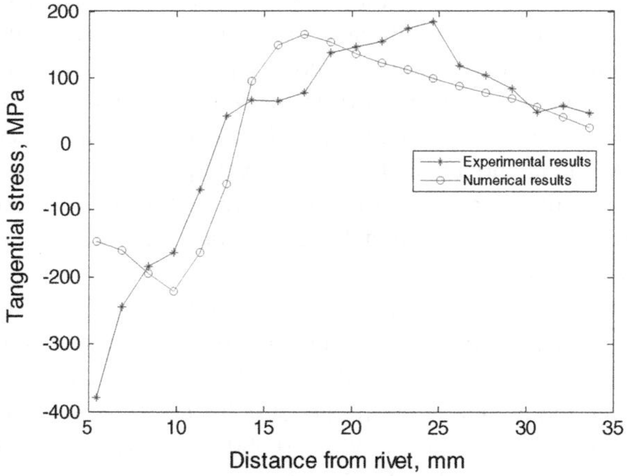

Fig. 7.12 Numerical and experimental [10] residual tangential stresses along the countersink line

- Project 2: Using the explicit FE described in Test Case 1 and Isight software, build a surrogate using Kriging method. The reader has the options to choose his own inputs and outputs of the Kriging method.
- Project 3: Steady-state analysis step is recommended to follow the explicit FE analysis, but it requires high computational power. Run the extra step and compare results.

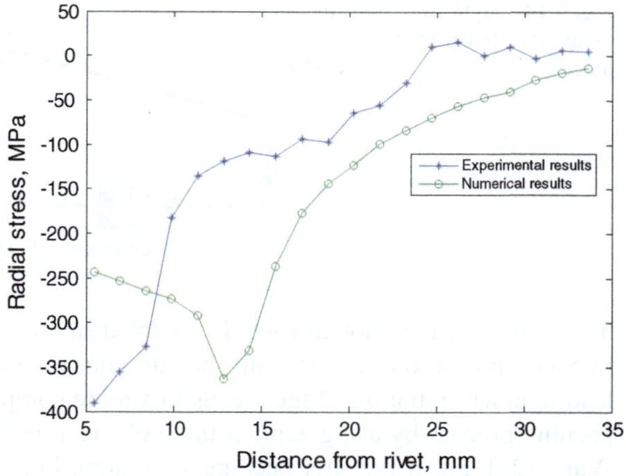

Fig. 7.13 Numerical and experimental [10] residual radial stresses along the countersink line

7.3 Shot Peening Simulation

Shot peening manufacturing process is used to improve structures' fatigue life or to form plates to follow a specific shape. Improving fatigue life of metallic aerospace structures is a mandatory due to the high maintenance cost. Shot peen is a cold cheap manufacturing process as there is no need for dies like riveting or following thermal treatment. So, how it works? it works by impacting the external surface of the structure part with a stream of hard spheres at definite velocity and impact angle. This creates a hardened layer with compressive residual stresses. This hardened layer will minimize the surface crack initiation which consequently improves fatigue life or cause the plate structure to spring back to form a specific shape. Many researchers have developed analytical theories [12–15] to study the effect of a single impact. However, the nonlinearity in material models and contact analysis during shot peening process, has led to the application of the finite element method [16–19]. The focus of this chapter is the shot peening for improving fatigue life of aerospace structures. The forming process using shot peening is mainly applied by the aircraft industry and it is out of scope of this chapter. Though the numerical technique applied here can be used in simulation of forming process, it will require longer computational time. Wang [20, 21] applied a surrogate approach to model the effect of shot peening on plates. Wang assumed each impact of the shot stream is acting independently, and their effects are distributed in a specific plastic layer. Then, each impact is replaced virtually by a thermal load that produces the same macroscopic effect. Shot peening stream density can then be simulated by multiple application of the calculated thermal load. This process can then be used to optimize the shot peen process parameters, especially for forming process, where the outcome of the process is the deformed final shape of the panel. Shot peening process usually is followed by the assembly process (riveting), while simulating the impact process with equivalent thermal load can model the global effect on the structure panel, but it will lack modeling the local

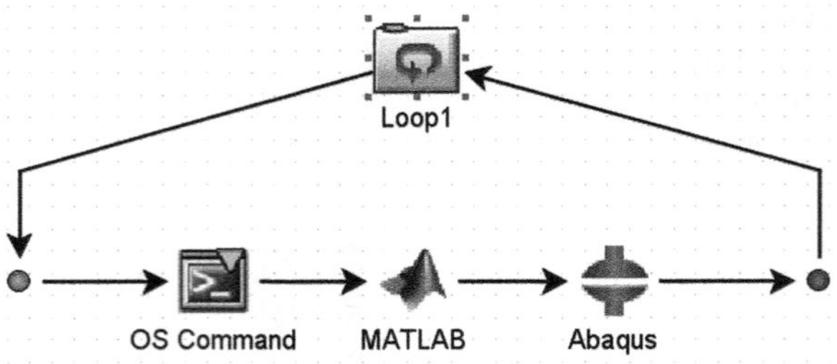

Fig. 7.14 Isight flowchart to model the shot peening process

details due to the shot impact. By local details, I mean the indentation due to impact, and the interaction with the subsequent manufacturing process that may lead to crack initiation. Which leads to a more complex scheme to model the shot peening process by using explicit finite element method, which was developed by Wang [21]. So, if the shot peening simulation intends for design optimization of the process parameters, then the thermal equivalent should be used. If the simulation intends to study the local effects of the process, such as crack initiation due to subsequent manufacturing process, then the explicit finite element scheme should be applied. In this section, the explicit finite element scheme is discussed in details using ABAQUS, MATLAB, and Isight. If the reader is interested in experimental study of shot peening and stress peen forming, then he should check the work done by H.Y. Miao [22]. They studied the relationships between the saturation, surface coverage, and roughness with respect to peening time, using aluminum Al2024 test strips. The influences of peening velocity and peening time on the resulting residual stress profiles have been experimentally presented. It would be an interesting project for the reader, if the finite element scheme described in the next section is applied on similar Al2024 samples and results are compared.

7.3.1 Finite Element Model

Wang [21] created the shot stream by modeling a series of impacts that covers $4^{(n-1)}$ impacts at a specific step n. This virtual coverage can model the progress of a single impact to a large number of multiple impacts. The first analysis step simulates the impact of one shot at the middle of the plate. Then, the second step analysis simulates the impact of additional $4^{(1)}$ impacts. Followed by other steps that simulate the impact of extra $4^{(n-1)}$ shots, until the number of impacts is covered. Simulation of shot peening is performed one by one, while including the residual stresses and deformations from previous impacts. A sequence of shot impacts uniformly distributed in both space and time is created. At each analysis step, the impact location begins from the smallest coordinates x and y, then increments y of the impact location while its x is fixed. Once the impact location reaches the edge of the panel, x is incremented and y starts from its smallest value.

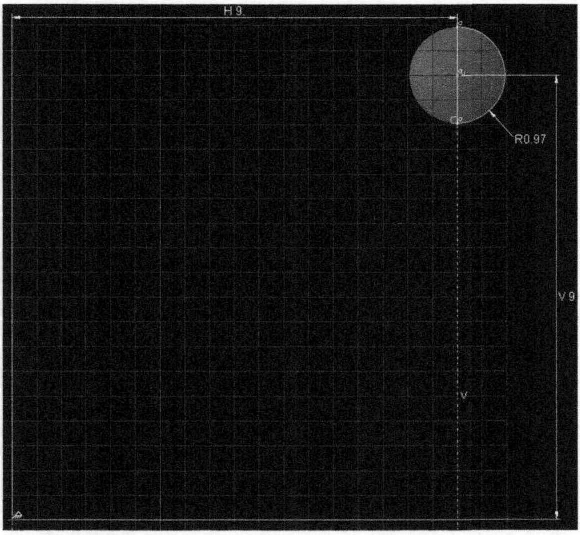

Fig. 7.15 Shot impact location defined for Isight

The simulation steps are tested in Test Case-1 and Test Case-2. So the core code will run the explicit finite element analysis, simulating impact of one sphere. While the second code generates the shot impact sequence, the third code controls the process and includes the previous induced residual stresses. The flowchart of the scheme is shown in Fig. 7.14 using Simulia Isight. The shot stream flow rate modeled is around 40 s, and the shot peening process is assumed to be 10 s for each 20 $\times$ 20 mm area. Following the explicit FE analysis, a static analysis step is performed to estimate the final stresses and deformations.

7.3.1.1 ABAQUS Module

It is the explicit finite element model that simulates the shot impact of a single hard sphere on the panel. The panel is fixed on a stand, which is assumed in the simulation that it is of similar material. The panel is aluminum alloy, 2024-T351, and the material is modeled with Johnson–Cook plastic model as defined in Eq. (7.1) and Table 7.1. The center of the spherical shot has to be defined with respect to fixed coordinates. Preferred with respect to the global coordinate system. So in ABAQUS sketch mode, two construction lines are sketched, and then fixed constraint is applied on them. The sphere center dimensions are defined with respect to these two lines. These two parameters (V9, H9) as shown on Fig. 7.15 can be used by Isight to change the shot location according to the programmed impact sequence.

The panel and stands are meshed with 3D Hex explicit stress elements with reduced integration, while the spheres are meshed with 3D Tet explicit stress elements. The meshed finite element model is shown in Fig. 7.16. The element

Fig. 7.16 Finite element model of single impact sphere

size for the panel is of size 0.1 mm near its top and gradually increasing to 0.5 mm at the bottom of the panel. Similar element size of 0.1 mm is chosen for the sphere, while the stand has element size of 1 mm. The sphere is modeled as rigid body, and a contact surface is defined with the top of the panel with frictionless contact. The bottom of the panel is constraint to the top of the stand. The perfect way to model the interaction between the panel and the stand is to add a contact surface. Adding contact surface between panel and stand is not necessary as the shot peening process for fatigue improvement is a surface process and will increase running time. It is recommended to add a contact surface between the panel and the stand in case of shot peening for forming. As shot peening for forming simulates the spring back of the panel due to stored kinetics and residual stresses, the interaction between the panel and the stand will has an effect.

Boundary conditions are shown in Fig. 7.17. The bottom of the stand is fixed in x, y, and z directions. In an attempt to reduce running time, only 10×10 mm of the panel is modeled with symmetry boundary conditions applied on three sides of the model. The symmetry boundary condition applied on each side is to add the effect of shot peening of the adjacent 10×10 mm area. If the modeled volume is at the edge of the panel, then the symmetry boundary condition is applied on three sides only.

The speed of shot impact is modeled in two analysis steps. In the first step, the velocity boundary condition is applied on the reference point of the spherical shot, and running the simulation for very short time (1e-9 s). Followed by the second step analysis, while canceling the velocity boundary condition and running simulation for (7e-6 s). Simulation time is not standard and it is a function of the shot

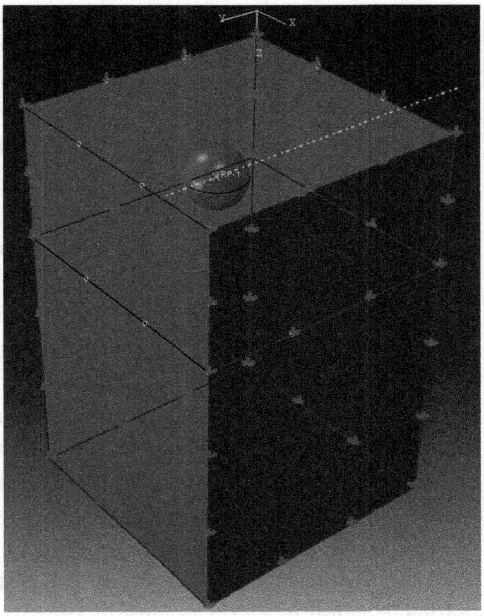

Fig. 7.17 Boundary conditions on panel

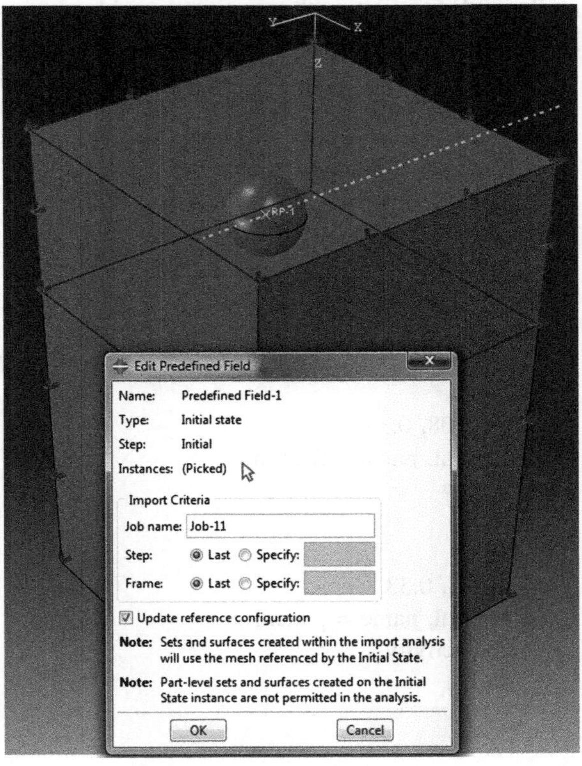

Fig. 7.18 Predefined field to include simulation results of previous shot impacts

size and shot peen parameters. The simulation parameters can be adjusted by trial and error.

The ABAQUS module is used by Isight for two objectives. First objective is the simulation of the single shot impact, and the second objective is to include in the simulation the previous impact results. This can be achieved by defining a Predefined Field in ABAQUS that reads material state results from previous simulation result file, as shown in Fig. 7.18 (for example Job-11). The content of the Job-11 file is updated with the new results, which is the task of Isight. Prior to integrating the ABAQUS module with Isight, a single shot impact is simulated without including the Predefined Field definition.

Substructure method is not available with explicit finite element analysis. Simulating shot peen process on large panel will require high computational power. Dividing the panel to multiple parts while applying symmetry boundary conditions on each part (depending to its location within the global panel), then assembling the numerical results later on the full panel, can reduce considerably the simulation running time. This scheme is still limited in ABAQUS 6.11, but it will be included in the future versions. Once this function is improved in the finite element software, it would be an interesting project to investigate the difference between running static assembly process of the multiple parts of the panel, and running a transient assembly scheme that follows the shot peen process parameters. A second good project is to simulate shot peen process on a panel, and then follow it by simulating the riveting assembly process. This project will investigate the effect of shot peening on the riveting process, and the final residual stresses that affect the assembly fatigue life. Figure 7.19 shows the contents of the ABAQUS module in Isight and the selected two parameters that Isight will change during simulation. These two parameters are the location of the sphere center that was defined in the Sketch module in ABAQUS.

ABAQUS input file is listed below,

```
** MATERIALS
**
*Material, name = Basem
*Density
2.7e-06,
*Elastic
2.2e + 08, 0.33
*Material, name = Steelm
*Density
7.5e-06,
*Elastic
2e + 08, 0.33
*Material, name = platem
*Conductivity
121000.,
*Density
2.78e-06,
```

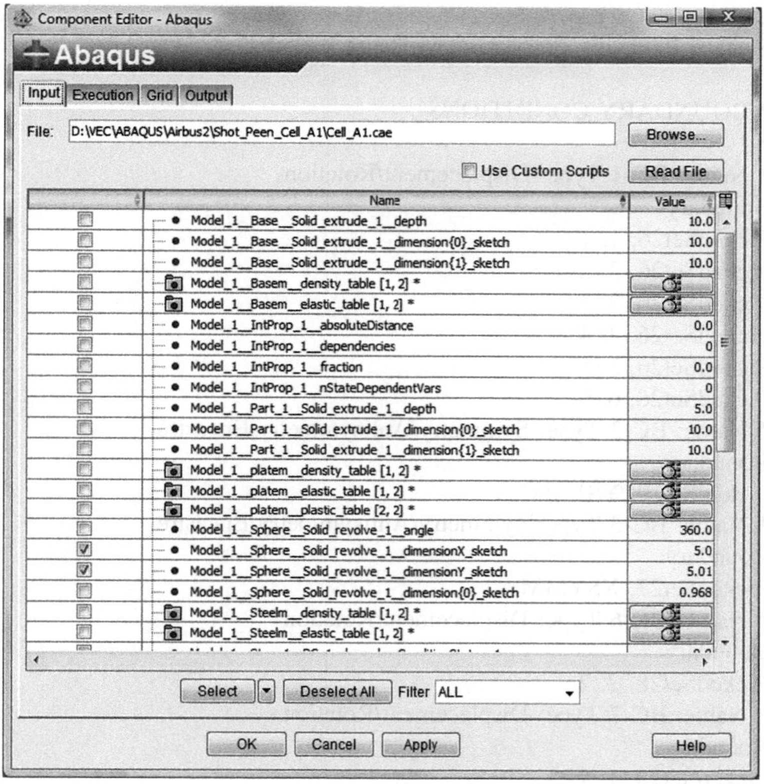

Fig. 7.19 ABAQUS module layout in Isight

```
*Elastic
6.9e + 07, 0.33
*Expansion
2.47e-05,
*Inelastic Heat Fraction
0.9,
*Plastic, hardening = JOHNSON–COOK
369000.,684000., 0.34, 1.7, 638., 25.
*Rate Dependent, type = JOHNSON–COOK
0.0083,1.
*Specific Heat
9.2e + 08,
**
** INTERACTION PROPERTIES
**
*Surface Interaction, name = IntProp-1
*Friction
```

```
0.,
*Surface Behavior, pressure-overclosure = HARD
**
** BOUNDARY CONDITIONS
**
** Name: BC-1 Type: Displacement/Rotation
*Boundary
_PickedSet26, 1, 1
_PickedSet26, 2, 2
_PickedSet26, 3, 3
_PickedSet26, 4, 4
_PickedSet26, 5, 5
_PickedSet26, 6, 6
** Name: BC-2 Type: Symmetry/Antisymmetry/Encastre
*Boundary
_PickedSet19, YSYMM
** Name: BC-3 Type: Symmetry/Antisymmetry/Encastre
*Boundary
_PickedSet27, XSYMM
** Name: BC-6 Type: Displacement/Rotation
*Boundary
_PickedSet28, 2, 2
** Name: BC-7 Type: Displacement/Rotation
*Boundary
_PickedSet29, 1, 1
**
** PREDEFINED FIELDS
**
** Name: Predefined Field-2 Type: Temperature
*Initial Conditions, type = TEMPERATURE
_PickedSet30, 25.
** ____________________________________
**
** STEP: Step-1
**
*Step, name = Step-1
*Dynamic, Explicit, adiabatic
, 1e-09
*Bulk Viscosity
0.06, 1.2
** Mass Scaling: Semi-Automatic
** Set-1
*Fixed Mass Scaling, elset = Set-1, factor = 25.
**
** BOUNDARY CONDITIONS
```

```
**
** Name: BC-5 Type: Velocity/Angular velocity
*Boundary, type = VELOCITY
_PickedSet22, 1, 1, 0.1
_PickedSet22, 2, 2, 0.1
_PickedSet22, 3, 3, 31000.
**
** INTERACTIONS
**
** Interaction: Int-1
*Contact Pair, interaction = IntProp-1, mechanical constraint = PENALTY,
cpset = Int-1
_PickedSurf14, _PickedSurf13
** Interaction: Int-2
*Contact Pair, interaction = IntProp-1, mechanical constraint = KINEMATIC,
cpset = Int-2
bases, plates
**
** OUTPUT REQUESTS
**
*Restart, write, number interval = 1, time marks = NO
**
** FIELD OUTPUT: F-Output-1
**
*Output, field
*Node Output
A, RF, U, V
*Element Output, directions = YES
EVF, LE, PE, PEEQ, PEEQVAVG, PEVAVG, S, SVAVG, TEMP
*Contact Output
CSTRESS,
**
** HISTORY OUTPUT: H-Output-1
**
*Output, history, variable = PRESELECT
*End Step
** ________________________________
**
** STEP: Step-2
**
*Step, name = Step-2
*Dynamic, Explicit, adiabatic
, 7e-06
*Bulk Viscosity
0.06, 1.2
```

```
**
** BOUNDARY CONDITIONS
**
** Name: BC-1 Type: Displacement/Rotation
*Boundary, op = NEW
_PickedSet26, 1, 1
_PickedSet26, 2, 2
_PickedSet26, 3, 3
_PickedSet26, 4, 4
_PickedSet26, 5, 5
_PickedSet26, 6, 6
** Name: BC-2 Type: Symmetry/Antisymmetry/Encastre
*Boundary, op = NEW
_PickedSet19, YSYMM
** Name: BC-3 Type: Symmetry/Antisymmetry/Encastre
*Boundary, op = NEW
_PickedSet27, XSYMM
** Name: BC-5 Type: Velocity/Angular velocity
*Boundary, op = NEW
** Name: BC-6 Type: Displacement/Rotation
*Boundary, op = NEW
_PickedSet28, 2, 2
** Name: BC-7 Type: Displacement/Rotation
*Boundary, op = NEW
_PickedSet29, 1, 1
**
** OUTPUT REQUESTS
**
*Restart, write, overlay, number interval = 1, time marks = NO
**
** FIELD OUTPUT: F-Output-1
**
*Output, field
*Node Output
A, RF, U, V
*Element Output, directions = YES
EVF, LE, PE, PEEQ, PEEQVAVG, PEVAVG, S, SVAVG, TEMP
*Contact Output
CSTRESS,
**
** HISTORY OUTPUT: H-Output-1
**
*Output, history, variable = PRESELECT
*End Step
```

7.3.1.2 MATLAB Module

The function of the MATLAB module in Isight is to define for ABAQUS the next shot impact location. As previously discussed in Sect. 7.3.1, the shot peening process is simulated on a number of steps to cover the total impacts with the panel. These shot peening steps are defined within the Matlab module and fid to ABAQUS for the next simulation. Matlab code is listed below,

```
a = [1; 9]; b = [1; 9];
x_min = min(a);
x_max = max(a);
y_min = min(b);
y_max = max(b);
mt = [2; 4; 8; 16]; nt = mt;
points (1,1) = 5.0; points(1,2) = 5.0;
indx = 2;
for ic = 1:4
m = mt(ic); n = m;
x_points = linspace(x_min,x_max,m);
y_points = linspace(y_min,y_max,n);
for i = 1:m
for j = 1:n
points(indx,1) = x_points(i);
points(indx,2) = y_points(j);
indx = indx + 1;
end
end
end
Model_1__Sphere__Solid_revolve_1__dimensionX_sketch = points (I1,1);
Model_1__Sphere__Solid_revolve_1__dimensionY_sketch = points (I1,2);
```

The (a, b) vectors describe the envelope of the shot impact, which in these simulations is shifted 1 mm from the edge of the panel.

7.3.1.3 OS Command Module

Seven ABAQUS files have to be updated prior to the next impact simulation. These files are:

*.abq, *. pac, *.mdl, *.odb, *.stt, *.res, *.prt.

These are the files required by ABAQUS to include the previous shot impact results. OS command module read the data from seven files of the current simulation and writes them on another seven files for the next impact simulation. The operation is shown in Fig. 7.20, assuming the current job analysis name is Job-1 and the Predefined Field file name is Job-11.

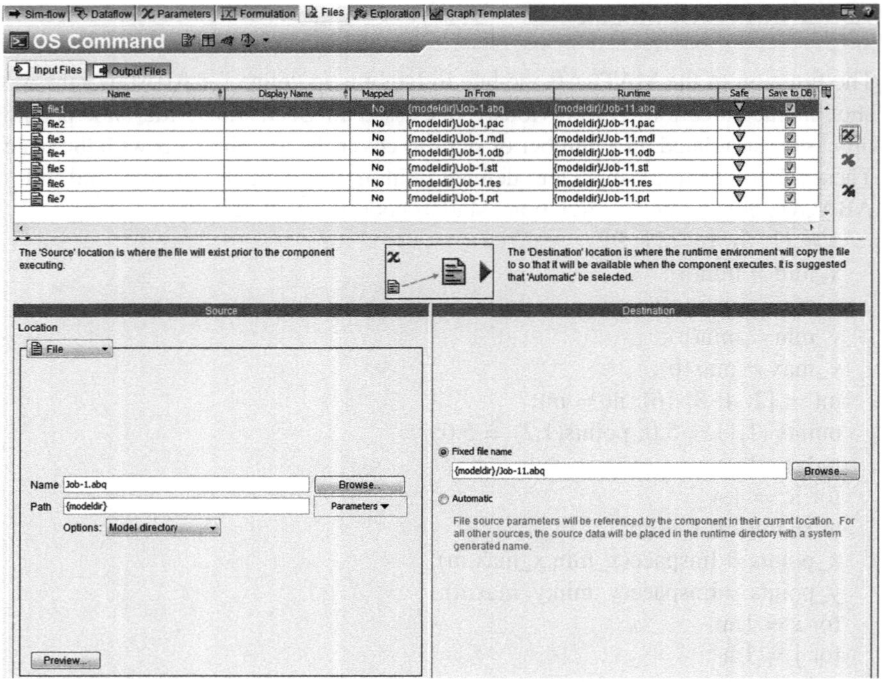

Fig. 7.20 OS command to prepare files for the next impact simulated

7.3.1.4 Loop1 Module

The Loop module is to control the shot impact sequence until the maximum number of shots is satisfied. It counts from two to the number of shots on the simulated area. The output of this module is called [I1] and its function can be seen in the MATLAB code listed above.

7.3.1.5 ABAQUS Static Step Analysis

The final result of the explicit finite element simulation of shot peen is followed by a static step analysis to estimate the final stresses and deformations.

7.3.2 Test Case-1

A single shot of a sphere of 0.97 mm radius at 31 m/s on a cylindrical plate (R 10 mm, H = 4 mm). The meshed finite element model is shown in Fig. 7.21. Due to symmetry only quarter of the assembly is modeled. The objective of the

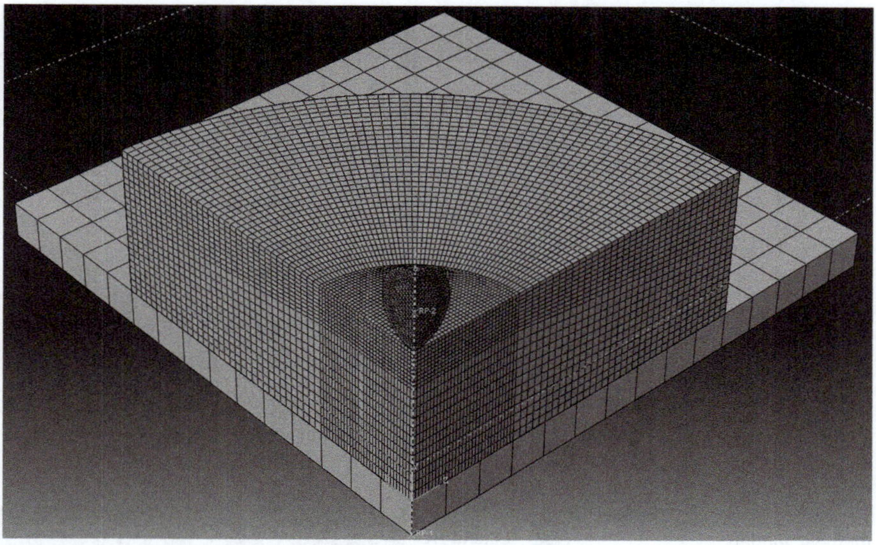

Fig. 7.21 Test Case-1 meshed finite element model of single shot

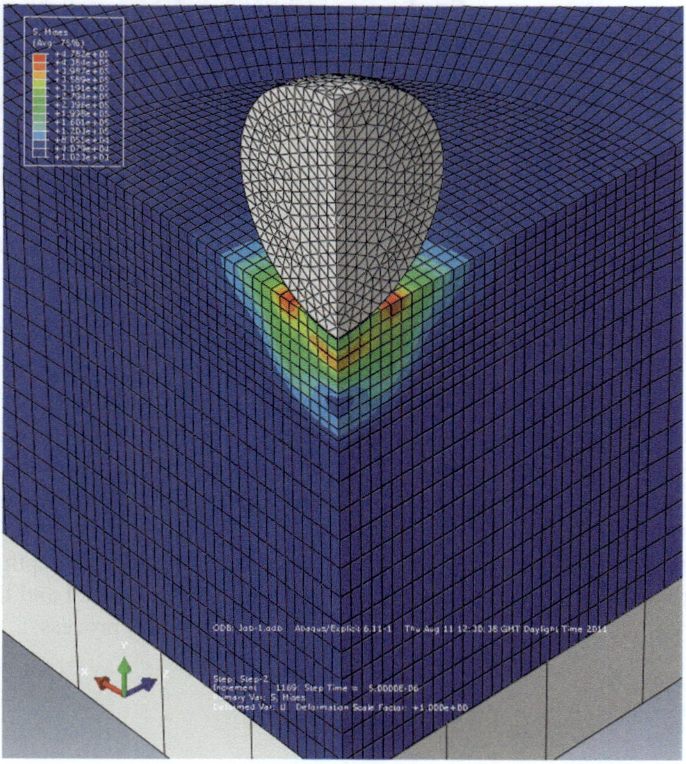

Fig. 7.22 Von Mises stress results

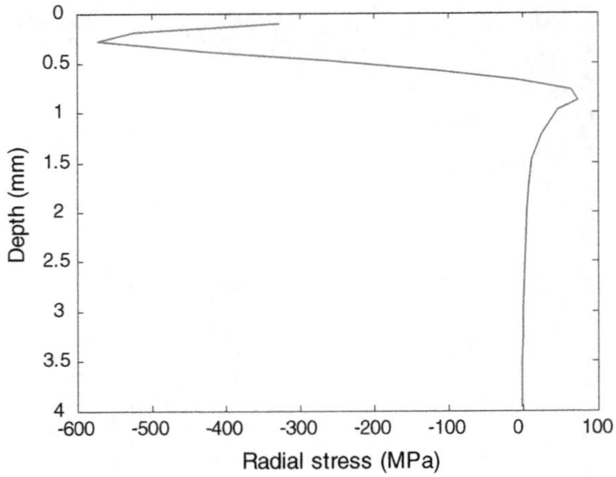

Fig. 7.23 Radial stress versus panel depth

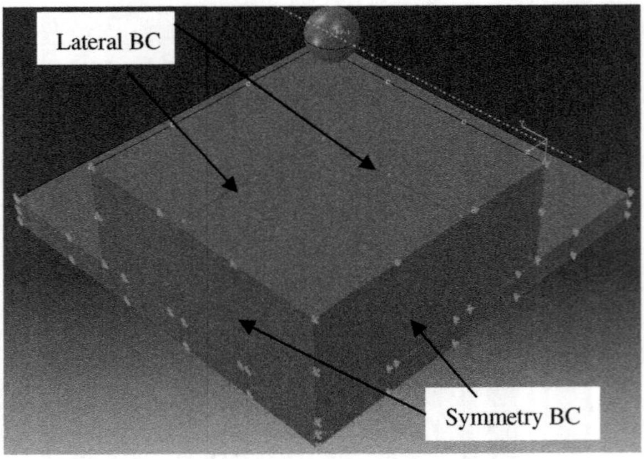

Fig. 7.24 Test Case-2 finite element model

single impact model is to adjust simulation parameters such as, step analysis time, and mesh density. It acts also as the initial point of the shot peening simulation. After running a couple of trials, the step analysis time is estimated to be 7e-6 s and the element size in the area of impact is chosen to be 0.1 and 0.2 mm elsewhere. For this test case, a contact surface is added between the panel and the stand. Simulation results are shown in Fig. 7.22 and the radial stress versus panel height is shown in Fig. 7.23.

As stress results show that the radial stress start decaying at depth of 2 mm, which proved that replacing the contact surface between the panel and the stand is necessary to accelerate simulation time of the shot peening process.

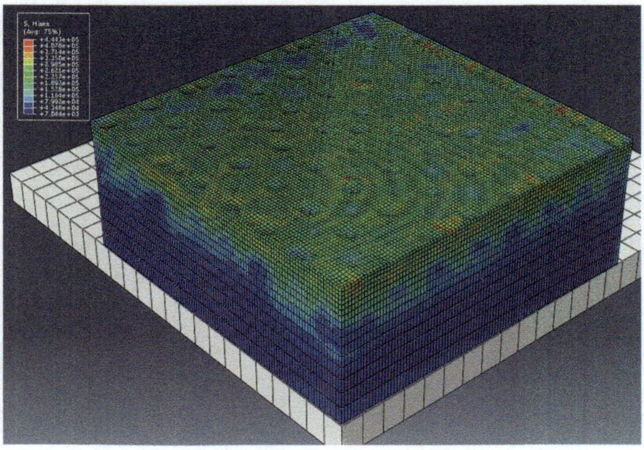

Fig. 7.25 Von Mises stresses

7.3.3 Test Case-2

The test setup shown in Fig. 7.24 is tested. It simulates the experimental shot peening condition of 0.28 MPa peening, 31 m/s impacts for 10 s on 20 × 20 and 4 mm thickness. Due to symmetry, only 10 × 10 mm is modeled. The explicit finite element model described above is used to study the impact of 100 (400/4) shots on the plate. This requires initial step with one shot in the middle of the plate, then use Isight program to simulate 4 Steps. The stand in this test case is modeled as a rigid body, through defining a reference point in ABAQUS and constraint it in six degrees of freedom. The plate is constraint to move in the lateral directions using tapes and similar boundary conditions are applied in the finite element model. Contact surfaces are defined between the plate and both the spherical shot and the stand. The contact surface is defined in this test case to allow the plate to bend and elongate while being shot peen. This is different from the Almen tests that are discussed in [22]. Stresses are shown in Fig. 7.25.

Von Mises stresses are calculated using explicit finite element followed by a static analysis step. During these analysis steps, the adiabatic heat effect is included to predict the amount of elastic energy converted to heat due to fast deformations. In my review to the research work in the area of shot peen, the effect of heat developed on the surface due to fast deformations is completely ignored. Thermal profiles developed on the plate during simulations are shown in Fig. 7.26. The maximum temperature reached at the end of simulations is 118 °C. This temperature is the result of 10 s shot peen at 31 m/s using 0.97 mm radius spherical shot. Running the shot peen simulations for 30 s leads to maximum temperature of 190 °C. Such temperature near top surface only of the plate will affect its forming process.

Equivalent plastic strain near the intersection of the two symmetry planes is shown in Fig. 7.27 and compared to experimental results [21].

Three projects can be suggested at this stage:

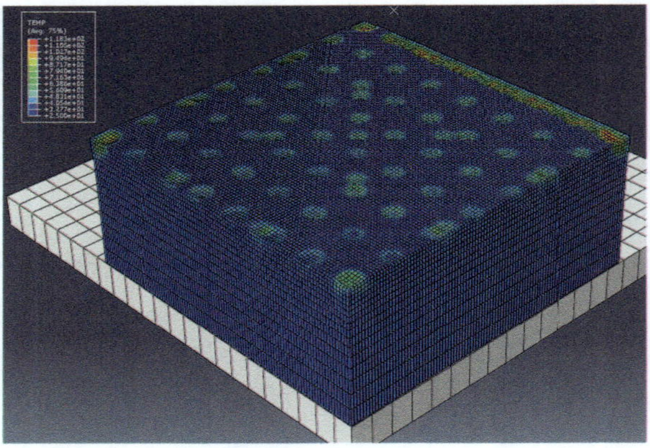

Fig. 7.26 Thermal profile due to shot peen simulations using Johnson–Cook material model

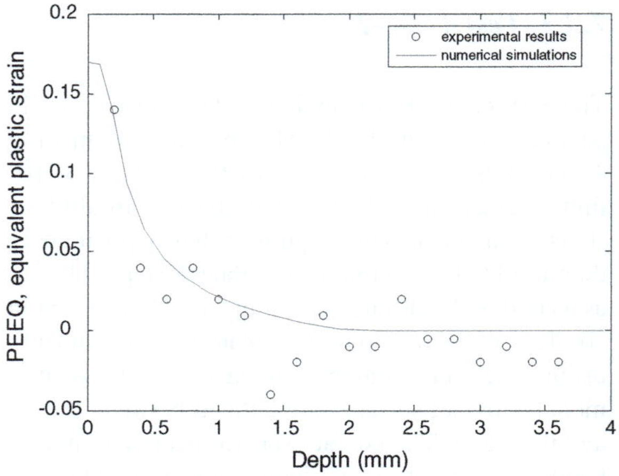

Fig. 7.27 PEEQ, equivalent plastic strain, numerical versus experimental [21]

- Project 1: Apply the shot peen simulation discussed above to validate the experimental results by Miao [22].
- Project 2: Effect of shot peen process on riveting.
- Project 3: Study the effect of heat developed during shot peen simulations on the finite element results and shot peen process for forming. This can be achieved by using coupled thermal stress explicit analysis to replace the stress explicit analysis discussed above.

7.4 Material Removal Simulation

Material removal is a machining process in which a hard turning tool removes material from a work-piece to produce the designed shape. Material removal is considered a secondary manufacturing process due to the high cost involved in the

Table 7.2 Selected milling cutting speeds for different materials. [http://www.southbaymachine.com/setups/cuttingspeeds.htm]

Material	Steel cutter, m/min	Carbide cutter, m/min
Alloy steel	12–20	45–75
Aluminum	150–300	300–600
Bronze	20–35	60–120
Cast iron	15–25	40–60
Stainless steel	10–25	30–90
Tool steel	18–20	40–60

process due to discarded material. The high accuracy in tolerances and surface finishes offered by the removal process makes it an attractive option for producing small quantities. Satellite structure components are produced in small quantities with high mounting accuracy requirements, which make material removal the perfect option. Assembly process of aerospace assemblies requires that each subassembly satisfies specific geometrical constraints. For example the aircraft wing assembly is preceded by the shot peen process and the riveting assembly of the wing panels to the stringers. These manufacturing processes sometimes produce unwanted deformations and require material removal to satisfy the geometrical tolerances, despite the high cost involved. The conventional material removal processes are to cut away small chips of material incrementally using a revolving cutting tool. Nonconventional machining processes may apply chemical or thermal process to remove material. Conventional machining processes are placed in three categories—single point cutting, multipoint cutting, and abrasive machining. Single point cutting is to use a stationary cutting tool with sharp edge to remove material from the work piece. Turning process is rotating the work piece while the cutting tool feeds into it, removing away material. Turning process is used for boring, grooving, or thread cutting. Multipoint material removal is to use a cutting tool with many sharp teeth that rotates against the work piece to remove material. Multipoint cutting processes are milling and drilling. Milling is to feed the work piece into the rotating cutting tool along defined paths to produce a variety of features. Milling process is used to create slots, chamfers, pockets, flat surfaces, and complex shapes. Drilling is to feed the rotating sharp tool vertically into the stationary work piece to create a hole. Drilling processes such as counter boring, counter sinking, and reaming can be used to create high accuracy holes, and threaded holes. Abrasive machining is to use a tool that has small abrasive particles to remove material from a work piece. Abrasive machining process is considered similar to milling or turning as each particle cuts into the workpiece removing a small chip of material. Abrasive process is used to improve the surface finish of a part, yet can still be used to form a work piece.

In this section, the milling process of metallic plates is simulated using finite element and discussed in details. Yet, the technique that to be discussed in the following section can be applied to simulate the other material removal processes. More practical details can be found in Workbook for Machine Tool Applications, Michael Bush [23]. Milling process is controlled by three parameters;

Table 7.3 Selected feed per tooth (CPT) using hard steel cutter. [http://skilledtradesmath.com/MachineTools.html]

Material	Face mills (mm)	Helical mills (mm)
Alloy steel	0.15	0.12
Aluminum	0.5	0.45
Brass and Bronze	0.3	0.28
Cast iron	0.3	0.25

Table 7.4 Selected feed per tooth (CPT) using carbide cutter. [http://skilledtradesmath.com/MachineTools.html]

Material	Face mills (mm)	Helical mills (mm)
Aluminum	0.5	0.4
Brass and Bronze	0.3	0.25
Cast iron	0.4	0.33

- Cutting speed.
- Feed rate.
- Depth of cut.

Cutting speed is defined in surface meters per minute of metal that can be removed efficiently. The cutting tool revolution per minute depends on cutting speed and tool diameter. Different parameters determine the cutting speed value, such as work-piece material, cutting tool material, and tool diameter. Milling machine cutting speeds for selected work-piece and tool material is shown in Table 7.2.

These values are based on practical experience [23, 24], and used to calculate the required revolutions per minute of the cutting tool,

$$\mathrm{Re\,v/\,min} = \frac{\text{cutting speed (m)} \times 320}{\text{Tool diameter (mm)}} \tag{7.2}$$

To improve cutting tool life,

- Apply lower cutting speed then increase gradually to the allowed maximum.
- Apply lower feed to get finer finish instead of increasing cutting speed.
- Apply coolant fluid during cutting operation [1, 23, 24].

Feed rate is the distance in mm per minute that work piece moves into cutting tool. The milling feed rate is dependent on chip size (chip per tooth), number of teeth in cutting tool, and revolution (r/min) of the cutter. Chip size is called CPT and it is the quantity of material to be removed by each tooth of the cutting tool. Factors that determine the feed rate are; cut dimensions, work-piece and tool materials, type of finish, and machine rigidity. Tables 7.3 and 7.4 show some examples of feed per tooth for different materials using either hard steel or carbide cutting tool. These values are for roughing cuts. Feed rate has to be reduced 50 % for finishing cuts.

These feed per tooth are based on practical experience [24]. Feed rate is calculated according to the following formula,

$$\text{Feed rate (mm/min)} = \text{No. of teeth} \times \text{CPT(mm)} \times \text{Revolution(r/min)} \quad (7.3)$$

Many researchers did work on machining simulation using finite element in an effort to optimize the process parameters (material—CPT—tool revolutions). Baker [25] developed a 2D, orthogonal metal-cutting process (turning) to study the influence of the cutting speed on the cutting force and the chip formation process. Ozel [26] simulated the orthogonal cutting process using finite element (Lagrangian analysis), by studying the effect of flow stress characteristics of work material at cutting regimes and friction characteristics mainly at the tool-chip interface. Arrazola and Ozel [27], studied on the effects of friction modeling in orthogonal cutting using the Arbitrary Lagrangian–Eulerian (ALE) fully coupled thermal stress analyses. They investigated the influence of limiting shear stress at the tool–chip contact on frictional conditions. Umbrello et al. [28] studied the effects of different sets of material constants of the Johnson–Cook constitutive equation in finite element modeling of orthogonal cutting of AISI 316L on the experimental and predicted cutting forces, chip morphology, temperature distributions, and residual stresses. Other researchers did some work to simulate the end milling cutting process. Soo et al. [29] used Lagrangian based code to simulate ball-nose end milling applying plasticity power law (independent from temperature). Chen et al. [30] developed a model for ball end milling with inclination angle using Johnson–Cook model (temperature dependent) and isotropic hardening rule to describe the properties of the work-piece material, and remeshing technology was adopted to obtain accurate results. Similar approach is discussed in this section, while using Johnson–Cook for both plasticity and damage progress.

7.4.1 Finite Element Model

The finite element model consists from three parts:

- Work-piece: 15 × 19 × 4 mm, made from Al2024-T351.
- End mill tool: 10 mm Tungsten Carbide End Mill.
- Holder to fix work piece and simulate feed rate.

7.4.1.1 Assembly Module

The tool is modeled as a rigid body using a reference point. The surface that will be in contact with the work piece is named in ABAQUS/CAE [Tool]. The height of this surface is 2 mm larger than the height of the work piece (1 mm from each side). So [29] meshed the tool with shell elements, in an attempt to reduce mesh

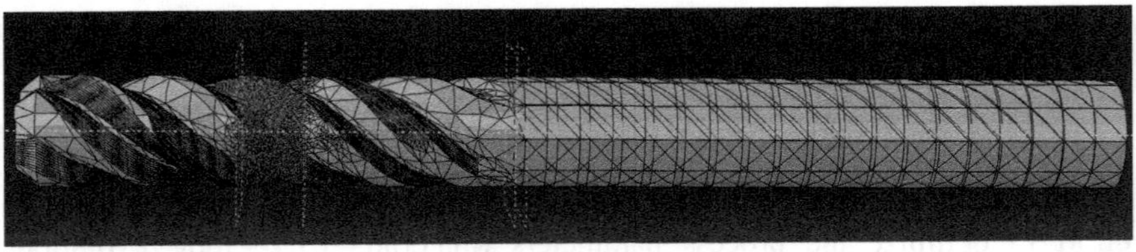

Fig. 7.28 Finite element model of the end mill tool

Fig. 7.29 Finite element model of the holder

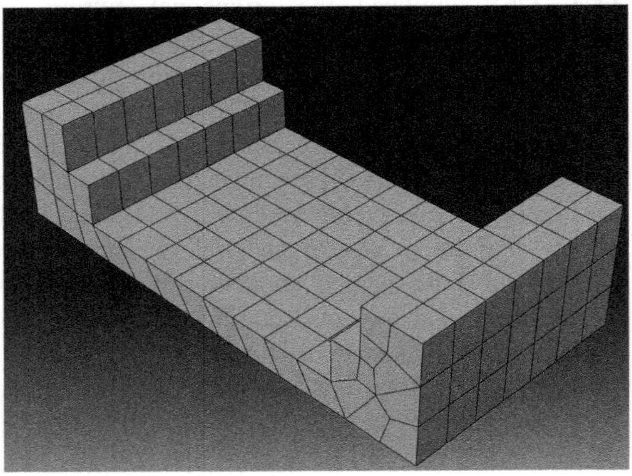

size. This requires adding virtual mass and rotation inertia to the 2D meshed model, as they affect the magnitude of force applied by the tool teeth on the work piece. Here, the tool is meshed with coarse (3 mm) C3D10 M explicit elements, except the part in contact with the work piece is meshed with 0.2 mm size elements. If the reader is interested in studying the forces and tool fatigue life, it is recommended to use more dense mesh. Increasing the number of elements of the tool does not affect the modeling technique to be discussed in this section, only requires more powerful pc. The meshed model of the tool is shown in Fig. 7.28.

The holder is also modeled as a rigid body using second reference point, and meshed with coarse mesh (1.5 mm element size). The function of the holder is to model the fixation of the work piece on the milling machine. The meshed model of the holder is shown in Fig. 7.29.

The work piece is modeled as a deformable body, and divided into two parts. The part to be machined with the end mill cutting tool is meshed with 0.2 mm C3D8R explicit elements, while the other part is meshed with 1 mm size elements. The depth of the part with dense mesh is double the cut depth. The work piece faces in contact with the holder are partitioned to create matching areas for the opposite ones on the holder. These partitions are to make the constraint or contact surface to be defined between the holder and the work piece more effective. The contact with the end mill cutting tool is defined as follows,

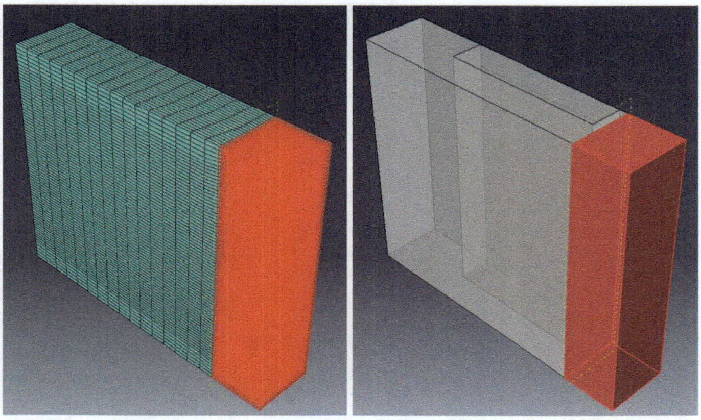

Fig. 7.30 General contact with the cutting tool

- Select the part with dense mesh and unhide the other one, then define in the part module a **Set** of elements and name it [plate]. In this set choose the whole elements of the part with dense mesh. Check Fig. 7.30.
- Define a Surface in the part module and select the outside surfaces of the part with dense elements, and name it [Surf-1].
- Edit the ABAQUS input file and change the Command line to define the Surf-1 to include the interior surfaces defined by the elements in Set [plate].

```
*Surface, type = ELEMENT, name = Surf-1
plate,
plate, interior
```

- Define the general contact between the tool and the work piece by selecting in the general contact definition the surfaces [Tool-1] and [Surf-1].
- Define the general contact properties, by defining the friction coefficient in the tangential direction and a hard contact with allowed separation in the normal direction.
- Define Tie constraints between the work piece and the holder as shown in Fig. 7.31.

7.4.1.2 Materials Module

Material response can be modeled as shown in Fig. 7.32. Segment 1–2 represents the elastic part of the material response, segment 2–3 represents the plastic yielding and strain hardening part, point 3 represents the damage initiation, and segment 3–4 represents the damage evolution and material stiffness degradation till fracture. The work piece flow stress as a function of strain, strain rate, and temperature is modeled using Johnson–Cook equation, Eq. (7.1) [segment 1–2–3]. Fracture of ductile materials can me modeled applying two different approaches.

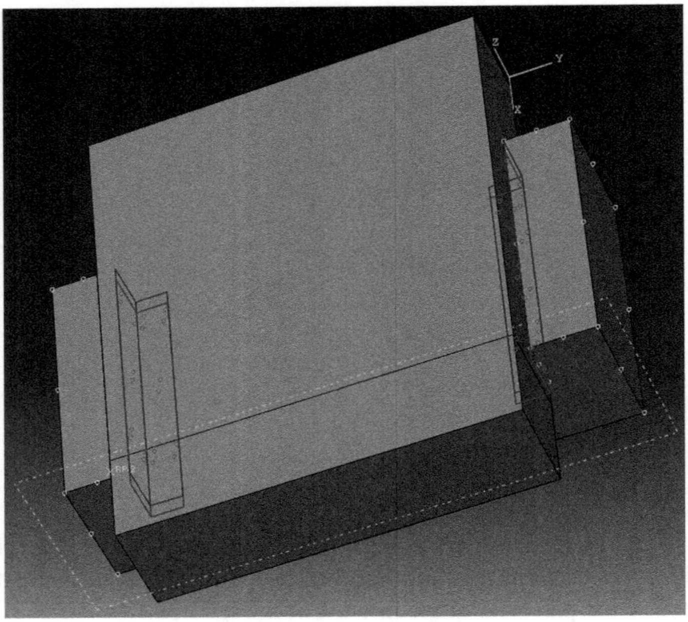

Fig. 7.31 Tie constraint definition between work piece and holder

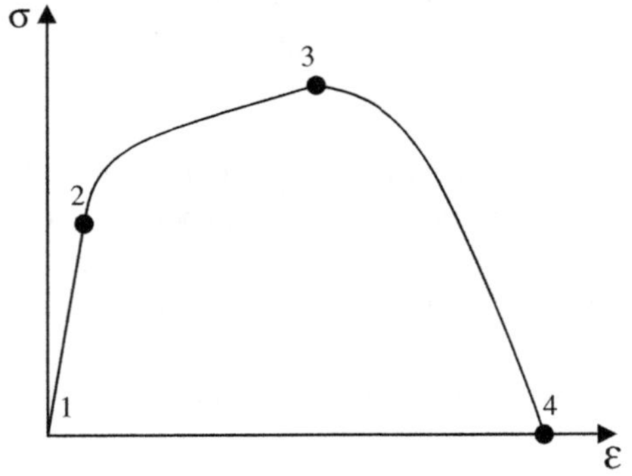

Fig. 7.32 Stress-Strain response of ductile material

One approach assumes fracture is due to nucleation, and growth of voids, while the equivalent plastic strain at the onset of damage is a function of stress triaxiality and strain rate. The other approach assumes fracture is due to shear band localization and the equivalent plastic strain at the onset of damage is a function of shear stress and strain rate. The approach adopted in this section is the ductile approach. This approach is more common in the literature, but the reader is free to apply the second approach and compare results.

The work piece flow stress as a function of strain, strain rate, and temperature is modeled using Johnson–Cook equation, Eq. (7.1). The damage initiation (*Damage Initiation in ABAQUS commands [31]) is simulated by using Johnson–Cook model, which is expressed in the following equation,

$$\bar{\varepsilon}_D^{pl} = [D1 + D2\ \exp(-D3.\eta)]\left[1 + D4.\ln\left(\frac{\dot{\bar{\varepsilon}}^{pl}}{\dot{\bar{\varepsilon}}_o}\right)\right]\left(1 + D5.\hat{\theta}\right) \tag{7.4}$$

The nondimensional temperature

$$\theta = \frac{T - T_{\text{room}}}{T_{\text{melt}} - T_{\text{room}}}$$

where T is the current temperature, T_{room} is the environment temperature, and T_{melt} is the material melting temperature. The effective plastic strain is defined as,

$$\bar{\varepsilon}^p = \int_0^t d\bar{\varepsilon}^p, \quad d\bar{\varepsilon}^p = \sqrt{\frac{2}{3} d\varepsilon_{ij}\, d\varepsilon_{ij}}$$

The effective stress is defined as,

$$\bar{\sigma} = \sqrt{\frac{3}{2}\sigma_{ij}\,\sigma_{ij}}$$

η is the ratio of pressure stress to effective stress,

$$\text{Pressure stress} = \text{Hydrostatic stress} = \frac{1}{3}\,\delta_{ij}\,\sigma_{kk}$$

$$\delta_{ij} = \text{Kornecker delta} = \{1 \text{ if } i=j\ ;\ 0 \text{ if } i \neq j\}$$

So how ABAQUS determines the damage initiation status? For each increment during the analysis, ABAQUS uses the Johnson–Cook failure equation and calculates the equivalent plastic strain at the onset of damage, $\bar{\varepsilon}_D^{pl}$ using the current stress triaxiality and strain rate. Then, ABAQUS calculate the scalar parameter, $\Delta\omega_D$,

$$\Delta\omega_D = \frac{\Delta\bar{\varepsilon}^{pl}}{\bar{\varepsilon}_D^{pl}\left(\eta,\ \dot{\bar{\varepsilon}}^{pl}\right)} \tag{7.5}$$

The damage initiation criteria is met when the following condition is satisfied,

$$\omega_D = \int \frac{d\bar{\varepsilon}^{pl}}{\bar{\varepsilon}_D^{pl}\left(\eta,\ \dot{\bar{\varepsilon}}^{pl}\right)} = 1 \tag{7.6}$$

Once the material damage criterion is initiated, the material stiffness starts degradation. The damage evolution law simulates the degradation rate of the material stiffness. For the ductile damage model, the material stiffness is modeled

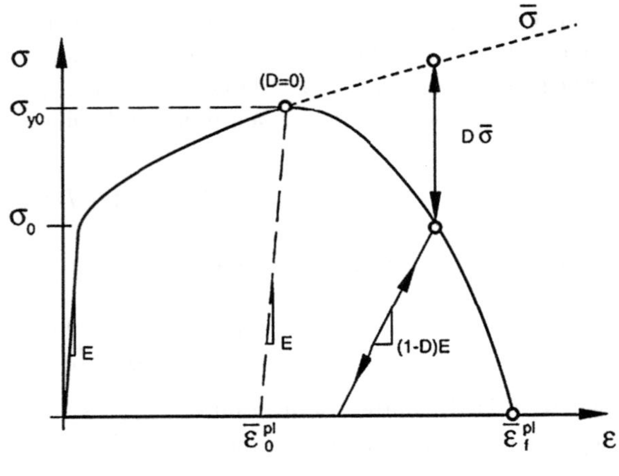

Fig. 7.33 Damage progress of ductile material (ABAQUS 6.11 manual [19])

with a scalar damage equation. During simulation, the stress tensor for the current material property is calculated according to the following equation,

$$\sigma = (1 - D)\,\bar{\sigma} \tag{7.7}$$

$\bar{\sigma}$ = effective stress tensor

σ = damaged stress tensor

D = Scalar damage variable

Figure 7.33 shows the material behavior under progressive damage. The onset of damage is at point $[\sigma_{\mathrm{yo}}, \bar{\varepsilon}_o^{pl}]$. $\bar{\varepsilon}_f^{pl}$ is the equivalent plastic strain at failure. The equivalent plastic strain $\bar{\varepsilon}^{pl}$ is mesh dependant, which makes estimating the damage scalar parameter D not practical. In an effort to resolve this issue, ABAQUS uses the equivalent plastic displacement $\bar{u}^{pl}$ instead, which is defined as,

$$\dot{\bar{u}} = (\text{Element characteristic length})\ \dot{\bar{\varepsilon}}^{pl} \tag{7.8}$$

Then damage evolution (D) as a function of the equivalent plastic displacement $\bar{u}^{pl}$ can be defined in linear or exponential form,

$$\dot{D} = \frac{\dot{\bar{u}}^{pl}}{\bar{u}_f^{pl}} \quad \text{or} \quad D = \frac{1 - \exp\left(-\alpha \frac{\bar{u}^{pl}}{\bar{u}_f^{pl}}\right)}{1 - \exp(-\alpha)} \tag{7.9}$$

where α is a material constant that can be adjusted to match experimental results. A second parameter that can be used to match experimental results is the maximum degradation parameter D_{max} at which ABAQUS deletes the elements from the mesh. The default value of this parameter is (0.99).

Then, the reader can ask a question, why do we need parameters to adjust finite element simulation results to match experimental ones? As in doing so it may reduces the value of finite element analysis!. The answer is no, and the finite element solution is as good as the material model used. The material models constants are evaluated based on experimental results like tensile or impact tests. Sometimes, the material behavior during the experimental testing for material parameter identification purpose is different from its behavior in a real test situation. The reasons for that could be the difference in strain rates or thermal conditions, in another term the stochastic nature of loads and material parameters. So, a parameter or two that are used to define the finite element model of the process are used to tune the numerical finite element results to better match experimental measurements. Once more, the reader asks himself, what is the purpose of doing finite element analysis then? To understand the purpose, assume an engineer is designing a mechanical part-A under specified loading condition. He then needs preliminary calculations for the part-A initial dimensions. This step can be performed based on experience or using finite element analysis. Then depending on loading conditions (static, dynamic, thermal), how many experimental tests he needs to perform to reach the final optimized design? while each iteration requires manufacturing a new part or assembly. So, it makes sense to perform one or two experimental testing, while tuning the finite element model parameters to match the experimental results. Once this finite element is developed, then the engineer will have the confidence to change any desired part dimensions or loading conditions using the FE model instead of the experimental testing. Once reached an acceptable solution to the problem in hand using the finite element model, the engineer then can perform another experimental testing with the new manufactured part to verify the results. This means cutting cost and increasing reliability.

Kay [32] published a report of how to estimate the Johnson–Cook parameters [D_1–D_5] using two different experimental tests. The first set of brackets in the Johnson–Cook fracture model represents the observation that the strain to fracture decreases as the hydrostatic tension increases. The second set of brackets represents the effect of an increased strain rate on the material ductility, while the third set of brackets represents the effect of thermal softening on the material ductility. Kay [20] used different approach to model damage evolution from the one used by ABAQUS. He did not use the equivalent plastic displacement to model damage evolution; instead he used Eq. (7.6) for damage evolution, while ABAQUS use the same equation for damage initiation, and then use the equivalent plastic displacement to model damage evolution. The value for equivalent plastic displacement at complete failure ($D = 1$) can be taken from ABAQUS example manual (Progressive failure of thin-wall aluminum extrusion). Clausen et al. [33] discussed in details using Split-Hopkinson to estimate Johnson–Cook plasticity and failure equations parameters of aluminum alloy AA5083. I assumed in the simulations that the failure equivalent plastic displacement is zero, which agrees with Kay [32].

7.4.1.3 Step Module

The tool is constraint in five degrees of freedom and only the rotation degree of freedom around the axis of revolution is kept free. Then applying velocity boundary condition, the tool revolution per second is defined. The unit system adopted in this analysis is [Kg–mm–s]. The holder is constraint in five degrees of freedom, and only the translation degree of freedom is kept free in the direction of feed rate. ABAQUS/explicit solver is used including the adiabatic heat effect, which simulates the softening of the material due to heat generated due to frictional force between cutting tool and work piece.

ABAQUS Input file is listed below,

```
**
*Elset, elset = plate, generate
1, 27000, 1
**
*Surface, type = ELEMENT, name = Surf-1
plate,
plate, interior
**
** Constraint: CP-1-Clamp-1-Workpiece-1
*Tie, name = CP-1-Clamp-1-Workpiece-1, adjust = no, no rotation, type =
SURFACE TO SURFACE
CP-1-Workpiece-1, CP-1-Clamp-1
** Constraint: CP-2-Clamp-1-Workpiece-1
*Tie, name = CP-2-Clamp-1-Workpiece-1, adjust = no, no rotation, type =
SURFACE TO SURFACE
CP-2-Workpiece-1, CP-2-Clamp-1
** Constraint: CP-3-Clamp-1-Workpiece-1
*Tie, name = CP-3-Clamp-1-Workpiece-1, adjust = no, no rotation, type =
SURFACE TO SURFACE
CP-3-Workpiece-1, CP-3-Clamp-1
** Constraint: CP-4-Clamp-1-Workpiece-1
*Tie, name = CP-4-Clamp-1-Workpiece-1, adjust = no, no rotation, type =
SURFACE TO SURFACE
CP-4-Workpiece-1, CP-4-Clamp-1
** Constraint: Constraint-1
*Rigid Body, ref node = _PickedSet19, elset = _PickedSet20, position =
CENTER OF MASS
** Constraint: Constraint-CL
*Rigid Body, ref node = _PickedSet45, elset = _PickedSet46
*End Assembly
**
** ELEMENT CONTROLS
**
*Section Controls, name = EC-1, ELEMENT DELETION = YES, MAX
DEGRADATION = 0.9
1., 1., 1.
**
** MATERIALS
```

```
**
*Material, name = AL2024-T351
*Conductivity
121000.,
*Damage Initiation, criterion = JOHNSON–COOK
0.112, 0.123, 1.5, 0.007, 0., 638., 25., 1.
*Damage Evolution, type = DISPLACEMENT
0.0,
*Density
2.78e-06,
*Elastic
7.31e + 07, 0.33
*Expansion
2.47e-05,
*Inelastic Heat Fraction
0.9,
*Plastic, hardening = JOHNSON–COOK
369000.,684000., 0.34, 1.7, 638., 25.
*Rate Dependent, type = JOHNSON–COOK
0.0083,1.
*Specific Heat
9.2e + 08,
*Material, name = "Tungsten Carbide"
*Density
1.58e-05,
*Elastic
6.9e + 08, 0.24
**
** INTERACTION PROPERTIES
**
*Surface Interaction, name = IntProp-1
*Friction
0.6,
*Surface Behavior, pressure-overclosure = HARD
**
** BOUNDARY CONDITIONS
**
** Name: BC-3 Type: Displacement/Rotation
*Boundary
_PickedSet24, 1, 1
_PickedSet24, 2, 2
_PickedSet24, 3, 3
_PickedSet24, 4, 4
_PickedSet24, 5, 5
** Name: BC-4 Type: Displacement/Rotation
```

```
*Boundary
_PickedSet47, 1, 1
_PickedSet47, 3, 3
_PickedSet47, 4, 4
_PickedSet47, 5, 5
_PickedSet47, 6, 6
**
** PREDEFINED FIELDS
**
** Name: Predefined Field-1 Type: Temperature
*Initial Conditions, type = TEMPERATURE
_PickedSet72, 25.
**
** INTERACTIONS
**
** Interaction: Int-1
*Contact, op = NEW
*Contact Inclusions
Tool-1.Tool, Workpiece-1.Surf-1
*Contact Property Assignment
,, IntProp-1
** ----------------------------------------------------------------
**
** STEP: Step-1
**
*Step, name = Step-1
*Dynamic, Explicit, adiabatic
, 0.0275
*Bulk Viscosity
0.06, 1.2
** Mass Scaling: Semi-Automatic
** Whole Model
*Variable Mass Scaling, dt = 5e-07, type = below min, frequency = 100
**
** BOUNDARY CONDITIONS
**
** Name: BC-2 Type: Velocity/Angular velocity
*Boundary, type = VELOCITY
_PickedSet23, 6, 6, 266.
** Name: BC-5 Type: Velocity/Angular velocity
*Boundary, type = VELOCITY
_PickedSet48, 2, 2, 424.
**
** OUTPUT REQUESTS
**
```

```
*Restart, write, number interval = 1, time marks = NO
**
** FIELD OUTPUT: F-Output-1
**
*Output, field, time interval = 0.0002
*Node Output
A, RF, U, V
*Element Output, directions = YES
EVF, LE, PE, PEEQ, PEEQVAVG, PEVAVG, S, SDEG, STATUS, SVAVG
*Contact Output
CSTRESS,
**
** HISTORY OUTPUT: H-Output-1
**
*Output, history, variable = PRESELECT
*End Step
** ________________________________
**
** STEP: Step-2
**
*Step, name = Step-2
*Dynamic, Explicit, adiabatic
, 0.0275
*Bulk Viscosity
0.06, 1.2
**
** OUTPUT REQUESTS
**
*Restart, write, number interval = 1, time marks = NO
**
** FIELD OUTPUT: F-Output-1
**
*Output, field, time interval = 0.0002
*Node Output
A, RF, U, V
*Element Output, directions = YES
EVF, LE, PE, PEEQ, PEEQVAVG, PEVAVG, S, SDEG, STATUS, SVAVG
*Contact Output
CSTRESS,
**
** HISTORY OUTPUT: H-Output-1
**
*Output, history, variable = PRESELECT
*End Step
```

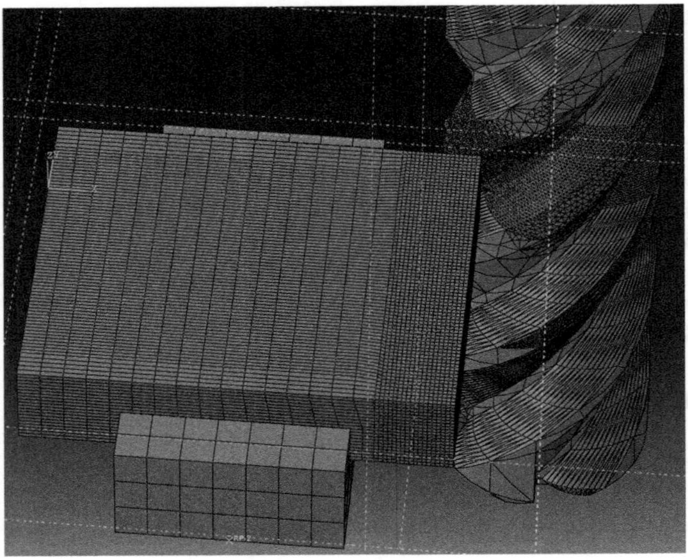

Fig. 7.34 Assembly module of the test case

7.4.2 Test Case-1

The assembly of the test case is shown in Fig. 7.34. The tool material is Carbide and the work piece is Al2024-T351. Selecting the milling cutting speed and CPT from Tables 7.2 and 7.3, and use Eqs. (7.2) and (7.3) to calculate cutting tool revolutions per second and work piece feed rate. The cutting tool revolutions per second is estimated to be 266 rev/s (1675 rad/s), and the feed rate is 424 mm/s. These values are based on tool cutting speed of 500 m/min and CPT of 0.4 mm.

The simulations results at different time steps are shown in Fig. 7.35. Thermal profile at the finished face is shown in Fig. 7.36. The maximum stress achieved during the cutting process is 550 MPa, and the maximum temperature is 120 °C with no coolant applied during the cutting process. Due to the dimensions of this part and the cutting process, running a steady state step to predict the final part shape will not be useful here, and there is no experimental result to validate the results. The following project is suggested to the reader:

- Project-1: Simulate the manufacturing process of the mounting plate of the satellite structure that was discussed in Chap. 2. Use the same material properties and tool dimensions that are used in this section. Predict final stresses and deformations using different cutting speeds with no coolant.
- Project-2: Manufacture an aircraft skin panel with stringers attached to it, using an end mill cutting process instead of the riveting assembly process. The skin panel dimensions are 250 × 250 × 5 mm, and the stringers are 30 mm height and 3 mm thickness. Start the simulations with a panel of 300 × 300 × 5 mm,

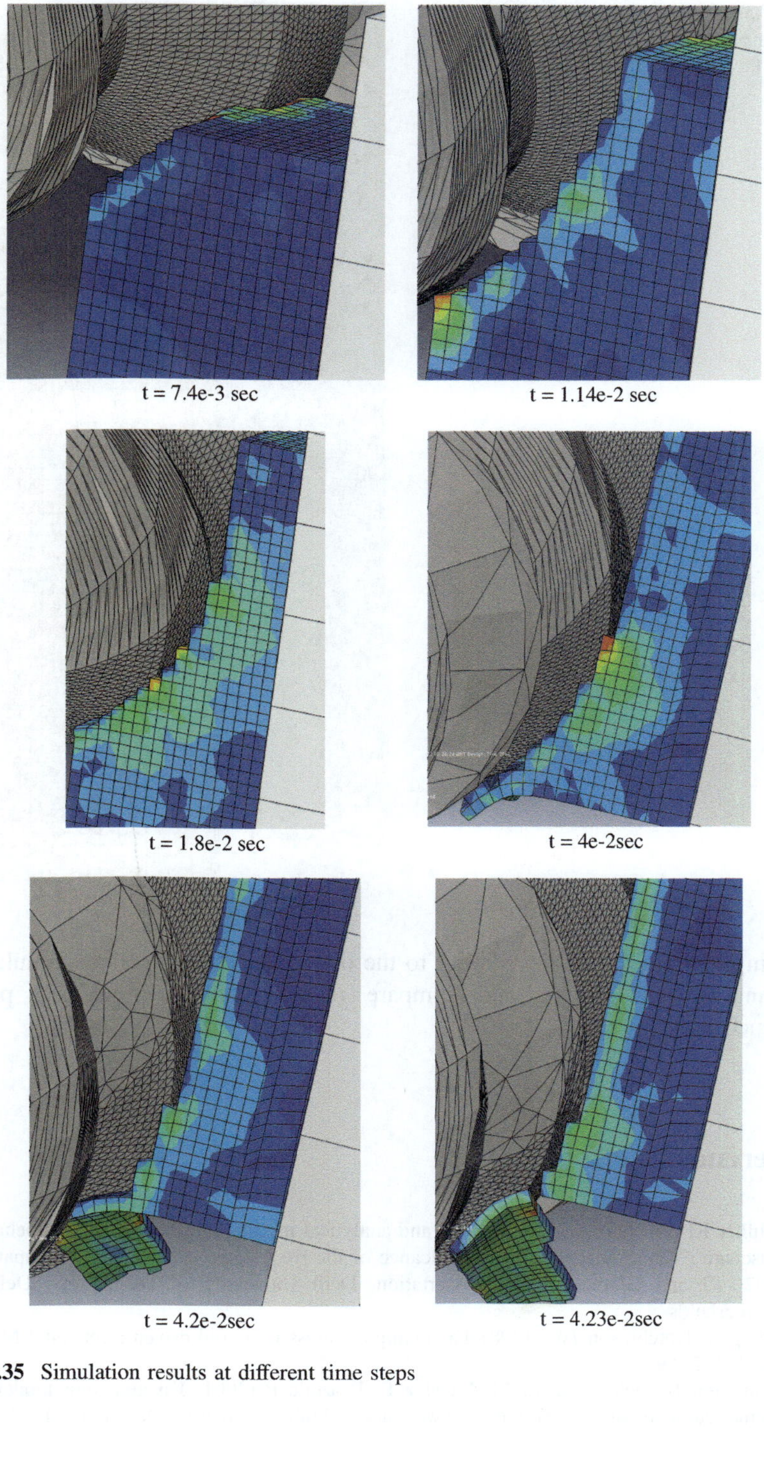

Fig. 7.35 Simulation results at different time steps

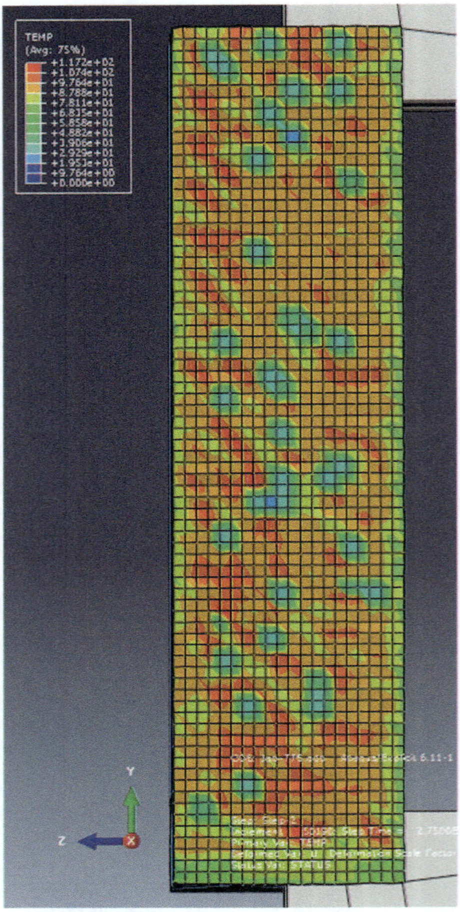

Fig. 7.36 Temperature profile on the finished work piece face

using material properties similar to the ones used in this section. Simulate the manufacturing process and compare results with the final real product dimensions.

References

1. Müller RPG (1995) An experimental and analytical investigation on the fatigue behavior of fuselage riveted lap joints: the significance of the rivet squeeze force and a comparison of 2024-T3 and Glare 3, Ph.D. Dissertation, Delft University of Technology, Delft, The Netherlands
2. Deng X, Hutchinson JW (1998) The clamping stress in a cold-driven rivet. Int J Mech Sci 40(7):683–694
3. Langrand B, Deletombe E, Markiewicz E, Drazetic P (2001) Riveted joint modeling for numerical analysis of airframe crashworthiness. Finite Elem Anal Des 38(1):21–44

4. Li G, Shi G (2004) Effect of the riveting process on the residual stress in fuselage lap joints. CASI J 50(2):91–105
5. Ryan L, Monaghan J (2000) Failure mechanism of riveted joint in fibre metal laminates. J Mater Process Technol 103(1):36–43
6. Szolwinski MP (1998) The mechanics and tribology of fretting fatigue with application to riveted lap joints, Ph.D. Dissertation, Purdue University, West Lafayette, IN
7. Szolwinski MP, Farris TN (2000) Linking riveting process parameters to the fatigue performance of riveted aircraft structures. J Aircraft 37(1):130–137
8. Rans C, Straznicky PV, Alderliesten R (2007) Riveting process induced residual stresses around solid rivets in mechanical joints. J Aircraft 44(1):323–329
9. Yunsu H (2006) Thermal contact conductance of nominaly flat surfaces. Springer J Heat Mass Transfer 43(1):1–5
10. Fox ME, Withers PJ (2007) Residual stresses in and around electromagnetically installed rivets measured using synchrotron and neutron diffraction. J Neutron Res 15(3–4):215–223
11. Repetto EA, Radovitzky R, Ortiz M, Lundquist RC, Sandstrom DR (1999) A finite element study of electromagnetic riveting. J Manuf Sci Eng 121:61–68
12. Johnson K L (1985) Contact mechanics. (Cambridge University Press, Cambridge)
13. Al-Hassani STS (1983) The shot peening of metals—mechanics and structures. SAE Report 821452, pp. 4513–4525 (Society of Automotive Engineers, Warrendale, Pennsylvania)
14. Wang T (2003) Numerical simulation and optimisation of shot peen forming processes. PhD Thesis, University of Cambridge
15. Al-Obaid YF (1990) A rudimentary analysis of improving fatigue life of metals by shot-peening. J Appl Mech 57:307–312
16. Follansbee PS, Sinclair GB (1984) Quasi-static normal indentation of an elasto-plastic half-space by a rigid sphere: 1. Int J Solids Structs 20(1):81–91
17. Sinclair GB, Follansbee PS, Johnson KL (1984) Quasi-static normal indentation of an elasto-plastic half-space by a rigid sphere—2: Results. Int J Solids Structs 21(2):865–888
18. Meguid SA, Shagal G, Stranart JC (1999) Finite element modelling of shot-peening residual stresses. J Mater Processing Technol. pp. 401–404
19. El Tobgy MS, Ng E, Elbestawi MA Three dimensional elastoplastic finite element model for residual stresses in the shot peening process. Proc
20. Wang T, Platts MJ, Wu J (2008) The optimization of shot peen forming process. J Mater Process Technol (206): 87–82
21. Kang X, Wang T, Platts J (2010) Multiple impact modelling for shot peening and peen forming. Part B J Eng Manuf 224:689–697
22. Miao HY, Demers D, Larose S, Perron C, Levesque M (2010) Experimental study of shot peening and stress peen forming. J Mater Process Technol 210:2089–2102
23. Michael Bush, Marc Warwahana (2001) Workbook for Machine Tool Applications: A Study Guide for the Skilled Trades. Kendall Hunt Publisher, Dubuque
24. Krar SF, Check AF (1998) Machine tool and manufacturing technology. Delmar Publishers, Albany. ISBN 0827363516
25. Baker M (2006) Finite element simulation of high-speed cutting forces. J Mater Process Technol 176(1–3):117–126
26. Ozel T (2006) The influence of friction models on finite element simulations of machining. Int J Mach Tools Mf 46(5):518–530
27. Arrazola PJ, Ö zel Tr (2010) Investigations on the effects of friction modeling in finite element simulation of machining. Int J Mech Sci 52(1):31–42
28. Umbrello D, Msaoubi R, Outeiro J (2007) The influence of Johnson–Cook material constants on finite element simulation of machining of AISI 316L steel. Int J Mach Tools Mf 47(3–4):462–470
29. Soo SL, Aspinwall DK, Dewes RC (2004) Three-dimensional finite element modeling of high-speed milling of Inconel 718. Part B J Eng Manuf 218:1555–1561
30. Chen XX, Zhao J, Li YE, Han SG, Cao QY, Li AH (2012) Numerical modeling of high-speed ball end milling with cutter inclination angle. Part B J Eng Manuf 226:606–616

31. ABAQUS Documentation. Version 6.11
32. Kay G (2002) Failure Modeling of Titanium-61-4 V and 2024-T3 Aluminum with the Johnson-Cook Material Model. Lawrence Livermore National Laboratory, U.S. Department of energy, http://www.doc.gov/bridge
33. Clausen A, Borvik T, Hopperstad OS, Benallal A (2004) Flow and flow and fracture characteristics of aluminum alloy AA5083–H116 as function of strain rate, temperature and triaxiality. J Mater Sci Eng A 364(1–2):260–272

Appendix A
Satellite Mechanical Loads

Mechanical loads can be static or dynamic. Static loads are constant or unchanging, and dynamic loads vary with time. Mechanical loads can also be external or self-contained. Types of loads for a satellite are:

- Static external loads: quasi-static loads due to inertia of supported components and structural modules from gravity or steady acceleration
- Static self-contained loads: mechanical preloads (intentional internal loads generated during assembly); thermo-elastic loads (generated by temperature changes)
- Dynamic external loads: engine thrust, sound pressure, and gusts of wind during launch; time-varying forces introduced by vibration during transportation to the launch site
- Dynamic self-contained loads: mass loading of a vibrating satellite during environmental testing or in space after the force that caused the excitation is removed
- Thermo-mechanical loads: on-orbit loads that arise from thermal cycles experienced by on-orbit environments
- Dynamic loads resulting from satellite maneuvers during orbit or attitude correction phases.

Every event in the life of a satellite introduces mechanical loads. Key events are manufacturing, handling, testing, transportation, prelaunch preparation, launch, satellite separation, and on-orbit operations. Launch generates the highest loads for most satellite structures, but any other event can be critical for specific parts of the structure. Mechanical load specifications for the satellite are presented below.

Two types of vibrations act on the satellite attachment points during transportation and LV flight. They are:

- Harmonic oscillations; and
- Random vibrations.

G. F. Abdelal et al., *Finite Element Analysis for Satellite Structures*,
DOI: 10.1007/978-1-4471-4637-7, © Springer-Verlag London 2013

Harmonic oscillations are characterized by the amplitude of vibroaccelerations and frequency. The parameters of harmonic oscillations are defined for sinusoidal vibrations during air transportation and axial and lateral sinusoidal vibrations during flight of the launch vehicle. Random vibrations are characterized by the spectral density of vibroaccelerations and duration of influence. The values of amplitude and spectral densities are given in the extreme octave points. The change of these values within the limits of each octave is linear in the logarithmic frequency scale. Random-vibration parameters are defined for road and rail transportation in three perpendicular directions individually. During launch, random vibrations act on the satellite/LV interface with approximately equal intensity of vibroaccelerations in each of the three randomly selected mutually perpendicular directions.

A.1 Mechanical Loads Before Integration with LV

Before integration with Dnepr launch vehicle, the satellite is subjected to the following mechanical environments:

- Quasi-static g-loads during handling;
- Quasi-static g-loads during rail transportation;
- Random-vibrations during rail transportation;
- Quasi-static g-loads during road transportation;
- Random-vibrations during road transportation.
- Quasi-static g-loads during air transportation;
- Sinusoidal vibrations during air transportation.

A.1.1 Handling

Various types of ground equipment are used to handle the satellite and its large components. Handling operations include loading flight articles in and out of transportation facilities, installing and removing them from integration and assembly place and test fixture. There are quasi-static g-loads during handling. The maximum quasi-static g-load is presented in Table A.1.

The load is along the direction of lifting (Z-direction) during handling case. During rail and road transportation, the X-axis of the vehicle is in the forward-motion direction, the Z-axis is vertically up, and the Y-axis makes the system right-handed. During ground operation in the horizontal position, n_x, n_y, and n_z are axial, lateral, and vertical g-loads, respectively.

Table A.1 Quasi-static g-loads

Strength-analysis case	n_x	n_y	n_z
Handling	–	–	−1.5
Rail transportation	±2	±0.7	−1±1
Road transportation	±2	±1.25	−1±2
Ground operation	±0.4	±0.5	−1±0.7

Table A.2 Random-vibration parameters during rail transportation

Band (Hz)					Duration (h)
0–10	10–20	20–40	40–60	60–100	
Vertical direction "down"					
$\frac{0.007}{0.187}$	$\frac{0.001}{0.1}$	$\frac{0.00025}{0.07}$	$\frac{0.00029}{0.075}$	$\frac{0.00011}{0.066}$	0.014 L
Lateral direction					
$\frac{0.00081}{0.09}$	$\frac{0.00023}{0.048}$	$\frac{0.00009}{0.042}$	$\frac{0.00018}{0.06}$	$\frac{0.000061}{0.05}$	0.004 L
Moving direction					
$\frac{0.0021}{0.15}$	$\frac{0.00031}{0.056}$	$\frac{0.000075}{0.039}$	$\frac{0.00009}{0.042}$	$\frac{0.000033}{0.036}$	0.002 L

Numerator—PSD (g^2 /Hz), denominator—R.M.S. acceleration (g)
L—transportation distance (km)

A.1.2 Ground Transportation

The satellite can be transported to the launch site by rail or by road. Without proper shipping containers, ground transportation can cause failure or fatigue damage to structural assemblies and electronic components. Structural loads are normally higher in transporting by road than by rail. There are two types of events that cause loads. The first is a transient event while the second is a random event. For ground transportation, both transient and random events are typically enveloped with a random-vibration power spectral density (PSD).

A.1.3 Rail Transportation

There are quasi-static g-loads and random vibrations during road transportation. The maximum quasi-static g-loads are presented in Table A.1. Table A.2 shows random-vibration parameters for the three mutually perpendicular directions. Random vibrations are equivalent to sinusoidal vibrations defined in Table A.3.

Table A.3 Equivalent sinusoidal vibrations during rail transportation

Direction	Band (Hz)					Duration (h)
	0–10	10–20	20–40	40–60	60–100	
Vibration-acceleration amplitude (g)						
Vertical "down"	0–0.434	0.164–0.227	0.113–0.153	0.165–0.195	0.12–0.144	0.014 L
Lateral	0–0.148	0.079–0.109	0.068–0.092	0.13–0.153	0.089–0.107	0.004 L
Along movement	0–0.238	0.091–0.126	0.062–0.084	0.092–0.108	0.066–0.079	0.002 L

L — transportation distance (km)
Variation of the vibration-acceleration amplitude within every band is linear at logarithmic scale of frequency

A.1.4 Road Transportation

There are quasi-static g-loads and random vibrations during road transportation. The maximum quasi-static g-loads are presented in Table A.1. Table A.4 shows random-vibration parameters for the three mutually perpendicular directions. Random vibrations are equivalent to sinusoidal vibrations defined in Table A.5.

A.1.5 Air Transportation

Similar to ground transportation, aircraft transportation environments consist of brief transients, such as landing and flight gusts, and random vibration produced by turbulent air. There are quasi-static g-loads and sinusoidal vibrations during the transportation. Table A.6 shows the maximum quasi-static g-loads during landing and flight. Table A.7 shows parameters of sinusoidal vibrations for the three mutually perpendicular directions.

The g-loads for every strength-analysis case act simultaneously. The X-axis of vehicle is in the forward-motion direction, the Z-axis is vertically up, and the Y-axis makes the system right-handed.

A.2 Mechanical Loads After Integration with LV and During Launch

After integration with Dnepr launch vehicle the satellite is affected by the following mechanical environments:

- Quasi static g-loads during ground operations;
- Quasi static g-loads during launch;
- Sinusoidal and random vibrations during launch;
- Acoustic environment during launch;
- Pyrotechnic shock environment.

Table A.4 Random-vibration parameters during road transportation

Direction	Band (Hz)								Duration (h)
	4–8	8–13	13–18	18–23	23–28	28–33	33–38	38–43	
Vertical "down"	$\frac{0.0039}{0.123}$	$\frac{0.0006}{0.055}$	$\frac{0.0007}{0.059}$	$\frac{0.0059}{0.168}$	$\frac{0.0039}{0.137}$	$\frac{0.0021}{0.1}$	$\frac{0.0009}{0.063}$	$\frac{0.0003}{0.038}$	0.021 L
Lateral	$\frac{0.00195}{0.087}$	$\frac{0.0003}{0.039}$	$\frac{0.00035}{0.042}$	$\frac{0.00295}{0.119}$	$\frac{0.00195}{0.097}$	$\frac{0.00105}{0.071}$	$\frac{0.00045}{0.045}$	$\frac{0.00015}{0.027}$	0.006 L
Along movement	$\frac{0.0011}{0.066}$	$\frac{0.0002}{0.032}$	$\frac{0.0002}{0.032}$	$\frac{0.0017}{0.092}$	$\frac{0.0011}{0.074}$	$\frac{0.0006}{0.055}$	$\frac{0.0002}{0.032}$	$\frac{0.0001}{0.022}$	0.003 L

Numerator—PSD (g^2 /Hz), denominator—R.M.S. acceleration (g)
L—transportation distance (km)

Table A.5 Equivalent sinusoidal vibrations during road transportation

Direction	Band (Hz)								Duration (h)
	4–8	8–13	13–18	18–23	23–28	28–33	33–38	38–43	
	Vibration-acceleration amplitude (g)								
Vertical "down"	0.208–0.291	0.114–0.144	0.155–0.181	0.525–0.586	0.477–0.52	0.382–0.41	0.281–0.292	0.165–0.173	0.021 L
Lateral	0.147–0.206	0.081–0.102	0.11–0.128	0.371–0.415	0.337–0.368	0.27–0.293	0.19–0.202	0.116–0.122	0.006 L
Along movement	0.11–0.155	0.066–0.083	0.083–0.0966	0.282–0.315	0.253–0.273	0.204–0.219	0.127–0.134	0.095–0.099	0.003 L

Variation of the vibration-acceleration amplitude within every band is linear at logarithmic scale of frequency
L—transportation distance (km)

Table A.6 Maximum quasi-static g-loads during air transportation

Strength-analysis case	n_x	n_y	n_z
Takeoff	±0.77	0	−3.93
Flight	0	±1	−1
	+0.2	0	+0.82
Landing	+1.37	0	−2.09

Table A.7 Sinusoidal vibrations for the three mutually perpendicular directions during air transportation

Band (Hz)	Vibration-acceleration amplitude (g)	Duration (min)
5–10	0.6	90
10–25	0.85	114
25–50	0.3	90
50–80	3.1	72
80–130	1.6	102
130–160	2.8	57
160–190	0.8	54
190–225	2.0	57
225–270	2.8	54
270–310	5.0	60

Note The duration corresponds with transportation distance of 10,000 km

Table A.8 Quasi static g-loads during launch

Loading case	Axial g-load n_z	Lateral g-load at SC/LV interface
Launch:	2.5±0.7	±0.3
Movement inside the launch canister	±1.0	±0.8
Movement beyond the launch canister		
Stage 1 operation:	3.0±0.5	0.5±0.5
Maximum dynamic pressure	7.5±0.5	0.1±0.5
Maximum axial acceleration		
Stage 2 operation:	7.8±0.5	0.2
Maximum axial acceleration		
Upper stage operation	−0.5	0.25

Note Lateral g-loads may act in any direction simultaneously with axial g-loads

A.2.1 Ground Operations

During launch preparation at the launch site, the satellite is integrated with Dnepr launch vehicle in the horizontal position. Therefore, the satellite needs special facilities to rotate it from the vertical position to the horizontal one. The launch vehicle "Dnepr" is transported in the horizontal position on a special road vehicle to the launch point. During this process, the satellite in the horizontal position is affected by quasi-static g-loads at the satellite/LV interface. These loads are presented in Table A.1.

Table A.9 Axial sinusoidal vibrations during flight of the launch vehicle

Band (Hz)	5–10	10–15	15–20
Vibration-acceleration amplitude (g)	0.5	0.6	0.5
Duration (s)	10	30	60

Note Variation of the vibration-acceleration amplitude within every band is linear at logarithmic scale of frequency

Table A.10 Lateral sinusoidal vibrations during flight of the launch vehicle

Band (Hz)	2–5	5–10	10–15
Vibration-acceleration amplitude (g)	0.2–0.5	0.5	0.5–1.0
Duration (s)	100	100	100

Note Variation of the vibration-acceleration amplitude within every band is linear at logarithmic scale of frequency

A.2.2 Launch

Launch starts when the booster engines ignite (lift-off) and ends with the separation of the satellite in its final orbit. A launch vehicle typically consists of stages. When the propellant from one stage is used up, the structure, storage tank, and engine of that stage separates from the launch vehicle, and the engine of the next stage ignites. Dnepr launch vehicle consists of two main propellant stages and an upper stage.

Table A.8 shows quasi-static g-loads for Dnepr launch vehicle. Accelerations in this table are represented in g's, or multiples of the earth's gravitational acceleration at sea level, which is about 9.81 m/s^2. Tables A.9 and A.10 show parameters of axial and lateral sinusoidal vibrations during flight of the launch vehicle. There are random vibrations in the three mutually perpendicular directions. Table A.11 shows parameters of these vibrations. Random vibrations are equivalent to sinusoidal vibrations defined in Table A.12.

There is an acoustic environment induced by Stage 1 engine operation and boundary layer noise. There is also pyrotechnic shock environment at fairing separation, upper stage separation, satellite separation, and deployment of solar panels. These loads act in the three mutually perpendicular directions. Values of the loads depend on the distance between the instrument and the adapter (at fairing separation and upper-stage separation) or on the distance between the instrument and the nearest pyrotechnic device (at satellite separation and deployment of solar panels).

Table A.11 Random vibrations in the three mutually perpendicular directions during launch

Band (Hz)	Loading case	
	Liftoff, LV flight segment where $M = 1$, q_{max} PSD (g^2/Hz)	1st stage burn (except for LV flight segment where $M = 1$, q_{max}), 2nd stage burn, 3rd stage burn
20…40	0.007	0.007
40…80	0.007	0.007
80…160	0.007–0.022	0.007
160…320	0.022–0.035	0.007–0.009
320…640	0.035	0.009
640…1280	0.035–0.017	0.009–0.0045
1280…2000	0.017–0.005	0.0045
R.M.S. acceleration (g) for 20–2,000 Hz band	6.5	3.6
Duration (s)	35	831

Variation of the vibration-acceleration amplitude within every band is linear at logarithmic scale of frequency

M means Mach number (equal to the vehicle speed divided by the local speed of sound), q_{max} means maximum dynamic pressure

Table A.12 Equivalent sinusoidal vibrations in the three mutually perpendicular directions during launch

Band (Hz)	Loading case	
	Liftoff, LV flight segment where M=1, q_{max} Vibration-acceleration amplitude (g)	1st stage burn (except for LV flight segment where M=1, q_{max}), 2nd stage burn, 3rd stage burn
20…40	0.6–0.812	0.6–0.812
40…80	0.812–1.063	0.812–1.063
80…160	1.063–2.351	1.063–1.326
160…320	2.351–3.489	1.326–1.769
320…640	3.489–3.882	1.769–1.969
640…1280	3.882–2.883	1.969–1.483
1280…2000	2.883–1.603	1.483–1.521
Duration (s)	35	831

Variation of the vibration-acceleration amplitude within every band is linear at logarithmic scale of frequency

A.3 Safety Factors

Safety factors for forces and bending moments must be taken as follows:

- Before integration with Dnepr launch vehicle:
- f = 1.5, for static g-loads,
- f = 2, for dynamic g-loads
- Air transportation: f = 2
- Ground operations after integration with Dnepr launch vehicle: f = 2
- During launch vehicle flight: f = 1.3

Appendix B
Load Analysis for Base-Driven Random Vibrations

This section shows how to predict a SDOF system's response to random vibration of its base or mounting structure. A mathematical concept called Fourier transform is used to solve the random-vibration problems. Moreover, it is used to obtain equivalent sinusoidal vibrations for random vibrations. This transform eases constructing transfer functions and helps in mathematically defining power spectral densities. Fourier transform can be considered as a Fourier series for a function having an infinite period. Mathematically, Fourier transform is

$$X(f) = \int_{-\infty}^{\infty} x(t)e^{-j2\pi ft}dt \tag{B.1}$$

where X is the transformed variable, as a function of frequency f for the variable x, which is a function of time t, and $j = \sqrt{-1}$. Given a Fourier transform, the time history can be recovered by the inverse Fourier transform:

$$x(f) = \int_{-\infty}^{\infty} X(f)e^{j2\pi ft}df \tag{B.2}$$

At first glance, these functions do not seem to relate to sine or cosine functions. Recall, however, that by Euler's theorem:

$$\begin{aligned} e^{jy} &= \cos y + j\sin y \\ e^{-jy} &= \cos y - j\sin y \end{aligned} \tag{B.3}$$

By using Fourier transform, an ordinary differential equation can be converted to an algebraic equation. The Fourier transform of a function's derivative is

$$\dot{X}(f) = j2\pi f\, X(f) \tag{B.4}$$

G. F. Abdelal et al., *Finite Element Analysis for Satellite Structures*,
DOI: 10.1007/978-1-4471-4637-7,

where $\dot{X}(f)$ is the Fourier transform of $\dot{x}(t)$, which is the first derivative of $x(t)$ with respect to time t, $X(f)$ is the Fourier transform of $x(t)$, and f is the frequency. Differentiating again:

$$\ddot{X}(f) = -(2\pi f)^2 X(f) \tag{B.5}$$

where $\ddot{X}(f)$ is the Fourier transform of $\ddot{x}(t)$, which is the second derivative of $x(t)$. To understand how to use the above equations, start with the equation of motion for an SDOF system:

$$m\,\ddot{x}(t) + 2m\omega_n\zeta\,\dot{x}(t) + k\,x(t) = F(t) \tag{B.6}$$

where m is the mass, ω_n is the natural circular frequency, ζ is the damping ratio, k is the stiffness, $x(t)$is the displacement, $\dot{x}(t)$ is the velocity, $\ddot{x}(t)$is the acceleration, F is force, and t is time. Dividing by mass and substituting $\omega_n = 2\pi f_n$,

$$\ddot{x}(t) + 4\zeta\pi f_n\,\dot{x}(t) + (2\pi f_n)^2 x(t) = \ddot{u}(t) \tag{B.7}$$

where f_n is the natural frequency and $\ddot{u}$ is the input acceleration, which is equal to F/m. Taking Fourier transform of the response $x(t)$ and the input acceleration $\ddot{u}(t)$, and using (B.4) and (B.5),

$$-(2\pi f)^2 X(f) + j2\pi f(4\zeta\pi f_n)X(f) + (2\pi f_n)^2 X(f) = \ddot{U}(f) \tag{B.8}$$

where $X(f)$ and $\ddot{U}(f)$ are the Fourier transforms of$x(t)$and$\ddot{u}(t)$. This expression can be rewritten in terms of acceleration Fourier transforms by substituting $\ddot{X}(f)$ for $X(f)$ per (B.5),

$$\ddot{X}(f) - j2\zeta\left(\frac{f_n}{f}\right)\ddot{X}(f) - \left(\frac{f_n}{f}\right)^2 \ddot{X}(f) = \ddot{U}(f) \tag{B.9}$$

By rearranging this expression,

$$\ddot{X}(f) = H(f)\,\ddot{U}(f) \tag{B.10}$$

where $\ddot{X}(f)$ is the output Fourier transform, which in this case is the Fourier transform of the time history of the response acceleration $\ddot{x}(t)$, $\ddot{U}(f)$ is the input Fourier transform, or in this case the Fourier transform of the time history of the input acceleration $\ddot{u}(t)$.$H(f)$ is the transfer function that relates the input and the output Fourier transform. When the force is applied to the mass, the transfer function is equal to:

$$H(f) = \frac{-(f/f_n)^2}{\left[1 - (f/f_n)^2\right] + j2\zeta(f/f_n)} \tag{B.11}$$

For base-driven loading,

$$H(f) = 1 + \frac{-(f/f_n)^2}{\left[1 - (f/f_n)^2\right] + j2\zeta(f/f_n)} \tag{B.12}$$

To assess a random forcing environment, the term root mean square is used, which is defined as:

$$\overline{x(t)}^2 = \frac{1}{T}\int_0^T x(t)x(t)dt \tag{B.13}$$

where $\overline{x(t)}^2$ is the mean square value of a response time history $x(t)$ and T is the duration. Substituting the inverse Fourier transform for one of the $x(t)$ terms per (B.2) produces:

$$\overline{x(t)}^2 = \frac{1}{T}\int_0^T x(t)\left[\int_{-\infty}^{\infty} X(f)e^{j2\pi ft}df\right]dt \tag{B.14}$$

By letting the duration T approach infinity, interchanging the order of integration, and taking advantage of symmetry of the Fourier transform, we finally arrive at Parseval's theorem:

$$\overline{x(t)}^2 = \int_0^{\infty}\left[\frac{2}{T}X(f)X(f)^*\right]df \tag{B.15}$$

where $X(f)^*$ is the complex conjugate of $X(f)$. The expression within brackets on the right-hand side of the equation is the power spectral density (PSD). Note that the product $X(f)\,X(f)^*$ is a real number and is equal to $|X(f)|^2$, which is the squared magnitude of the Fourier transform.

Random vibration is characterized with a power spectral density (PSD) curve. The term "power" is a general term that can represent acceleration, velocity, displacement, force, strain, etc., depending on the parameter needed to be described. An acceleration PSD at frequency f, which is designated W(f), is the mean square acceleration within a selected frequency band (whose center is f) divided by the bandwidth. PSD is commonly expressed in units of g^2/Hz, where g is the acceleration of Earth's gravity at sea level.

Equation (B.7) establishes that the Fourier transform of the output is equal to the product of the transfer function and the Fourier transform of the input. Based on Parseval's theorem [Eq. (B.12)],

$$\ddot{x}_{rms} = \sqrt{\int_0^{\infty} \left[|H(f)|^2 W_{\ddot{X}}(f) \right] df} \tag{B.16}$$

where $\ddot{x}_{rms}$ is the root mean square (rms) of the response acceleration, f is the frequency, $|H(f)|$ is the gain of the complex transfer function, and $W_{\ddot{X}}(f)$ is the input acceleration PSD. For practical applications, $W_{\ddot{X}}(f)$ is defined over a finite frequency range, and the integral in Eq. B.12 is evaluated numerically.

For an SDOF system or any mounted structure with a single dominant response frequency, Eq. (B.13) can be approximated with the following formula, which is used to convert the acceleration PSD into root-mean-square vibration acceleration:

$$\ddot{x}_{rms} = \sqrt{\frac{\pi f_n W_{\ddot{X}}(f_n)}{10 + 0.05 f_n}} \tag{B.17}$$

where $\ddot{x}_{rms}$ is the rms response acceleration, f_n is the natural frequency, $W_{\ddot{X}}(f_n)$ is the input acceleration PSD at frequency f_n.

By itself, the rms values of the response are not sufficient for design. For about 68 % of the duration of the environment, the absolute value of the acceleration will be less than the rms value. Peak response will be quite a bit higher. Thus, an appropriate probability level must be established for the design load, and then multiply the rms acceleration by the factor necessary to obtain that probability level.

From satellite mechanical load requirements, the satellite structure should be designed according to the third standard deviation, so the limit acceleration is calculated from the following formula:

$$A = 3\sqrt{\frac{\pi f_n W_{\ddot{X}}(f_n)}{10 + 0.05 f_n}} \tag{B.18}$$

This formula is used to calculate the vibration-acceleration amplitude of the equivalent sinusoidal harmonic vibration for road and rail transportation and launch vehicle flight.

9781447146360